Recent Developments in Electronic Materials and Devices

Related titles published by The American Ceramic Society:

Dielectric Materials and Devices
Edited by K.M. Nair, Amar S. Bhalla, Tapan K. Gupta, Shin-Ichi Hirano, Basavaraj V. Hiremath, Jau-Ho Jean, and Robert Pohanka
©2002, ISBN 1-57498-118-8

Boing-Boing the Bionic Cat and the Jewel Thief
By Larry L. Hench
©2001, ISBN 1-57498-129-3

The Magic of Ceramics
By David W. Richerson
© ISBN 1-57498-050-5

Boing-Boing the Bionic Cat
By Larry L. Hench
©2000, ISBN 1-57498-109-9

Electronic Ceramic Materials and Devices (Ceramic Transactions, Volume 106)
Edited by K.M. Nair and A.S. Bhalla
©2000, ISBN 1-57498-098-X

Ceramic Innovations in the 20th Century
Edited by John B. Wachtman Jr.
©1999, ISBN 1-57498-093-9

Dielectric Ceramic Materials (Ceramic Transactions, Volume 100)
Edited by K.M. Nair and A.S. Bhalla
©1997, ISBN 0-57498-066-1

Advances in Dielectric Ceramic Materials (Ceramic Transactions, Volume 88)
Edited by K.M. Nair and A.S. Bhalla
©1996, ISBN 1-57498-033-5

Hybrid Microelectronic Materials (Ceramic Transactions, Volume 68)
Edited by K.M. Nair and V.N. Shukla
©1995, ISBN 1-57498-013-0

Ferroic Materials: Design, Preparation, and Characteristics (Ceramic Transactions, Volume 43)
Edited by A.S. Bhalla, K.M. Nair, I.K. Lloyd, H. Yanagida, and D.A. Payne
©1994, ISBN 0-944904-77-7

Grain Boundaries and Interfacial Phenomena in Electronic Ceramics (Ceramic Transactions, Volume 41)
Edited by Lionel M. Levinson and Shin-Ichi Hirano
©1994, ISBN 0-944904-73-4

For information on ordering titles published by The American Ceramic Society, or to request a publications catalog, please contact our Customer Service Department at 614-794-5890 (phone), 614-794-5892 (fax),<customersrvc@acers.org> (e-mail), or write to Customer Service Department, 735 Ceramic Place, Westerville, OH 43081, USA.

Visit our on-line book catalog at <www.ceramics.org>.

Volume 131

Recent Developments in Electronic Materials and Devices

Proceedings of the Advances in Dielectric Materials and Multilayer Electronic Devices symposium, held at the 103rd Annual Meeting of The American Ceramic Society April 22–25, 2001, in Indianapolis, Indiana.

Edited by

K.M. Nair
E.I. duPont de Nemours & Company

A.S. Bhalla
The Pennsylvania State University

S.I. Hirano
Nagoya University

Published by

The American Ceramic Society
735 Ceramic Place
Westerville, Ohio 43081
www.ceramics.org

Proceedings of the Advances in Dielectric Materials and Multilayer Electronic Devices symposium, held at the 103rd Annual Meeting of The American Ceramic Society April 22–25, 2001, in Indianapolis, Indiana.

Cover photo: "SEM microstructure of as-cast surfaces of $SrTiO_3$" is courtesy of Stephen J. Lombardo, Rajesh V. Shende, and Daniel S. Krueger, and appears as figure 2 (right) in their paper "High Breakdown Strength and High Dielectric Constant Capacitors in the Strontium Zirconate and Strontium Titanate Solid Solution System," which begins on page 67.

Library of Congress Cataloging-in-Publication Data

A CIP record for this book is available from the Library of Congress.

For information on ordering titles published by The American Ceramic Society, or to request a publications catalog, please call 614-794-5890.

4 3 2 1–05 04 03 02

ISSN 1042-1122
ISBN 1-57498-145-5

Contents

POISSON'S RATIOS IN HIGH COUPLING FERROELECTRIC CERAMICS

Arthur Ballato
US Army Communications-Electronics Command
Fort Monmouth, NJ 07703-5201

ABSTRACT

Poisson's ratio is a function of orientation in anisotropic bodies. Moreover, the value in piezoceramics depends also upon imposed electrical conditions. We discuss both forms of variation: angular changes, and piezoelectric modification by electrical loads.

INTRODUCTION

Poisson's ratio, ν, is defined for isotropic media as the quotient of lateral contractive strain to longitudinal extensive strain when a simple terminal tractive stress is applied to a vanishingly thin bar or rod [1], [2]. The ratio finds application in a number of areas of applied elasticity and solid mechanics, for example, as indication of the mechanical coupling between various vibrational modes of motion [3]. For poled electroceramics, Poisson's ratio is a function of bar orientation with respect to polarization axis. This type of variability is to be expected. What is surprising is the size of the variability that can be produced via the piezo effect by altering electrical boundary conditions. One may use this second source of adjustment to provide means of tuning novel MEMS (microelectromechanical structures) and MOEMS (microoptoelectromechanical structures) filter and resonator devices for future applications such as advanced cellular communications involving mechanically resonant microstructures integrated with electronic and optical circuitry [4], [5]. These devices require extension of Poisson's ratio considerations to a variety of crystalline and polycrystalline substances.

In most materials, the dimensionless number ν is positive. In crystals and poled electroceramics, ν takes on different values, depending on the chosen directions of stress and strain. The maximum value of ν = +1/2 is obtained in the incompressible medium limit, where volume is preserved; see the Appendix. For ordinary materials, values of +1/4 to +1/3 are typical, but in crystals ν may vanish, or even take on negative values. In order to provide a synoptic, yet relatively uncomplicated picture, Fig. 1 sketches the bounds on ν as function of the traditional Lamé constants of an isotropic medium. Table I relates various elastic measures for substances or conditions indicated in the figure [6], [7]. Analytical formulas for Poisson's ratio are expressed in

terms of elastic moduli. For the case of crystals of general anisotropy, these expressions are quite unwieldly, but for substances in the hexagonal system the symmetry elements reduce the complexity considerably. Piezoceramics, and many of the materials under consideration for future microdevices are characterized by hexagonal symmetry.

Table I. Relations among isotropic elastic parameters

SUBSTANCE OR CONDITION	ν	λ	μ	Y	κ	v_{shear}/v_{long}
Ideal fluids	½	λ	0	0	λ	0
Many metals	1/3	2μ	μ	$8\mu/3$	$8\mu/3$	½
Poisson relation	¼	μ	μ	$5\mu/2$	$5\mu/3$	$1/\sqrt{3}$
Pure rigidity	0	0	μ	2μ	$2\mu/3$	$1/\sqrt{2}$
Perfect compressibility	– 1	$-2\mu/3$	μ	0	0	$\sqrt{3}/2$
Incompressible liquids	½	∞	0	0	∞	0
Incompressible solids	---	∞	∞	∞	∞	---

[v_{shear}/v_{long}] is ratio of shear to longitudinal wave velocities

The hexagonal crystal system comprises seven point groups (6/m mm, 6/m, 6, 6mm, 6-bar, 6-bar m2, and 622), and includes a number of the binary semiconductor systems, and their alloys [2], [8]. These have the piezoelectric 6mm wurtzite structure; examples are GaN and AlN. The family of poled electroceramics, including $BaTiO_3$, PZT, PLZT, PMN, PMN-PT, and related alloys are characterized by symmetry ∞mm, that is, they are transversely isotropic. However, this symmetry is equivalent to 6mm for all tensor properties up to and including rank five [9]; this includes elasticity. All hexagonal groups have the same elastic matrix scheme, so the purely elastic portion of Poisson's ratio is the same for all of these groups. The piezoelectric portion depends on the class; classes 6/m and 6/m mm are not piezoelectric, and the remaining five groups each have different matrix schemes. We will consider only the 6mm piezoceramic case; a more general treatment is given in Ref. [10].

HEXAGONAL STIFFNESS AND COMPLIANCE RELATIONS

Relations for Poisson's ratio are most simply expressed in terms of the elastic compliances [$s_{\lambda\mu}$]. It is often the case, however, that the most accurate determinations of the elastic constants (resonator and transit-time methods) yield values for the stiffnesses [$c_{\lambda\mu}$] directly [11] – [17]; the conversions are given below. For the hexagonal system, the elastic stiffness and compliance

matrices have identical form. Referred to the x_k axes as defined in the IEEE Standards [17], the matrices are:

$$\begin{matrix} c_{11} & c_{12} & c_{13} & 0 & 0 & 0 \\ c_{12} & c_{11} & c_{13} & 0 & 0 & 0 \\ c_{13} & c_{13} & c_{33} & 0 & 0 & 0 \\ 0 & 0 & 0 & c_{44} & 0 & 0 \\ 0 & 0 & 0 & 0 & c_{44} & 0 \\ 0 & 0 & 0 & 0 & 0 & c_{66} \end{matrix} \qquad \begin{matrix} s_{11} & s_{12} & s_{13} & 0 & 0 & 0 \\ s_{12} & s_{11} & s_{13} & 0 & 0 & 0 \\ s_{13} & s_{13} & s_{33} & 0 & 0 & 0 \\ 0 & 0 & 0 & s_{44} & 0 & 0 \\ 0 & 0 & 0 & 0 & s_{44} & 0 \\ 0 & 0 & 0 & 0 & 0 & s_{66} \end{matrix}$$

Stiffness and compliance are matrix reciprocals; the five independent components of each are related by:

$$(c_{11} + c_{12}) = s_{33} / S \qquad (c_{11} - c_{12}) = 1 / (s_{11} - s_{12}) \qquad (1, 2)$$

$$c_{13} = - s_{13} / S \qquad c_{33} = (s_{11} + s_{12}) / S \qquad (3, 4)$$

$$c_{44} = 1 / s_{44} \qquad S = s_{33} (s_{11} + s_{12}) - 2 s_{13}^{2} \qquad (5, 6)$$

Additionally, one has $s_{66} = 2(s_{11} - s_{12})$. The compliances are given in terms of the stiffnesses simply by interchange of symbols, but with $c_{66} = (c_{11} - c_{12})/2$. The equality of the 11 and 22 components together with the given relations between the 66, 11 and 12 components imply transverse isotropy; that is, all directions perpendicular to the unique six-fold axis, (i.e., in the basal plane), are elastically equivalent.

POISSON'S RATIO FOR NONPIEZOELECTRIC CRYSTALS

Poisson's ratio for crystals is given in general as $\nu_{\alpha\beta} = - s_{\alpha\beta}'/s_{\beta\beta}'$, where x_β is the direction of the longitudinal extension, x_α is the direction of the accompanying lateral contraction, and the $s_{\alpha\beta}'$ and $s_{\beta\beta}'$ are the appropriate rotated elastic compliances referred to this right-handed axial set [2]. This relation follows from a definition of the ratio as negative quotient of lateral to longitudinal strains for a single imposed longitudinal stress. It suffices to take x_1 as the direction of longitudinal extension; then two Poisson's ratios are defined by the orientations of the lateral axes x_2 and x_3: $\nu_{21} = - s_{12}'/s_{11}'$ and $\nu_{31} = - s_{13}'/s_{11}'$. Application of the definition requires specification of the orientation of the rotated x_k coordinates with respect to the crystallographic directions, and transformation of the compliances accordingly. For completeness, we note that in crystals the Young's modulus Y is a function of orientation, and is defined by $Y = 1/s_{11}'$. The Lamé moduli λ, μ for isotropic elastic bodies are defined as $\lambda = c_{12} = c_{13} = c_{23}$, and $\mu = c_{44} = c_{55} = c_{66}$. The remaining nonzero isotropic stiffnesses are $(\lambda + 2\mu) = c_{11} = c_{22} = c_{33}$.

ROTATED HEXAGONAL COMPLIANCES – GENERAL

The unprimed compliances are referred to a set of right-handed orthogonal axes related to the crystallographic axes in the manner defined by the IEEE Standard [17]. Direction cosines a_{mn} relate the transformation from these axes to the set specifying the directions of the applied longitudinal extension (x_1), and the resulting lateral contractions (x_2 and x_3). General expressions for the transformed hexagonal compliances that enter the formulas for ν_{21} and ν_{31} are:

$$s_{11}' = s_{11}[a_{11}^2 + a_{12}^2]^2 + s_{33}[a_{13}^4] + (s_{44} + 2s_{13})[a_{13}^2][a_{11}^2 + a_{12}^2] \quad (7)$$

$$\begin{aligned} s_{12}' = {} & s_{11}[a_{11}a_{21} + a_{12}a_{22}]^2 + s_{33}[a_{13}^2 a_{23}^2] + \\ & s_{44}[a_{13}a_{23}][a_{11}a_{21} + a_{12}a_{22}] + s_{12}[a_{11}a_{22} - a_{12}a_{21}]^2 + \\ & s_{13}[a_{23}^2[a_{11}^2 + a_{12}^2] + a_{13}^2[a_{21}^2 + a_{22}^2]] \end{aligned} \quad (8)$$

$$\begin{aligned} s_{13}' = {} & s_{11}[a_{11}a_{31} + a_{12}a_{32}]^2 + s_{33}[a_{13}^2 a_{33}^2] + \\ & s_{44}[a_{13}a_{33}][a_{11}a_{31} + a_{12}a_{32}] + s_{12}[a_{11}a_{32} - a_{12}a_{31}]^2 + \\ & s_{13}[a_{33}^2[a_{11}^2 + a_{12}^2] + a_{13}^2[a_{31}^2 + a_{32}^2]] \end{aligned} \quad (9)$$

TRANSFORMATION MATRIX FOR GENERAL ROTATIONS

Poisson's ratio for the most general case of 6mm point symmetry may be derived by considering the transformation matrix for a combination of two coordinate rotations: a first rotation about x_1 by angle θ, (x_1 is any axis in the basal plane, i.e., perpendicular to the polar axis), and a second rotation about the resulting x_2 by angle ψ. When these angles are set to zero, the x_1, x_2, x_3 axes coincide respectively with the reference crystallographic directions. For nonzero angles, the direction cosines a_{mn} are as follows, with the abbreviations c(θ) and s(θ) for cos(θ) and sin(θ), etc.:

$c(\psi)$	$s(\theta)s(\psi)$	$-c(\theta)s(\psi)$
0	$c(\theta)$	$s(\theta)$
$s(\psi)$	$-s(\theta)c(\psi)$	$c(\theta)c(\psi)$

Substitution of these a_{mn} into the expressions for s_{11}' , s_{12}' , and s_{13}' , and thence into the formulas $\nu_{21} = -s_{12}'/s_{11}'$ and $\nu_{31} = -s_{13}'/s_{11}'$ formally solves the problem, for specified values of θ and ψ, in the absence of the piezoelectric contribution. The condition of transverse isotropy, stated above, renders all results independent of the azimuthal angle in the plane normal to the polar axis.

BULK MODULUS

The bulk modulus, or compressibility, κ, is often associated with considerations requiring use of the Poisson's ratio. It is defined as the hydrostatic pressure required to bring about a unit relative change of volume of a substance; it is always positive. For solids of the most general anisotropy, κ is found from the relation

$$[(s_{11} + s_{22} + s_{33}) + 2\,(s_{23} + s_{13} + s_{12})] = 1 / \kappa. \qquad (10)$$

For the hexagonal system this reduces to:

$$[(2\,s_{11} + s_{33}) + 2\,(2\,s_{13} + s_{12})] = 1 / \kappa. \qquad (11)$$

PIEZOELECTRIC CONSTITUTIVE EQUATIONS

The four linear sets of piezoelectric constitutive matrix equations, according to the IEEE conventions [17], are comprised of two sets containing mixed intensive and extensive variables:

$$T = c^E\,S - e_t\,E \qquad D = e\,S + \varepsilon^S\,E \qquad (12, 13)$$

$$S = s^D\,T + g_t\,D \qquad E = -\,g\,T + \beta^T\,D \qquad (14, 15)$$

and two sets containing intensive and extensive variables arranged homogeneously:

$$S = s^E\,T + d_t\,E \qquad D = d\,T + \varepsilon^T\,E \qquad (16, 17)$$

$$T = c^D\,S - h_t\,D \qquad E = -\,h\,S + \beta^S\,D \qquad (18, 19)$$

Symbols in these equations are: elastic stress (T), elastic strain (S), dielectric displacement (D) and electric field (E), elastic stiffness (c), elastic compliance (s), dielectric permittivity (ε), dielectric impermeability (β), and four varieties of piezoelectric coefficients (e, g, d, and h). Superscripted material constants (e.g., c^E) are those values obtained when the superscripted quantity is held constant; subscript "t" denotes transposed matrix. Because the Poisson ratio is defined in terms of a single applied stress, the sets containing piezo "d" or "g" coefficients are preferred, as they contain fewer terms.

MATERIAL COEFFICIENTS FOR WURTZITE STRUCTURE – 6mm

Piezoceramics are characterized in their elastic, piezoelectric, and dielectric properties by the phenomenological term scheme for 6mm symmetry. Using the "d" set of equations (16) – (17), this symmetry yields the 9x9 Van Dyke matrix given in Table II referred to an unrotated axial set [17].

Table II. Unrotated Van Dyke matrix for class 6mm

s_{11}	s_{12}	s_{13}	0	0	0	0	0	d_{31}
s_{12}	s_{11}	s_{13}	0	0	0	0	0	d_{31}
s_{13}	s_{13}	s_{33}	0	0	0	0	0	d_{33}
0	0	0	s_{44}	0	0	0	d_{15}	0
0	0	0	0	s_{44}	0	d_{15}	0	0
0	0	0	0	0	s_{66}	0	0	0
0	0	0	0	d_{15}	0	ε_{11}	0	0
0	0	0	d_{15}	0	0	0	ε_{11}	0
d_{31}	d_{31}	d_{33}	0	0	0	0	0	ε_{33}

PIEZOELECTRIC CONTRIBUTION TO POISSON'S RATIO

We take the simple case of a bar of 6mm piezoceramic with polar axis in the x_3 (Z) direction, with tensile stress applied in the x_1 direction, and electrodes on the planes having normals parallel to x_3. This is the situation depicted in Fig. 2 when angle θ equals zero. Poisson's ratio is properly defined only in the case where lateral dimensions approach zero; use of finite lateral dimensions is an acceptable approximation in many engineering applications, provided these dimensions are small compared to bar length. It is easiest to use either the "g" or "d" constitutive sets, since these have stress as independent mechanical variable; we will use the "d" set.

From Table II, application of stress T_1 gives the following equations:

$$S_1 = s_{11}^E T_1 + d_{31} E_3 \tag{20}$$

$$S_2 = s_{12}^E T_1 + d_{31} E_3 \tag{21}$$

$$S_3 = s_{13}^E T_1 + d_{33} E_3 \tag{22}$$

$$D_3 = d_{31} T_1 + \varepsilon_{33}^T E_3 \tag{23}$$

At this point, the terminal electric boundary condition (BC) must be specified. We assume that an external load imposes the immittance condition

$$D_3 = -\varepsilon' E_3. \tag{24}$$

Therefore, (23) becomes

$$d_{31} T_1 + (\varepsilon_{33}^T + \varepsilon') E_3 = 0, \tag{25}$$

which permits elimination of E_3:

$$E_3 = (-d_{31} T_1)/(\varepsilon_{33}^T + \varepsilon'). \tag{26}$$

This is inserted into (20)-(22) to yield

$$S_1 = [s_{11}^E - d_{31}^2/(\varepsilon_{33}^T + \varepsilon')]T_1 \tag{27}$$

$$S_2 = [s_{12}^E - d_{31}^2/(\varepsilon_{33}^T + \varepsilon')]T_1 \tag{28}$$

$$S_3 = [s_{13}^E - (d_{31}\, d_{33})/(\varepsilon_{33}^T + \varepsilon')]T_1. \tag{29}$$

One could impose mechanical BCs as well, if desired; we discuss only the purely electrical BC given. This BC can be realized in some cases by use of another (external) crystal with electrodes, or, more generally, by attachment of an external active or passive electrical network. The ε' can be a complex number. We will consider below only the simple case where the electrical BC is imposed by an external load capacitor C'. This corresponds to the depiction of Fig. 2 with closed switch.

POISSON'S RATIO FOR PIEZOELECTRIC MATERIALS

Two Poisson's ratios, ν_{21} and ν_{31}, are defined as:

$$\nu_{21} = -S_2/S_1 \text{ and } \nu_{31} = -S_3/S_1. \tag{30}$$

From (27) – (29), one obtains, in terms of the piezo "d" coefficients:

$$\nu_{21} = -[s_{12}^E - d_{31}^2/(\varepsilon_{33}^T + \varepsilon')]/[s_{11}^E - d_{31}^2/(\varepsilon_{33}^T + \varepsilon')] \tag{31}$$

$$\nu_{31} = -[s_{13}^E - (d_{31}\, d_{33})/(\varepsilon_{33}^T + \varepsilon')]/[s_{11}^E - d_{31}^2/(\varepsilon_{33}^T + \varepsilon')] \tag{32}$$

ALTERATION OF ν BY VARIATION OF ELECTRICAL BC

The electrical immittance condition (24) imposed by an external load is a generalization of the short-circuit (SC) and open-circuit (OC) limiting cases usually quoted. In order to characterize attachment of an external load capacitor C' to the piezoceramic, we assign a value C to the capacitance appearing across the piezoceramic's electrode system, and relate ($\varepsilon + \varepsilon'$) to (C + C'). To go from the point relation ($\varepsilon + \varepsilon'$) to the global relation (C + C') we consider that C = ε times G, where G is a factor that depends solely on geometry (for a parallel-plate geometry, G = A/t, where "A" is lateral area, and "t" is thickness). Then both C and C' are normalized to G: G = (C/ε) = (C'/ε'); thus $\varepsilon' = \varepsilon$ (C'/C), and

$$(\varepsilon + \varepsilon') = \varepsilon\,(1 + C'/C). \quad (33)$$

Thus, the externally attached C' affects the total permittivity ($\varepsilon + \varepsilon'$) through the quantity (1 + C'/C) = (1 + ζ). When C' $\rightarrow$ 0, one has OC conditions; when 1/C' $\rightarrow$ 0, one has SC conditions; any other value of C' yields an intermediate condition. One can use an inductor, resistor, or generally, any immittance in place of C', including an active device.

When a load capacitor is placed in parallel with the piezoceramic bar, one simply replaces ε_{33}^T by ε_{33}^T (1 + C'/C), where C' is the load capacitance value, and C is the capacitance of the electrode system on the piezobar. Letting ζ = C'/C ($0 \le \zeta \le \infty$), one has

$$\nu_{21} = -\,[s_{12}^E - d_{31}^2/(\varepsilon_{33}^T\,(1+\zeta))]/[s_{11}^E - d_{31}^2/(\varepsilon_{33}^T\,(1+\zeta))] \quad (34)$$

$$\nu_{31} = -\,[s_{13}^E - (d_{31}\,d_{33})/(\varepsilon_{33}^T\,(1+\zeta))]/[s_{11}^E - d_{31}^2/(\varepsilon_{33}^T\,(1+\zeta))] \quad (35)$$

LIMITING CASES: SHORT- AND OPEN-CIRCUIT CONDITIONS

In the limit $1/\varepsilon' \rightarrow 0$, one obtains the short-circuit (SC) relations:

$$\nu_{21}^{(SC)} = -\,(s_{12}^E/s_{11}^E) \qquad \nu_{31}^{(SC)} = -\,(s_{13}^E/s_{11}^E). \quad (36, 37)$$

The limit $\varepsilon' \rightarrow 0$ yields the open-circuit (OC) equations:

$$\nu_{21}^{(OC)} = -\,(s_{12}^E - d_{31}^2/\varepsilon_{33}^T)/(s_{11}^E - d_{31}^2/\varepsilon_{33}^T) \quad (38)$$

$$\nu_{31}^{(OC)} = -\,(s_{13}^E - (d_{31}\,d_{33})/\varepsilon_{33}^T)/(s_{11}^E - d_{31}^2/\varepsilon_{33}^T). \quad (39)$$

Piezocoupling factor $k_{31}^2 = d_{31}^2/(s_{11}^E\,\varepsilon_{33}^T)$ relates the SC and OC ratios:

$$\nu_{21}^{(OC)} = [\nu_{21}^{(SC)} + k_{31}^2]/(1 - k_{31}^2) \quad (40)$$

$$\nu_{31}^{(OC)} = [\nu_{21}^{(SC)} + k_{31}^2 (d_{33}/d_{31})]/(1 - k_{31}^2). \quad (41)$$

In terms of the piezo "g" coefficients, a parallel derivation yields

$$\nu_{21} = -[s_{12}^D + g_{31}^2/(\beta_{33}^T + \beta')]/[s_{11}^D + g_{31}^2/(\beta_{33}^T + \beta')] \quad (42)$$

$$\nu_{31} = -[s_{13}^D + (g_{31}\, g_{33})/(\beta_{33}^T + \beta')]/[s_{11}^D + g_{31}^2/(\beta_{33}^T + \beta')] \quad (43)$$

In the limit $\beta' \to 0$, one obtains the short-circuit (SC) relations:

$$\nu_{21}^{(SC)} = -(s_{12}^D + g_{31}^2/\beta_{33}^T)/(s_{11}^D + g_{31}^2/\beta_{33}^T) \quad (44)$$

$$\nu_{31}^{(SC)} = -(s_{13}^D + (g_{31}\, g_{33})/\beta_{33}^T)/(s_{11}^D + g_{31}^2/\beta_{33}^T) \quad (45)$$

The limit $1/\beta' \to 0$ yields the open-circuit (OC) equations:

$$\nu_{21}^{(OC)} = -(s_{12}^D/s_{11}^D) \qquad \nu_{31}^{(OC)} = -(s_{13}^D/s_{11}^D). \quad (46, 47)$$

The Poisson's ratios computed via the "d" and the "g" coefficients are equal. Thus we obtain for the SC condition:

$$-\nu_{21}^{(SC)} = (s_{12}^E/s_{11}^E) = (s_{12}^D + g_{31}^2/\beta_{33}^T)/(s_{11}^D + g_{31}^2/\beta_{33}^T) \quad (48)$$

$$-\nu_{31}^{(SC)} = (s_{13}^E/s_{11}^E) = (s_{13}^D + (g_{31}\, g_{33})/\beta_{33}^T)/(s_{11}^D + g_{31}^2/\beta_{33}^T) \quad (49)$$

Similarly, for the OC condition one has:

$$-\nu_{21}^{(OC)} = (s_{12}^D/s_{11}^D) = (s_{12}^E - d_{31}^2/\varepsilon_{33}^T)/(s_{11}^E - d_{31}^2/\varepsilon_{33}^T) \quad (50)$$

$$-\nu_{31}^{(OC)} = (s_{13}^D/s_{11}^D) = (s_{13}^E - (d_{31}\, d_{33})/\varepsilon_{33}^T)/(s_{11}^E - d_{31}^2/\varepsilon_{33}^T) \quad (51)$$

ROTATED COORDINATES – OBLIQUE POLAR AXIS

In a coordinate system where the polar axis is normal to the x_1 axis (direction of applied stress), yet is oblique to the x_2 and x_3 axes (Fig.2 with $\theta \neq 0$; rotated Y cuts), one needs the quantities s_{11}' ; s_{12}' ; s_{13}' ; d_{31}' ; d_{32}' ; d_{33}' ; and ε_{33}'; this is seen from Table III, which recasts the 6mm term scheme of Table II for this rotation.

Table III. Van Dyke matrix for class 6mm, about axis normal to polar axis

s_{11}'	s_{12}'	s_{13}'	s_{14}'	0	0	0	d_{21}'	d_{31}'
s_{12}'	s_{22}'	s_{23}'	s_{24}'	0	0	0	d_{22}'	d_{32}'
s_{13}'	s_{23}'	s_{33}'	s_{34}'	0	0	0	d_{23}'	d_{33}'
s_{14}'	s_{24}'	s_{34}'	s_{44}'	0	0	0	d_{24}'	d_{34}'
0	0	0	0	s_{55}'	s_{56}'	d_{15}'	0	0
0	0	0	0	s_{56}'	s_{66}'	d_{16}'	0	0
0	0	0	0	d_{15}'	d_{16}'	ε_{11}'	0	0
d_{21}'	d_{22}'	d_{23}'	d_{24}'	0	0	0	ε_{22}'	ε_{23}'
d_{31}'	d_{32}'	d_{33}'	d_{34}'	0	0	0	ε_{23}'	ε_{33}'

Values for the required rotated compliances s' are given in (7) – (9). The necessary rotated piezoelectric and dielectric relations are as follows:

$$d_{31}' = (d_{15} - d_{33})(a_{11}a_{31} + a_{12}a_{32})\, a_{13} + d_{31}\, (a_{11}^2 + a_{12}^2)\, a_{33} \quad (52)$$

$$d_{32}' = (d_{15} - d_{33})(a_{21}a_{31} + a_{22}a_{32})\, a_{23} + d_{31}\, (a_{21}^2 + a_{22}^2)\, a_{33} \quad (53)$$

$$d_{33}' = (d_{31} + d_{15})(a_{31}^2 + a_{32}^2)\, a_{33} + d_{33}\, a_{33}^3 \quad (54)$$

$$\varepsilon_{33}' = \varepsilon_{11}\, (a_{31}^2 + a_{32}^2) + \varepsilon_{33}\, a_{33}^2 \quad (55)$$

Rotated constitutive equations are:

$$S_1' = s_{11}^{E'}\, T_1' + d_{31}'\, E_3' \quad (56)$$

$$S_2' = s_{12}^{E'}\, T_1' + d_{32}'\, E_3' \quad (57)$$

$$S_3' = s_{13}^{E'}\, T_1' + d_{33}'\, E_3' \quad (58)$$

$$D_3' = d_{31}'\, T_1' + \varepsilon_{33}^{T'}\, E_3' \quad (59)$$

with external load

$$D_3' = -\varepsilon'\, E_3'. \quad (60)$$

Finally, the Poisson's ratios for the rotated case are:

$$\nu_{21}' = -\, [s_{12}^{E'} - d_{31}'^2/(\varepsilon_{33}^{T'} + \varepsilon')]/[s_{11}^{E'} - d_{31}'^2/(\varepsilon_{33}^{T'} + \varepsilon')] \quad (61)$$

$$\nu_{31}' = -[s_{13}^{E'} - (d_{31}' d_{33}')/(\varepsilon_{33}^{T'} + \varepsilon')]/[s_{11}^{E'} - d_{31}'^2/(\varepsilon_{33}^{T'} + \varepsilon')] \quad (62)$$

With an external load capacitor, $(\varepsilon^{T'} + \varepsilon')$ is again replaced by $\varepsilon^{T'}(1 + \zeta)$. A parallel derivation yields the two Poisson's ratios for rotated piezoceramics in terms of the piezo "g" coefficients. Limiting cases are as follows:

SC ($1/\varepsilon' \rightarrow 0$):

$$\nu_{21}' = -(s_{12}^{E'}/s_{11}^{E'}) \quad (63)$$

$$\nu_{31}' = -(s_{13}^{E'}/s_{11}^{E'}). \quad (64)$$

OC ($\varepsilon' \rightarrow 0$):

$$\nu_{21}' = -[s_{12}^{E'} - d_{31}'^2/\varepsilon_{33}^{T'}]/[s_{11}^{E'} - d_{31}'^2/\varepsilon_{33}^{T'}] \quad (65)$$

$$\nu_{31}' = -[s_{13}^{E'} - (d_{31}' d_{33}')/\varepsilon_{33}^{T'}]/[s_{11}^{E'} - d_{31}'^2/(\varepsilon_{33}^{T'}]. \quad (66)$$

INPUT VALUES FOR NUMERICAL EXAMPLES

Bechmann [18] measured c^E, e, and ε^S values for $BaTiO_3$. The remaining material coefficients have been computed using the following matrix relations [19]:

$$[s^E] = [c^E]^{-1} ; [d] = [e][s^E]; [\beta^S] = [\varepsilon^S]^{-1} ; [h] = [\beta^S][e]; [\varepsilon^T] = [\varepsilon^S] + [e][d_t];$$

$$[\beta^T] = [\varepsilon^T]^{-1} ; [c^D] = [c^E] + [h_t][e]; [s^D] = [c^D]^{-1} ; [g] = [\beta^T][d].$$

Tables IV-VI contain the pertinent numerical values for this substance. In some entries additional digits have been retained to avoid round-off error.

Table IV. Elastic constants of $BaTiO_3$

$\lambda\mu \rightarrow$	11	33	13	44	66	[12]
$c_{\lambda\mu}^E$	166.	162.	77.5	42.9	44.8	76.4
$c_{\lambda\mu}^D$	167.54	189.54	70.98	54.88	44.8	77.94
$s_{\lambda\mu}^E$	8.552	8.893	-2.843	23.31	22.32	-2.609
$s_{\lambda\mu}^D$	8.180	6.734	-1.947	18.22	22.32	-2.980

units: c in [GPa]; s in $[TPa]^{-1}$ Mass density is $\rho = 5.72$ [Mg/m^3].

Table V. Piezoelectric constants of $BaTiO_3$

$\lambda\mu\rightarrow$	15	31	33
e	11.6	-4.4	18.6
h	1.033	-0.3503	1.481
g	18.82	-4.705	11.34
d	270.4	-79.04	190.4

units: e in [C/m^2]; h in [V/nm]; g in [m^2/kC]; d in [pm/V]

Table VI. Dielectric constants of $BaTiO_3$

$\lambda\mu\rightarrow$	11	33
ε^S	11.23	12.56
ε^T	14.37	16.80
β^S	890.5	796.2
β^T	696.1	595.3

ε in [nF/m]; β in [m/nF]

APPLICATION TO PIEZOCERAMICS

Using the values in Tables IV – VI for barium titanate, (36) – (39) yield the entries in Table VII for the unrotated case, where the angle θ in Fig. 2 equals zero.

Table VII. Poisson's ratios for unrotated $BaTiO_3$

Electrical BC	ν_{21}	ν_{31}
OC	0.3643	0.2380
SC	0.3050	0.3325
Ratio (OC/SC)	1.1945	0.7160

For barium titanate, Poisson's ratio ν_{21} increases by 19%, while ν_{31} decreases by 28%, as the electrical condition is varied from SC to OC. These are quite large changes, yet the barium titanate coupling $|k_{31}| = 20.9\%$ is not a very large value. The changes brought about by BC changes will be even more dramatic in higher coupling (e.g., single-crystal) materials. The quotient $R_{21} = \nu_{21}^{(OC)}/\nu_{21}^{(SC)} > 1$, while $R_{31} = \nu_{31}^{(OC)}/\nu_{31}^{(SC)} < 1$; this is because d_{31} and d_{33} have opposite signs.

Variation of electrical BCs on piezoelectrics may be used to alter acoustic delay times [20], and, in conjunction with external circuits, used to adjust frequencies [21] and temperature coefficients [22] of piezoresonators. These applications differ fundamentally from alterations of Poisson's ratio by means of electrical BC variations discussed here, although they share the piezoelectric mechanism in common.

In the following figures, all θ rotations are about an axis in the basal plane, as shown in Fig. 2; the material is barium titanate [18]. Fig. 3 is a plot of the piezocoupling factor $|k_{31}'(\theta)|$. Figures 4 and 5 show the Poisson's ratios for SC and OC conditions, respectively, while Figure 6 depicts the variations in R_{21} and R_{31}. Figures 7 and 8 respectively portray the variations in $\nu_{21}'(\zeta)$ and $\nu_{31}'(\zeta)$ versus ratio of capacitances, ζ, with $\theta = 0$. When $\zeta = 0$, one has the OC condition; in the limit of large ζ, one approaches the SC condition.

Table VIII gives some representative examples, for short-circuit conditions, of rotated orientations for additional materials; entries are computed from data in Refs. 23 and 24.

Table VIII. Poisson's ratio, Young's modulus, and compressibility of selected piezoceramics under short-circuit boundary conditions.

COMPOSITION	$\theta°, \psi°$	ν_{21}	ν_{31}	Y	κ
$Ba\,Ti\,O_3$	0, 0	0.305	0.333	117.0	106.3
PZT-4	45, 0	0.380	0.380	81.3	92.9
PZT-5A	0, 45	0.380	0.392	58.6	89.1
PZT 52/48	45, 45	0.401	0.354	63.7	93.5
PZT 65/35	0, 0	0.290	0.395	110.3	95.6
$Pb_{0.76}\,Ca_{0.24}\,Ti\,O_3$	45, 0	0.399	0.399	136.1	69.6
$Pb_{0.96}\,La_{0.04}\,Ti\,O_3$	0, 45	0.232	0.163	147.2	81.4
$Pb_{0.89}\,Nd_{0.11}\,Ti\,O_3$	45, 45	0.235	0.262	149.4	94.0

[Y and κ in GPa]

CONCLUSIONS

Poisson's ratio, with respect to rotated coordinate axes, is derived for piezoceramics. The effects of orientation and piezoelectricity are separately described, and it is shown that values for the ratio may be altered by changing the electrical boundary conditions. Numerical examples are provided.

REFERENCES

[1]A. E. H. Love, "*A Treatise on the Mathematical Theory of Elasticity*," 4th Ed., Dover, New York, 1944, p. 13.

[2]W. G. Cady, *Piezoelectricity*, McGraw-Hill, New York, 1946; Dover, New York, 1964.

[3]M. Onoe and H. F. Tiersten, "Resonant frequencies of finite piezoelectric ceramic vibrators with high electromechanical coupling," IEEE Trans. Ultrasonics Engineering, vol. UE-10, pp. 32-39, July 1963.

[4]M. A. Stroscio, K. W. Kim, S. Yu, and A. Ballato, "Quantized acoustic phonon modes in quantum wires and quantum dots," J. Appl. Phys., vol. 76, pp. 4670-4675, 1994.

[5]S. Yu, K. W. Kim, M. A. Stroscio, G. J. Iafrate, and A. Ballato, "Electron-acoustic-phonon scattering rates in rectangular quantum wires," Phys. Rev. B, vol. 50, pp. 1733-1738, 1994.

[6]A. Ballato, "Poisson's ratio for tetragonal, hexagonal, and cubic crystals," IEEE Trans. Ultrason., Ferroelec., Freq. Control, vol. 43, no. 1, pp. 56-62, January 1996.

[7]A. Ballato, "Poisson's ratio for poled electroceramics," Technical Report ARL-TR-432, US Army Research Laboratory, Fort Monmouth, NJ, August, 1995, 18 pp. ADA299045.

[8]J. F. Nye, *Physical Properties of Crystals*, Clarendon Press, Oxford, 1957; Oxford University Press, 1985.

[9]H. J. Juretschke, *Crystal Physics*, W A Benjamin, Reading, MA, 1974.

[10]A. Ballato, "Poisson's ratio for piezoceramics," Proc. 10th European Frequency and Time Forum, pp. 127-131, Brighton, UK, March 1996. IEE Conference Pub. 418.

[11]"IRE Standards on Piezoelectric Crystals, 1949," Proc. IRE, vol. 37, no. 12, pp. 1378-1395, December 1949.

[12]"IRE standards on piezoelectric crystals - the piezoelectric vibrator: definitions and methods of measurements, 1957," Proc. IRE, vol. 45, no. 3, pp. 353-358, March 1957.

[13]"IRE standards on piezoelectric crystals: determination of the elastic, piezoelectric, and dielectric constants - the electromechanical coupling factor, 1958," Proc. IRE, vol. 46, no. 4, pp. 764-778, April 1958. (IEEE Standard no. 178).

[14]"IRE standards on piezoelectric crystals: measurements of piezoelectric ceramics, 1961," Proc. IRE, vol. 49, no. 7, pp. 1161-1169, July 1961. (IEEE Standard no. 179).

[15]"Standard definitions and methods of measurement for piezoelectric vibrators," IEEE Standard no. 177, IEEE, New York, May 1966.

[16]"IEEE Standard on Piezoelectricity," IEEE Standard 176-1978, IEEE, New York. Reprinted in IEEE Trans. Sonics Ultrason., vol. SU-31, no. 2, Part 2, 55pp., March 1984.

[17]"IEEE Standard on Piezoelectricity," ANSI/IEEE Standard 176-1987, The Institute of Electrical and Electronics Engineers, New York, 10017.

[18]R. Bechmann, "Elastic, piezoelectric, and dielectric constants of polarized barium titanate ceramics and some applications of the piezoelectric equations," J. Acoust. Soc. Amer., vol. 28, pp. 347-350, 1956.

[19]A. Ballato, "Piezoelectricity: Old effect, new thrusts," IEEE Trans. Ultrason., Ferroelec., Freq. Control, vol. 42, no. 5, pp. 916-926, September 1995.
[20]A. Ballato, "Variable Delay Line," US Patent 3,286,205, issued 15 November 1966.
[21]A. Ballato, "Method and Apparatus for Testing Crystal Elements," US Patent 4,158,805, issued 19 June 1979.
[22]A. Ballato, "Adjustment of the Frequency-Temperature Characteristics of Crystal Oscillators," US Patent 4,607,239, issued 19 August 1986.
[23]B. Jaffe, W. R. Cook, Jr. and H. Jaffe, *Piezoelectric Ceramics*, Academic Press, New York, 1971. ISBN 0-12-379550-8.
[24]Landolt-Börnstein, Numerical Data and Functional Relationships in Science and Technology, New Series, Group III: Crystal and Solid State Physics, Volumes III/1, 1966; III/11, 1979; and III/29a, 1992. Springer-Verlag, Berlin, New York.

APPENDIX

POISSON'S RATIO FOR AN INCOMPRESSIBLE MEDIUM

An incompressible medium is defined as one in which volume is preserved. For a long thin isotropic rod comprised of such a medium, a longitudinal stress (T_1) will produce a longitudinal strain (S_1), and a corresponding contraction of the lateral cross-section. Without loss of generality, one may assume a rectangular cross-section. Initial dimensions are L_1, W_2, and H_3; volume is $V = (L_1\ W_2\ H_3)$. In the deformed state, the new dimensions are $[L_1\ (1 + S_1)]$, $[W_2\ (1 + S_2)]$, and $[H_3\ (1 + S_3)]$. Isotropy implies that contractile strains are equal: $S_2 = S_3$. Volume conservation requires that

$$V = (L_1\ W_2\ H_3) = \{[L_1\ (1 + S_1)]\ [W_2\ (1 + S_2)]\ [H_3\ (1 + S_3)]\}, \qquad \text{(A1)}$$

so that

$$(1 + S_1)\ (1 + S_2)^2 = 1. \qquad \text{(A2)}$$

Poisson's ratio, ν, is defined as the ratio $[-\ S_2/S_1]$ in the small deformation limit. One therefore obtains in this limit $(S_1 + 2S_2) = 0$, or $\nu = +\ ½$.

FIGURE CAPTIONS

Figure 1. Limits on the Lamé constants of isotropic solids. Symbols employed are: V, vacuum; G, gas; R, rubber; L, liquid; IL, incompressible liquid; M, metal, ceramic; P, plastic; C, cement; GL, glass.
Figure 2. Electroded piezoceramic bar oriented obliquely to the polar axis, with applied stress field, and provision for external capacitive load.
Figure 3. Piezocoupling $|k_{31}'(\theta)|$ of rotated orientations of barium titanate.
Figure 4. Rotated Poisson's ratios for short-circuited conditions.
Figure 5. Rotated Poisson's ratios for open-circuited conditions.
Figure 6. Effect of piezoelectricity on rotated Poisson's ratios.
Figure 7. Poisson's ratio $\nu_{21}'(\zeta)$ versus ratio of load to piezobar capacitances.
Figure 8. Poisson's ratio $\nu_{31}'(\zeta)$ versus ratio load to piezobar of capacitances.

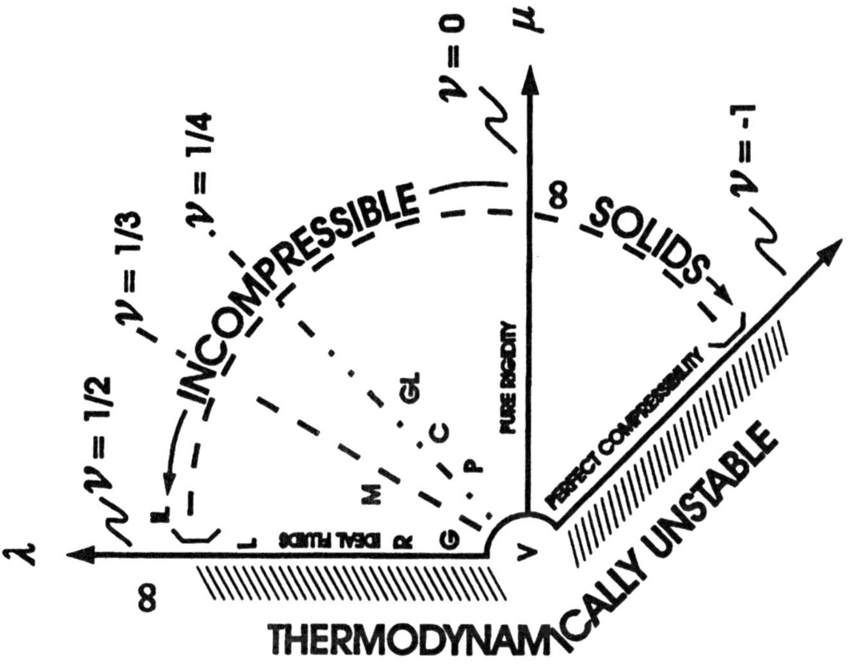

λ
μ
$\nu = 1/2$
$\nu = 1/3$
$\nu = 1/4$
$\nu = 0$
$\nu = -1$
INCOMPRESSIBLE
SOLIDS
PURE RIGIDITY
PERFECT COMPRESSIBILITY
IDEAL FLUIDS
THERMODYNAMICALLY UNSTABLE
L
R
G
V
M
C
P
GL

Figure 1.

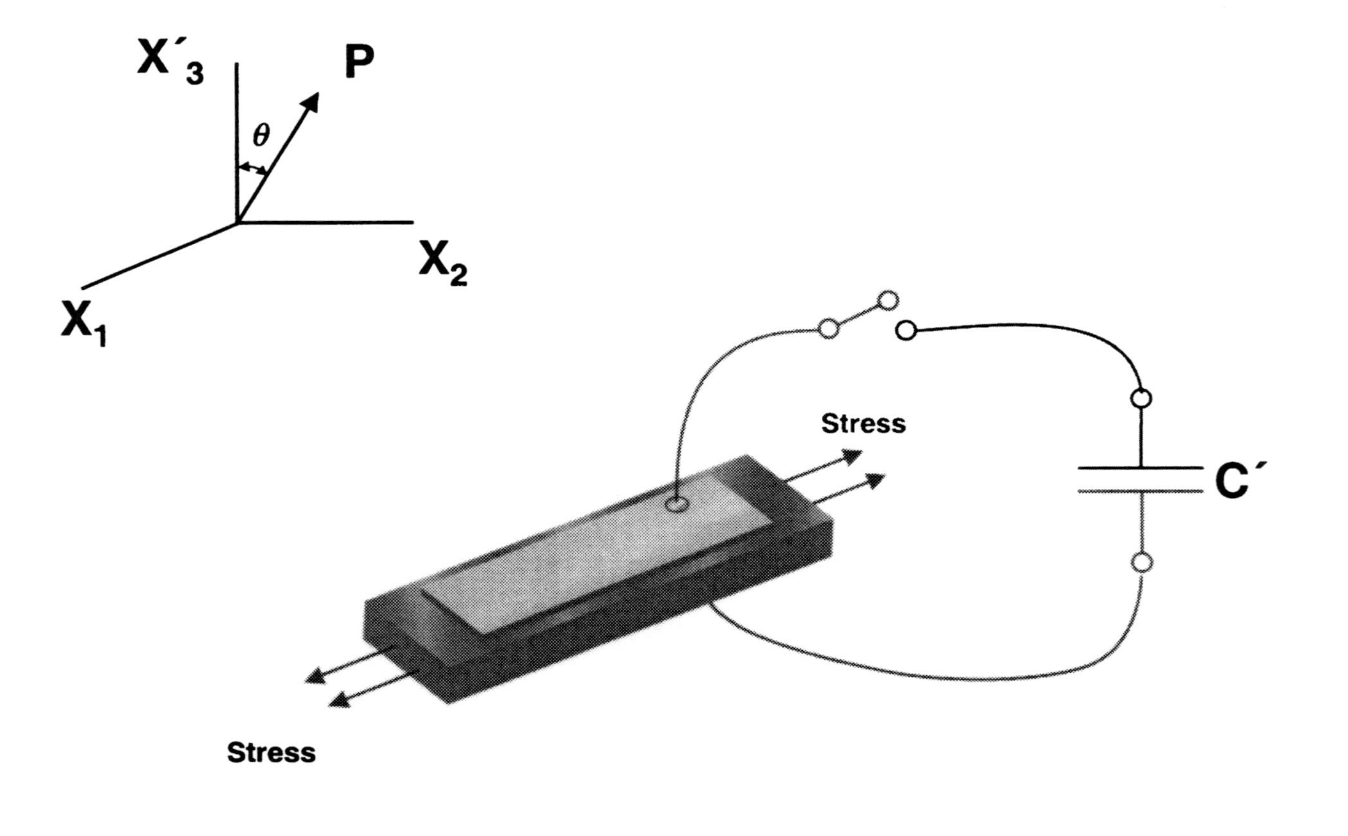

Figure 2.

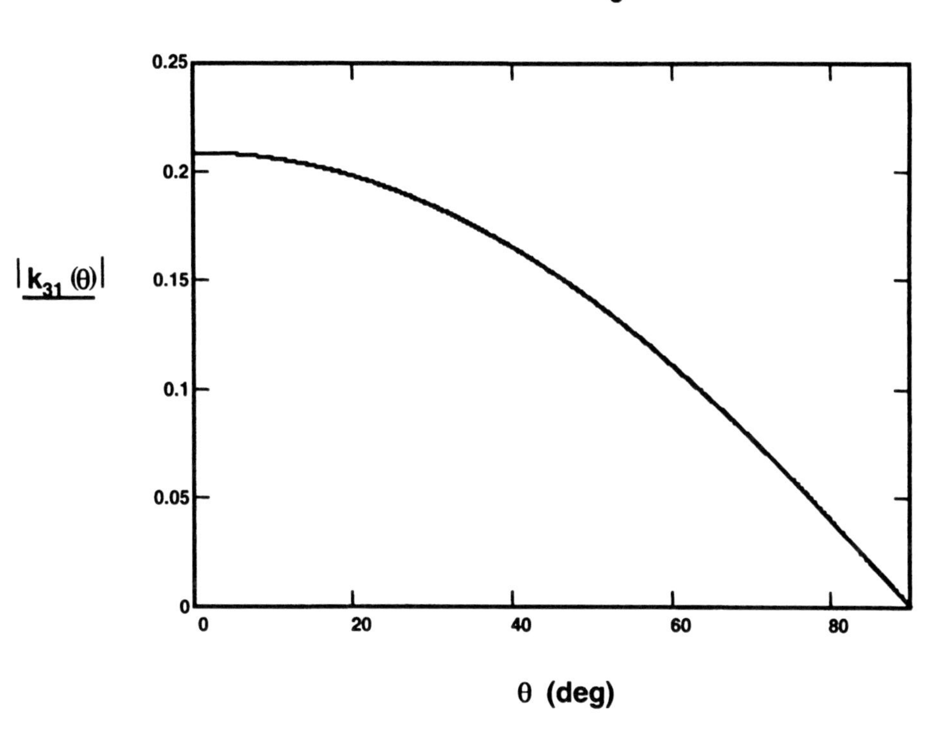
Rotated Piezocoupling
Ba Ti O3
|k31 (θ)|
0.25
0.2
0.15
0.1
0.05
0
0
20
40
60
80
θ (deg)

Rotated Poisson's Ratio for Short – and Open – Circuited Conditions

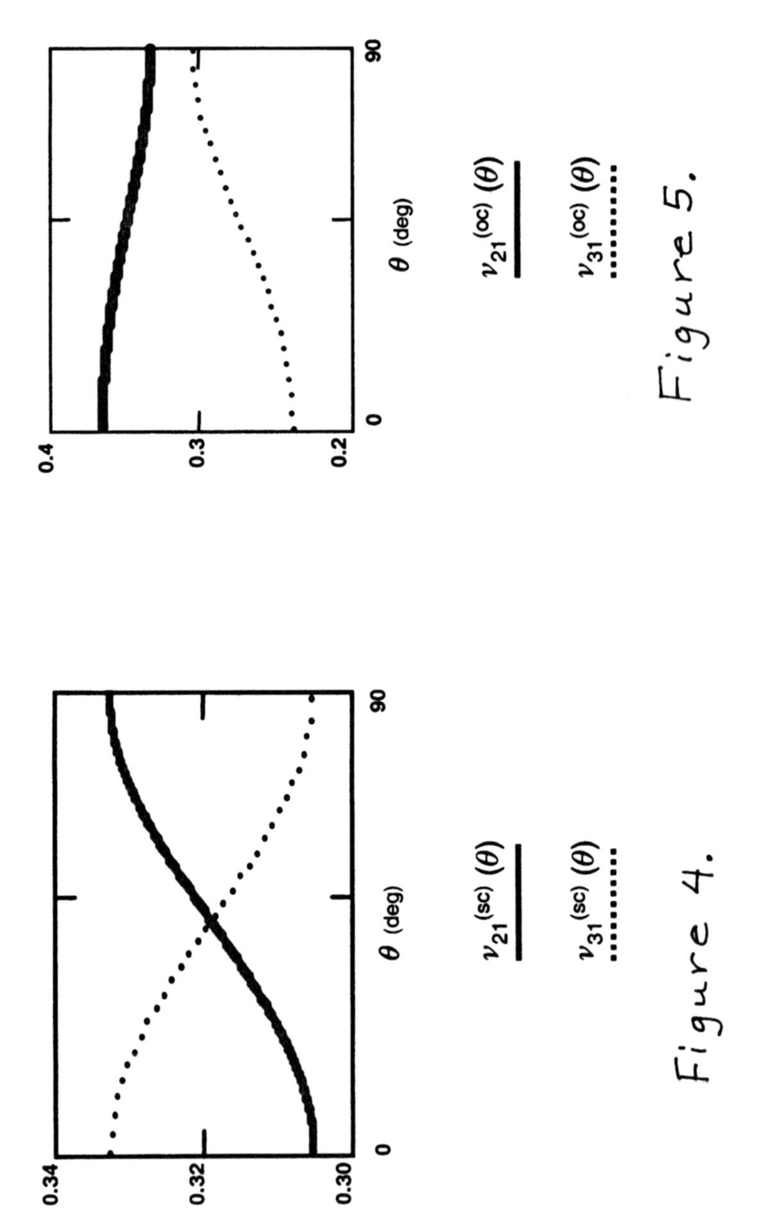

Figure 4.

Figure 5.

Effect of Piezoelectricity on Poisson's Ratios

Ba Ti O_3

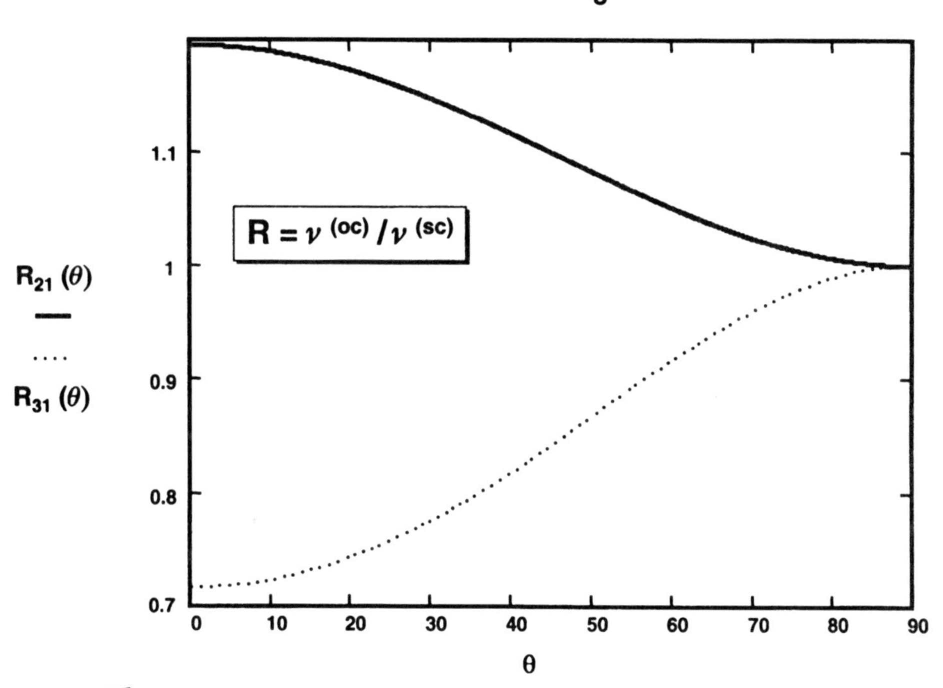

Figure 6.

Effect Of Load Capacitor

Ba Ti O_3

$$\zeta = \left(\frac{\text{Load Capacitance}}{\text{Bar Capacitance}}\right) \qquad \theta = 0$$

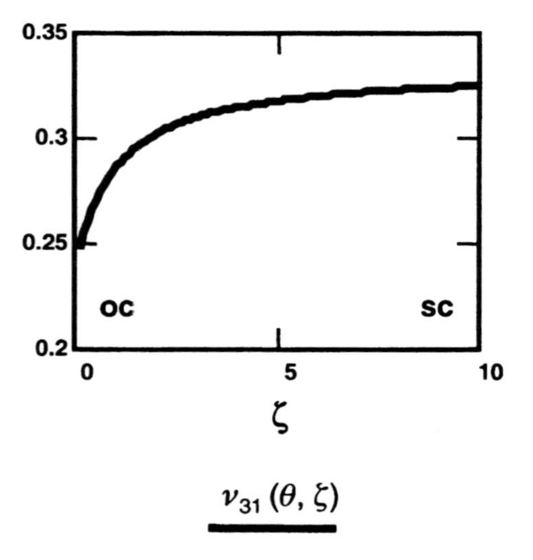

Figure 8.

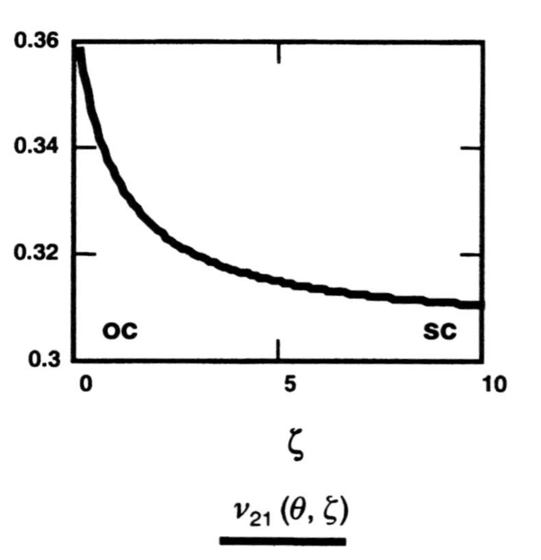

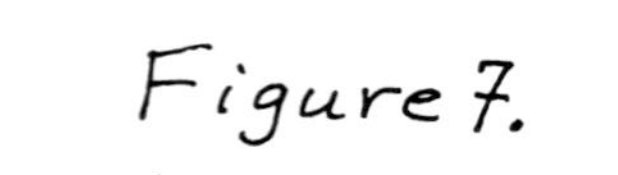

Figure 7.

DETERMINATION OF BINDER DECOMPOSITION KINETICS FOR PVB-$BaTiO_3$-Pt MULTILAYER CERAMIC CAPACITORS

Stephen J. Lombardo and Rajesh V. Shende
Department of Chemical Engineering
University of Missouri
Columbia, MO 65211, USA

ABSTRACT

The development of binder burnout cycles to maximize yield of ceramic components is often a trial-and-error procedure. In this work, we show how the activation energy and preexponential factor for binder decomposition can be determined from data obtained from thermogravimetric analysis for barium titanate capacitors with poly(vinyl butyral) (PVB) as the binder. The kinetic parameters can then be incorporated into a coupled transport and kinetic model to predict the buildup of pressure within a ceramic green body. With this approach, one can then specify heating rates and hold conditions for binder burnout cycles.

INTRODUCTION

During binder burnout [1], the decomposition products of polymer degradation [2,3] lead to an increase in pressure within the pore space of the ceramic body. As a consequence, a number of failure modes [4-6] may occur such as cracking, fracture, and blistering. Although a number of models [5-12] with differing viewpoints have been developed to describe the buildup of pressure, a general feature of such models is that they contain a source term, which accounts for the rate of binder decomposition. The rate expression is represented in an Arrhenius form, which consists of an activation energy, a preexponential factor, and a concentration dependence. Values for the first two quantities are often determined by first assuming a kinetic model and then fitting the decomposition data obtained from a thermogravimetric analyzer (TGA).

In this work, we evaluate binder decomposition data obtained with a TGA and show how the activation energy and preexponential factor can be obtained by the use of integral kinetic expressions. We next see to what extent the simulated kinetics agree with the data obtained in the experiments. The kinetic parameters determined by such analyses can then be incorporated into transport models for predicting the buildup of pressure within a ceramic body. This methodology can be used to provide guidelines for specifying heating rates and hold temperatures and times for binder burnout cycles in the processing of multilayer ceramic capacitors (MLCs).

EXPERIMENTAL

The samples used in the work were prepared from barium titanate powder (Tamtron X7R412H, TAM Ceramics Inc., Niagara Falls, NY), Pt-electrode paste, and a poly(vinyl butyral) (PVB) binder solution (B73305 Ferro Corp., San Marcos, CA). The binder fraction was approximately 10% of the total sample weight. The weight loss experiments were conducted with a Perkin-Elmer Model 1020 Series TGA 7, the details of which have been described elsewhere [12]. Typically, 50 mg of sample were loaded into the TGA, and the flow rate of air was set at 90 cm^3/min.

RESULTS AND DISCUSSION

The rate expression for a thermally activated process can be represented in terms of the conversion, α, as

$$\frac{d\alpha}{dt} = A\exp\left[\frac{-E}{RT}\right]f(\alpha) \tag{1}$$

where E is the activation energy, A is the preexponential factor, T is the temperature, and R is the gas constant. The effect of the conversion on the rate depends on the mechanism of binder decomposition and commonly used forms of $f(\alpha)$ [13,14] are for zero-, first-, and second-order decomposition. For complex systems in which a porous medium is present during binder decomposition, the apparent weight loss may be diffusion controlled and models have been developed for this case as well [14]. In Table I, we present the forms of $f(\alpha)$ for some of the commonly assumed mechanisms.

Table I. Kinetic models of polymer decomposition and their integrated forms.

Kinetic Model	$f(\alpha)$	$F(\alpha)$
Zero order	$(1-\alpha)^0$	α
First order	$(1-\alpha)^1$	$-ln(1-\alpha)$
Second order	$(1-\alpha)^2$	$\alpha/(1-\alpha)$
Diffusion (Valensi)	$[-ln(1-\alpha)]^{-1}$	$(1-\alpha)\,[ln(1-\alpha)]+\alpha$

To obtain values for A and E from TGA weight loss data, Eq. 1 can be integrated between two times, t_1 and t_2, which correspond to two temperatures, T_1 and T_2, and to two degrees of conversion, α_1 and α_2. Because the exponential function with temperature dependence in the denominator cannot be integrated analytically, either numerical methods or approximate expressions must be used. Lee and Beck [13] have developed an approximate method of high accuracy to determine the activation energy and preexponential factor from TGA weight loss data:

$$\ln\left[\frac{F(\alpha)}{T^2}\right] = \ln\left[\frac{AR}{\beta(E+2RT)}\right] - \frac{E}{RT} \tag{2}$$

where β is the linear heating rate used in the experiment. The function $F(\alpha)$ is the integrated form of $f(\alpha)$ and is given by

$$\mathrm{F}(\alpha) = \int \frac{1}{\mathrm{f}(\alpha)} d\alpha \tag{3}$$

Expressions for $F(\alpha)$ corresponding to the different kinetic models are also given in Table I. To obtain values of A and E, $F(\alpha)$ is computed from TGA data at each temperature and then the left-hand-side of Eq. 2 is plotted versus $1/T$. From the slope of the graph, the activation energy can be determined and this value can then be substituted back into Eq. 2 to determine the preexponential factor.

The results of TGA weight loss experiments for pure PVB at 10°C/min and for PVB-$BaTiO_3$-Pt MLCs at heating rates of 5 and 10 °C/min are presented in Fig. 1. Two important observations can be made from this data. First, in the presence of the ceramic and Pt-electrode materials, the polymer decomposes more rapidly as compared to the pure polymer. Second, the degree of weight loss is seen to depend upon the heating rate used in the experiments. For the faster heating rate, a given degree of conversion occurs at higher temperature as compared to the slower heating rate.

The data in Fig. 1 for pure PVB and for the PVB-$BaTiO_3$-Pt MLCs were analyzed by some of the kinetic models listed in Table I. Figure 2 is a representative plot obtained by using the integral method of Eq. 2 with the Valensi diffusion model and shows three regions of linear behavior as a function of temperature. Stage I correspond to conversions of the first 2% of binder and thus do not constitute a significant portion of the data. The stage III region at high temperatures corresponds to roughly 20% of the data, whereas stage II corresponds to the 78% of the conversion range.

In Table II are listed the values of A and E for the stage II region along with the regression coefficients. Although the regressions coefficients corresponding to all of the kinetic models are high (>0.98), the values of A and E are quite different. For PVB-$BaTiO_3$-Pt MLCs, the preexponential factors vary over nearly seven orders of magnitude and the activation energies vary by a factor of two. The two quantities vary in a similar manner [15,16], however, with smaller values of A arising with the smaller values of E.

Table II. Activation energies and preexponential factors determined by the integral methods of analysis for weight loss measured at β=10°C/min.

Model	A (1/s)	E (kJ/mol-K)	r^2 (-)	1-α (-)
Pure PVB				
First order	7.65×10^2	66.9	0.99	0.99-0.20
Valensi diffusion	1.36×10^8	135.3	0.99	0.99-0.36
PVB-BaTiO$_3$-Pt				
Zero order	9.37×10^2	58.0	0.99	0.99-0.32
First order	8.19×10^3	65.8	0.99	0.99-0.21
Second order	6.04×10^4	72.9	0.98	0.99-0.32
Valensi diffusion	2.98×10^9	127.7	0.99	0.98-0.21

To assess the accuracy of the values of *A* and *E*, the kinetic parameters in Table II were used to simulate the TGA data obtained in the experiments. Figure 3 compares the predicted kinetics with the experimental data for pure PVB at a heating rate of 10°C/min. It can be observed that both the first order and Valensi diffusion models describe the data very well, in spite of the fact that the kinetic parameters obtained from the two models differ substantially.

Figure 4 shows a comparison between experimental and the predicted weight loss profiles for PVB-BaTiO$_3$-Pt MLCs. It shows that all of the kinetic models provide a reasonable representation of the experimental data at low temperature. At higher temperature, however, the models deviate to some extent from the data, especially for second-order kinetics. The data in Fig. 4 do indicate, however, that very different values for the kinetic parameters can provide reasonable descriptions of the rate of binder burnout, but that no conclusions can be made on the decomposition mechanism of the binder.

The kinetic parameters in Table II from TGA data taken at a heating rate of 10°C/min for PVB-BaTiO$_3$-Pt MLCs were also used to simulate the experimental data recorded at 20°C/min. The results in Fig. 5 illustrate that under these circumstances, the level of agreement between the simulated weight loss and the experimental data is poorer. This indicates that using the kinetic parameters determined from TGA data at one heating rate for prediction of the weight loss at another heating rate may not be accurate.

Although not shown here, the kinetic parameters determined here have been used in a coupled transport and kinetic model for describing the buildup of pressure within ceramic green bodies [12,16]. By using this approach, binder removal cycles can then be specified in a quantitative manner.

CONCLUSIONS

The kinetics of binder weight loss have been measured at different heating rates, and the features in the weight loss profiles are seen to depend on the heating rates used in the experiments. The weight loss data were analyzed by four different mechanisms for zero order, first order, second order, and diffusion-controlled kinetics. For pure PVB degradation, the first order and Valensi diffusion models both represent the experimental data very well. For PVB-$BaTiO_3$-Pt MLCs, the kinetic models provide a reasonable representation of the experimental data, especially at lower temperatures. Although use of integral expressions with the different models leads to very different values for the activation energy and preexponential factor, all of the values were capable of predicting the experimental data within ±10%.

REFERENCES

1. J. A. Lewis, "Binder Removal From Ceramics," *Annual Rev. Mater. Sci.*, **27**, 147-73 (1997).
2. H.H.G. Jellinek, "Degradation and Depolymerization Kinetics," pp. 1-37 in *Aspects of Degradation and Stabilization of Polymers*. Edited by H.H.G. Jellinek, Elsevier, New York, 1978.
3. C. David, Ch. 1 in *Comprehensive Chemical Kinetics-Degradation of Polymers*, Vol. 14. Edited by C H. Gamford and C.F.H. Tipper, Elsevier, New York, 1975.
4. J.R.G. Evans, M.J. Edirisinghe, J.K. Wright, and J. Crank, "On the Removal of Organic Vehicle from Molded Ceramic Bodies," *R. Soc. London A*, **432**, 321 (1991).
5. P. Calvert and M. Cima, "Theoretical Models for Binder Burnout," *J. Am. Ceram. Soc.*, **73**, 575-79 (1990).
6. B. Peters and S.J. Lombardo, "Optimization of Multi-Layer Ceramic Capacitor Geometry for Maximum Yield during Binder Burnout," to appear in *J. Mat. Sci.: Mat. Electr.* (2001).
7. M. R. Barone and J. C. Ulicny, "Liquid-Phase Transport During Removal of Organic Binders in Injection-Molded Ceramics," *J. Am. Ceram. Soc.*, **73**, 3323-33 (1990).
8. R.M. German, "Theory of Thermal Debinding," *Int. J. Powder Metall.*, **23**, 237-45 (1987).
9. G.Y. Stangle, and I.A. Aksay, "Simultaneous Momentum, Heat and Mass Transfer With Chemical Reaction in A Disordered Porous Medium: Application to Binder Removal From A Ceramic Green Body," *Chem. Eng. Sci.*, **45**, 1719-31 (1990).
10. D. S. Tsai, "Pressure Buildup and Internal Stresses During Binder Burnout: Numerical Analysis, " *AIChE J.*, **37**, 547-54 (1991).

11. A.C. West and S.J. Lombardo, "The Role of Thermal and Transport Properties on The Binder Burnout of Injection Molded Ceramic Components," *Chem. Eng. J.* **71**, 243-52 (1998).
12. L.C.-K. Liau, B. Peters, D.S. Kruger, A. Gordon, D.S. Viswanath, and S.J. Lombardo, "The Role of Length Scale on Pressure Increase and Yield of PVB-$BaTiO_3$-Pt Multi-Layer Ceramic Capacitors During Binder Burnout, " *J. Am. Ceram. Soc.*, **83**, 2645-53 (2000).
13. T.V. Lee and S.R. Beck, "A New Integral Approximation Formula for Kinetic Analysis of Nonisothermal TGA Data, " *AIChE J.*, **30**, 517-19 (1984).
14. T.C.K. Yang, W.H. Chang and D.S. Viswanath, "Thermal Degradation of Poly(Vinyl Butyral) in Alumina, Mullite and Silica Composites, "*J. Therm. Anal.*, **47**, 697-13 (1996).
15. S.J. Lombardo and A.T. Bell, "A Review of Theoretical Models of Adsorption, Diffusion, Desorption, and Reaction of Gases on Metal Surfaces," *Surface Science Reports*, **13**, 1-72 (1991).
16. R.V. Shende and S.J. Lombardo, "Determination of Binder Decomposition Kinetics for Specifying Heating Parameters in Binder Burnout Cycles, " submitted to *J. Am. Ceram. Soc.* (2001)

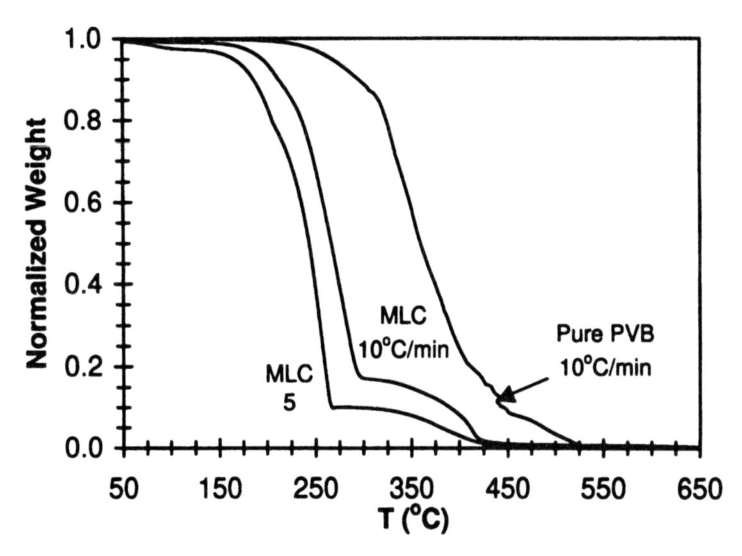

Figure 1. Weight loss profiles of pure PVB at 10°C/min and PVB- $BaTiO_3$-Pt MLCs at different heating rates during TGA experiments in air.

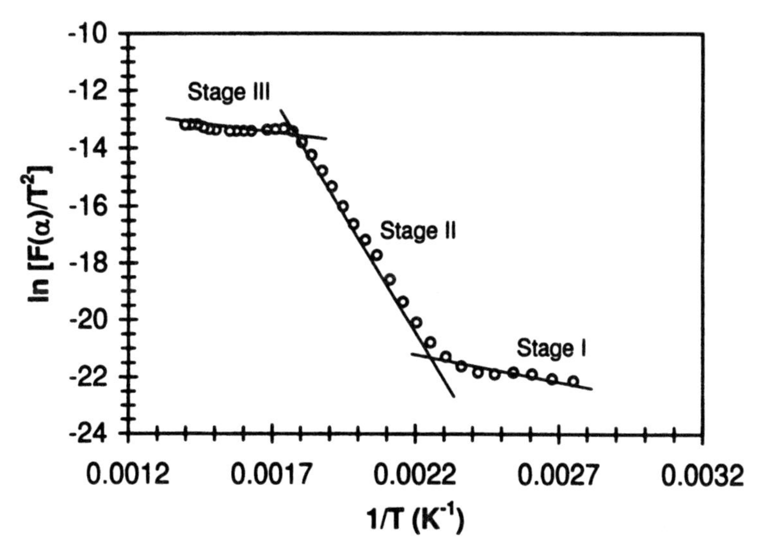

Figure 2. Integral method of analysis of TGA data taken at 10°C/min using the Valensi diffusion model for PVB-$BaTiO_3$-Pt MLC.

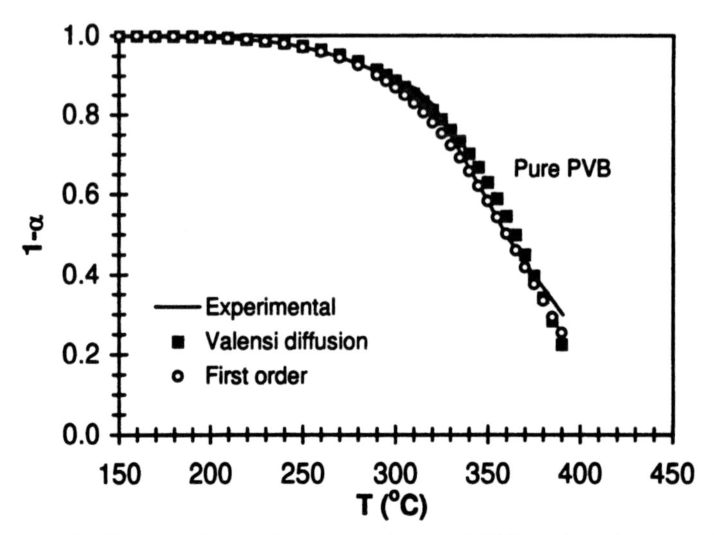

Figure 3. Comparison of an experimental TG weight loss profile (solid line) recorded at 10°C/min with the kinetics simulated by first order and Valensi diffusion models using the parameters obtained by analyzing TGA data taken at 10°C/min.

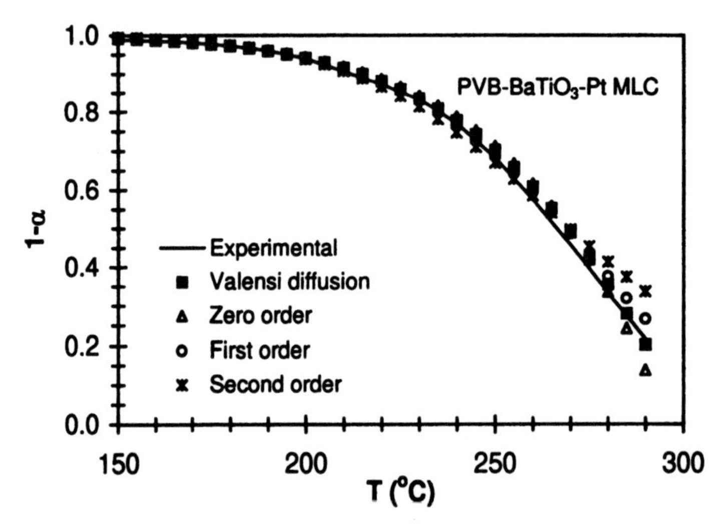

Figure 4. Comparison of an experimental TG weight loss profile (solid line) recorded at 10°C/min with the kinetics simulated by the four models using the parameters obtained by analyzing TGA data taken at 10°C/min.

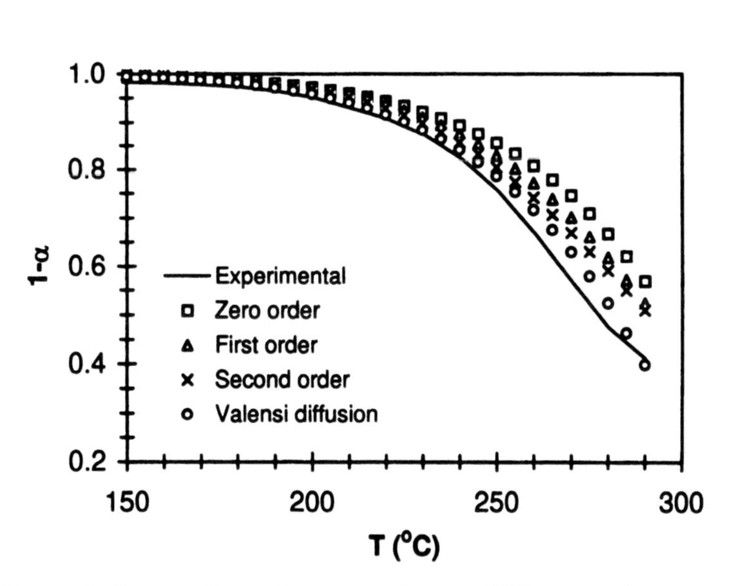

Figure 5. Comparison of an experimental TG weight loss profile (solid line) recorded at 20°C/min with the kinetics simulated by the four models using the parameters obtained by analyzing TGA data taken at 10°C/min.

CHARACTERIZATION OF THE SOL-GEL DERIVED PZT THICK FILMS ON METAL SUBSTRATES

Jinrong Cheng*, Wenyi Zhu, Nan Li and L.Eric Cross
Materials Research Laboratory
The Pennsylvania State University
University Park, PA 16803, USA

Zhongyan Meng
School of Materials Science and Engineering
Shanghai University
Shanghai 201800, P.R.China

ABSTRACT

The ferroelectric lead zirconate titanate (PZT) films with various thicknesses were deposited on commercial stainless steel (SS) substrates by using the modified 2-methxyethanol (2-MOE) based sol-gel spin-on procedures. Rapid sintering and quick cooling routes were utilized to reduce the diffusion and oxidation reactions in the crystallization process. A thin layer of $PbTiO_3$ (PT) was introduced between the PZT films and metal foils to provide optimal nucleation sites for the PZT films, and promote the PZT films were well crystallized at the low temperature of 550°C. The phase structure and dielectric properties of the PZT films were characterized in detail. A ferroelectric hysteresis loop with the remnant polarization of 35 $\mu C/cm^2$ and the coercive field of 99 kv/cm were obtained for the PZT films with PT seeding layer deposited on the SS substrates.

INTRODUCTION

Ferroelectric PZT films have attracted considerable attention for integrated actuator and ultrasonic applications [1,2]. Most PZT films are currently deposited on platinized silicon wafers due to the compatibility with Integrated Circuit Technology, however, limiting some potential applications. To deposit PZT films on metal substrates, such as stainless steel substrates or other alloy foils is of great interest in combining the excellent sensor and actuator properties of PZT films

* To whom correspondence should be addressed

with the easy integrability of these substrates into engineering systems. It will permit a broader range of applications of the PZT films. Nevertheless, only few papers report the PZT films deposited on metal substrates and give some attractable electrical properties [3,4].

Deposition of the PZT films onto metal substrates is usually hindered by high film growth temperature (typically above 600°C). The high temperature and reactive ambient during film growth promote interdiffusion and substrate degradation. Some metal oxide electrodes, such as RuO_2 and $SrRuO_3$, were often considered to hinder the diffusion between PZT films and platinized silicon substrates [5,6]. However, the preparation of metal oxide electrodes resulted in complex processing and high cost. The lead titanate (PT) thin films have the similar perovskite structure to the PZT films and the low crystallization temperature around 500°C, and can be prepared by using sol-gel methods. In order to lower the growth temperature of PZT films, it may be a nature choice using PT as a seeding layer. The idea proved feasible for PZT films deposited on platinized silicon substrates. The author once introduced PT seeding layer between the PZT films and NiTi alloy substrates [7]. However, as a seeding layer of PZT films, the PT thin films were not used on the SS substrate yet. This paper focuses on preparing the PZT films with improved dielectric and ferroelectric properties on commercial SS substrates by using the PT seeding layer and the low temperature annealing process.

Experimental Procedure

Both PZT and PT films were prepared by 2-MOE based sol-gel spin-on techniques respectively, the detail of which has been previously reported [7]. Precursors were prepared by the inverse mixing order (IMO) methods. The major starting materials include anhydrous lead acetate, zirconium n-propoxide and titanium tetrobutoxide, which were dissolved in 2-MOE solvents. The PZT precursor has the Zr/Ti ratio of 52/48. The mole concentrations of the PZT and PT are 0.65 M and 0.25 M respectively. 304 SS sheets (#8 polish) were ordered from Penn Stainless Products, Inc. PT seeding layer was first spin coated onto SS substrates and crystallized at 550°C for 0.2 hours. Then, the PZT films were repeatedly coated on the top of the PT thin films up to desired thickness. The final annealing was carried out at 500°C and 550°C for 0.5 hours. PZT films were also directly deposited on SS substrates and annealed at different temperatures of 550°C, 600°C and 650°C. Rapid sintering and cooling processes were used to reduce the diffusion effects at the interface of SS substrate-PZT films. After film deposition, the platinum top electrodes with 1.6 and 0.8 mm in diameter were sputtered on the PZT films through a mask. Phase structures of PZT films were identified using XRD techniques. The film thickness was examined by using

Alpha-step 500 profiler. Dielectric properties, leakage current densities and hysteresis loops were measured by using the HP 4274A LCR meter, Model 6517 electrometer/high resistance meter and the modified Sawyer-Tower circuit.

RESULTS AND DISCUSSION

Crack free PZT films with different thicknesses were coated onto SS substrates. The thickness of PZT films is linearly dependent on coating layers, as shown in Fig.1. The 3.427 μm-thick PZT films were prepared with 30 coating layers. PZT films were well combined with SS substrates. The SS substrate has larger thermal expansion coefficient than that of PZT films. In the cooling process, the mismatch of the thermal expansion coefficients made the PZT films burden pressing stress, which was beneficial to prohibiting PZT films from cracking.

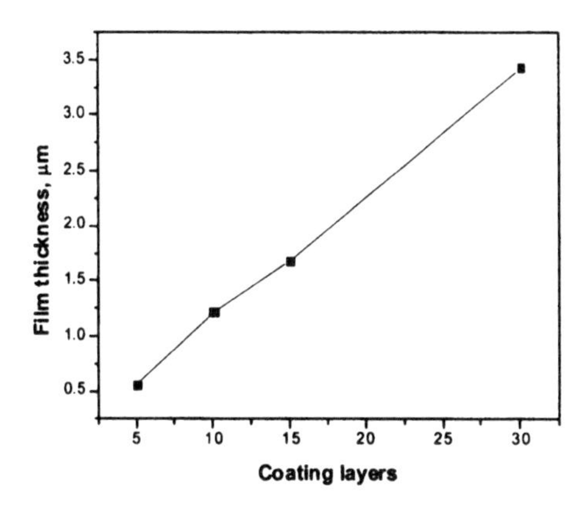

Fig.1: The dependence of the thick ness of PZT films on coating layers

Phase evolution of the PZT films on SS substrates: Fig.2 (a) shows the XRD patterns of PZT films deposited on SS substrates with a PT seeding layer. It is observed that the PZT films can be well crystallized after annealing at 550°C for 0.5 hours. All films consist of the perovskite phase with no detectable pyrochlore phase and other metal oxide phases. Three sharp peaks with high intensity occur at 2θ of 21.4531, 43.7006 and 30.6144 corresponding to the diffraction planes of (001), (002) and (101) for perovskite phases respectively. The 1.679 μm-thick PZT films reveal (00h) preferred orientations to some extent. However, the PZT films keep amorphous state with a few of pyrochlore phases after annealing at 500°C. Seen from Fig.2 (b), the XRD patterns of PZT films directly deposited on

SS substrates, the major phase of the PZT films are pyrochlore phase after annealing at 550°C for 0.5 hours. Nevertheless, though the crystallinity of the PZT films is somewhat improved due to increasing the annealing temperature to 650°C, the peaks are widened and pyrochlore phases are found indicating the deterioration of the film quality and substrates. In addition, the $PbZrO_3$ phases are found separated from PZT phases.

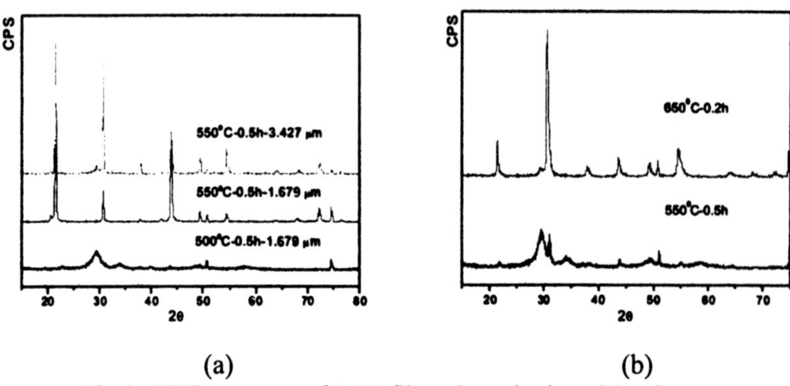

(a) (b)

Fig.2: XRD patterns of PZT films deposited on SS substrates (a) with PT seeding layer, (b) without PT seeding layer

In fact, it is difficult to directly coat PZT films onto SS substrate. Applying low annealing temperature, PZT films either were still in the amorphous state or had pyrochlore phases, which were easily stable in low temperatures. The high annealing temperature resulted in strong diffusion and oxidation reactions so that the Zr-rich phases were formed due to the easy oxidation of titanium element. The formation of perovskite PZT films is controlled by nucleation because the active energy of nucleation is much higher than that of growth [7]. The perovskite PT thin films have similar crystal structure with PZT, which can be crystallized at the low temperature of 500°C. Therefore, introducing the PT thin film on SS substrates before the deposition of PZT films provided easy nucleation sites so that the PZT films could nucleate easily and grow at low annealing temperature of 550°C with improved crystallinity. The PT seeding layer could also compensate the loss of titanium element prohibiting the separation of perovskite PZT phases.

Dielectric and ferroelectric properties: Fig.3 and Fig.4 give dielectric constants and dissipation of PZT films deposited on SS substrates with and without PT the seeding layer respectively. Dielectric constants decrease with increasing frequencies, however, increases with increasing annealing temperature. Under the frequency of 1 KHz, the room dielectric constant and dissipation of the

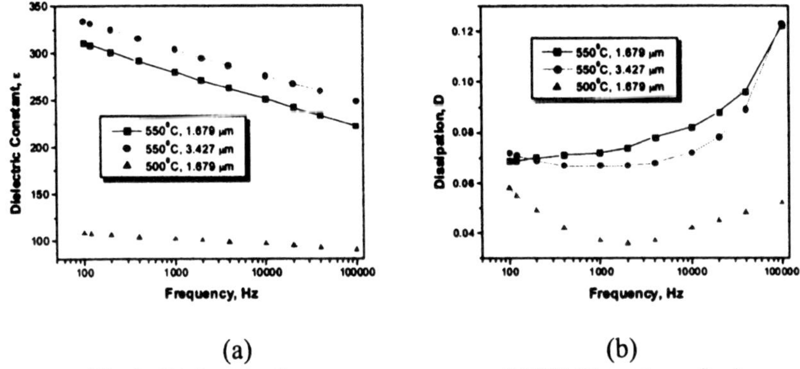

(a) (b)

Fig.3: Dielectric frequency spectrum of PZT films deposited on SS substrates with PT seeding layer
(a) Dielectric constant- frequency relationships
(b) Dissipation- frequency relationships

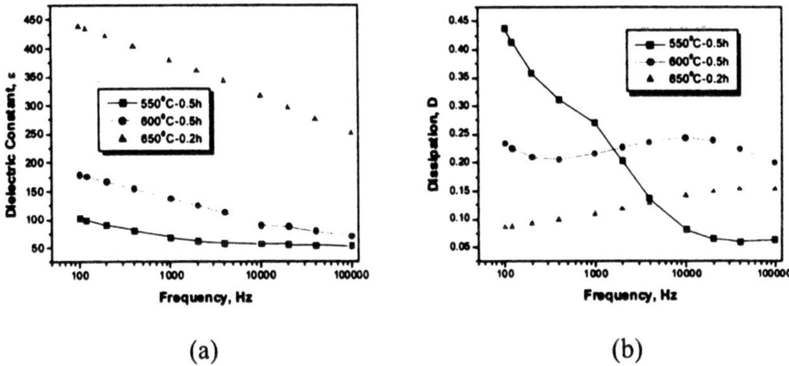

(a) (b)

Fig.4: Dielectric frequency spectrum of PZT films deposited on SS substrates without PT seeding layer
(a) Dielectric constant- frequency relationships
(b) Dissipation- frequency relationships

1.679 μm-thick PZT films with and without the PT seeding layer is 280, 0.072 and 68, 0.27 respectively. It proved that the PT seeding layer improved the dielectric properties of PZT films on SS substrates, which corresponding to the improvement of phase purity. The dielectric constants are not very high because that the PT thin films have low dielectric constant compared with PZT films. On the basis of the low-permittivity interface layer model [8], the measured relative permittivity ε of the film is given by

$$1/\varepsilon = 1/\varepsilon_f + (d_s / \varepsilon_s + d_i / \varepsilon_i)(1/d) \qquad (d >> d_s, d_I)$$

where ε_s, ε_f and ε_i are the relative permittivity of the surface layer containing fine grains on the top of the film, the intrinsic PZT film and the interface layer on the bottom electrode, respectively, d, d_i, and d_s are the thickness of the film, the interface layer and the surface layer. Even though the thickness of low-permittivity interface layer was much thinner, it would greatly reduce the effective dielectric constants. Therefore, the thickness of the PT films should be restricted as thin as possible.

The leakage current density of PZT films on SS substrates was shown in Fig. 5. PZT films still have low leakage current densities after coupled with SS substrates.With PT seeding layer, the leakage current density of PZT films is in the range of 10^{-9}~10^{-7} μC/cm^2 under drive field of 100 kv/cm, which is lower than that of PZT films without PT seeding layer.

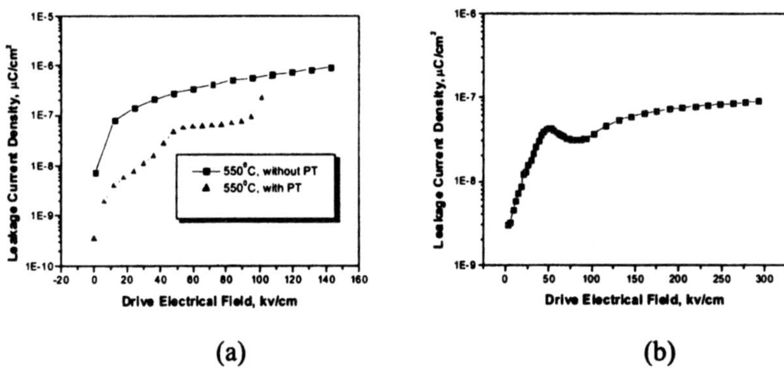

Fig.5: Leakage current density of PZT films deposited on SS substrates with different thickness (a) 1.679 μm, (b) 3.427 μm

For the 3.427 μm-thick PZT films, the leakage current densities are linearly dependent on electric field at low electric fields (Fig.5 b) indicating that ohmic conduction at the film-SS substrate interface is predominant factor in determining the current flowing through the metal-insulator-metal (MIM) structure. In this region, the film-SS substrates contact barrier limits current injection, thermally excited electrons are believed to be the major source of the current. A transition region is evident in which the current decreases. This may be due to electron trapping, which generates a depletion region of negative space charges, which lowers the effective electric field at the cathode (SS substrates) and, hence, reduces the injection current. However, this transition region diminishes as decreasing the PZT film thickness, as shown in Fig.5 (a). With further increasing the electric field beyond a critical onset level, electrons become free from the trap centers, a quadratic behavior of I-V characteristic occurs as shown. This mechanism is generally described as space-charge-limited conduction [9]. The I-V characteristics and the conduction mechanism of PZT films on metal substrate are similar to that on normal platinized silicon substrates.

Fig.6 (a) shows ferroelectric hysteresis loops with good shape for the PZT films deposited on SS substrates with PT seeding layers indicating that the excellent ferroelectric properties were obtained. The 1.679 μm-thick PZT films have well-saturated loops with the remnant polarization (Pr) and coercive field (Ec) of 35 $\mu C/cm^2$ and 99 kv/cm respectively. The value of Pr is even higher than that of some PZT films deposited on platinized silicon substrates. The large coercive field may be due that partial electrical volts were applied on the low permittivity interface layer, which reduce the effective fields applied on the PZT layers. The PZT films with large coercive field can endure the applications in relatively high electrical field. However, without PT seeding layer, it is hard to obtain good hysteresis loops for PZT films on SS substrates. As shown in Fig.7 (b), all loops have round end indicating the PZT films are much leaky. The dielectric dissipation was so high that no ideal hysteresis loops could be obtained for these PZT films directly deposited on SS substrates.

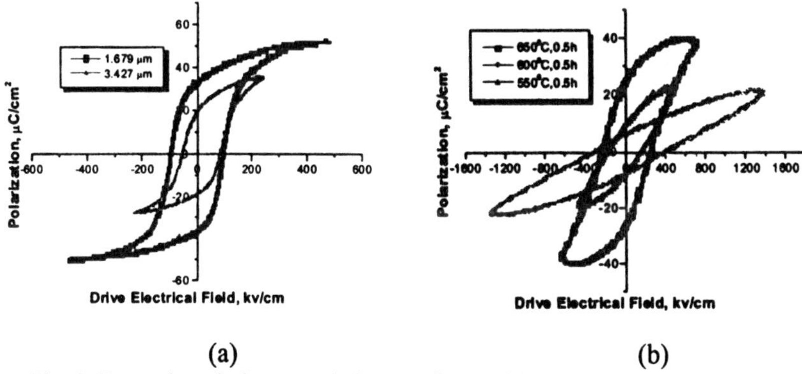

(a) (b)

Fig.6: Ferroelectric hysteresis loops of PZT films deposited on SS substrates (a) with PT seeding layer, (b) without PT seeding layer

CONCLUSIONS

Crack free PZT thick films with more than 3 μm in thickness were prepared on SS substrates. The PT seeding layer improved the crystallinity of PZT films at low annealing temperature of 550°C. Well-developed perovskite phases were obtained, and the diffusion and oxidation reactions between PZT films and SS substrates were effectively restricted. The dielectric dissipation of PZT films deposited on SS substrates was greatly reduced by introducing PT seeding layer, which contributed to the improved dielectric and ferroelectric properties of PZT films. Though the dielectric constants were not very high, it could satisfy some applications which high permittivity were not specially required. An almost ideal hysteresis loop with the remnant polarization of 35 $\mu C/cm^2$ and coercive field of 99 kv/cm revealed excellent ferroelectricity of the 1.679 μm-thick PZT films and its potential device applications.

ACKNOWLEDGEMENT

We are pleased to acknowledge support from the office of Naval Research (ONR) under contract No. N 00014-99-1-1011.

REFERENCE

[1]D.L.Polla,L.F.Francis, "Processing and Characterization of Piezoelectric Materials and Integration into Microelectro Mechanical System," *Annu. Rev. Mater. Sci.* 28, 563-97 (1998).

[2]David Liu, J.P.Mevissen, "Thick Layer Deposition of Lead Perovskite Using Diol-based Chemical Solution Approach," *Integrated Ferroelectrics* 18, 263-74 (1997).

[3]M. Giersbach, S. Seifert, D. Sporn, T. Hauke and H. Beige, "Piezoelectric Properties of PZT Thin Films on Metallic Substrates," Ferroelectrics, 241, 175-82 (2000).

[4]Q. Zou, H. E. Ruda and B. G. Yacobi, "Dielectric Properties of Lead Zirconate Thin Films Deposited on Metal Foils," Appl. Phys. Lett., 77 [7] 1038 (2000).

[5]Kuo-Shung Liu and Tzu-Feng Tseng, "Improvement of $(Pb_{1-x}La_x)(Zr_yTi_{1-y})_{1-x/4}O_3$ Ferroelectric Thin Films by Use of $SrRuO_3$/Ru/Pt/Ti Bottom Electrodes," Appl. Phys. Lett., 72 [10] 1182 (1998).

[6]J.H.Cho and K.C.Park, "Comparison of Epitaxial Growth of $PbZr_{0.53}Ti_{0.47}O_3$ on $SrRuO_3$ and $La_{0.5}Sr_{0.5}CoO_3$,"Appl. Phys. Lett., 75 [4] 549 (1999).

[7]Jinrong Cheng, Zhongyan Meng, "Synthesis and properties of PT/PZT thin film multilayers on shape memory alloys," Proc.9th CIMTEC-World Ceramics Congress and Forum on New Materials, in Advances in Sci. and Tech. 25, 61 (1999).

[8]K.Sumi, H.Giu, M.Hashimoto,"Thickness Dependence of Structural and Ferroelectric properties of Sol-gel $Pb(Zr_{0.56T}Ti_{0.44})_{0.90}(Mg_{1/3}Nb_{2/3})_{0.10}O_3$ Films", Thin Solid Films, 330, 183-189 (1998).

[9]Xixin Qu, "Thin Film Physics (in Chinese)," Shanghai Sci. and Technology Press, (1986) 124.

A STUDY ON HOT-PRESSED 0.3PZN-0.7PZT PIEZOELECTRIC CERAMICS

Yongli Xu and Donglu Shi
University of Cincinnati
497 Rhodes PO Box 210012
Cincinnati OH45221-0012

Shangping Li, Peng Wang and Shi Tian
Beijing University of Aero. & Astro.
37 Xueyuan Road
Beijing 100083, P.R. China

ABSTRACT

0.3PZN-0.7PZT piezoelectric ceramics were made by hot-press (HP) and annealing. SEM image shows that there is almost no grain growth during hot-press. However the grain size increases more than six times during annealing. X-Ray diffraction shows that two non-perovskite phases $Pb_3Nb_2O_8$ and pure Pb phase come out in the hot-pressed samples, which disappears after annealing at 850°C and 1250°C respectively. The piezoelectric coefficient d_{33} reaches maximum with the Pb content 1.5% less than the stoichiometry after having annealed at 1250°C for 7 hrs. With the optimized processing parameters, the piezoelectric properties are d_{33} =715 pC/N, K_p=0.65, tgδ=1.9%, ε_r =3457, and Q_m=50.1.

INTRODUCTION

In recent years, research on actuators and related materials, for example piezoelectric materials, is very active. The properties of piezoelectric ceramics, such as density, piezoelectric coefficient, electric fatigue, and toughness, are key factors for the long lasting devices. However low piezoelectricity, low density, high pores, and short life in high frequency electric fatigue frustrates the wide use of the conventional piezoelectric ceramics for actuators. Although the relaxation based ferroelectric's single crystals show superior electric properties over polycrystalline ceramics[1] and great potential for the future devices, the high cost, small size, and technical problems prohibit their extensive applications.

Compared with conventional sintering process, hot-press possesses many advantages such as low sintering temperature, high density, good homogeneity of piezoelectricity, controllable grain size and high toughness, and good electric fatigue properties[2]. Recent research results have revealed that the microstructure dominates the fatigue properties[3,4]. The smaller the grain is, the better the fatigue properties will be. As of the small grain size, hot-pressed PZT-PYW ceramics possess better piezoelectricity and higher toughness[5,6]. In this paper, the piezoelectric properties of PZN-PZT system were studied using hot-press method for the potential application in solid-state scanner.

1. EXPERIMENTAL PROCEDURE

The piezoelectric ceramics used in this study have the following formulation: $Pb_{0.97}La_{0.03}(Zn_{1/3}Nb_{2/3})_{0.3}Zr_{0.37}Ti_{0.33}O_3$. Reagent grade of Pb_3O_4, TiO_2, La_2O_3, $5ZnO{\cdot}2CO_2{\cdot}4H_2O$, and Nb_2O_5 were used as the row materials. Superfine ZrO_2 powder with surface area $30m^2/g$ was made by chemical method. After weighing, the powder was wet ball milled for 10hrs. in zirconia media. The slurry was dried in an oven at 120°C overnight. Then pressed into one large pellet and calcined at 850°C for 3hrs. The calcined pellet was broken and wet ball-milled for 24hrs. After dried the powder was ready for hot-press. The powder was first compacted into pellets 20mm in diameter and 10mm in height in a steel die at pressure of 50MPa. The pellet was then transferred into the graphite die and put into the hot-press furnace with the following process parameters: temperature: 900~1000°C, dwell time 1hr., pressure: 7~49MPa. The hot-pressed samples were annealed at various temperatures and time.

The hot-press pellets were cut into 1mm thickness, with silver paint applied on both sides, calcined at 720°C for 20min, and then poled at 120°C with 30KV/cm for 15min in silicon oil. The piezoelectric properties were measured 24hrs after the poling.

The density of sintered bodies was measured by immersion method. Microstructure was observed using a Scanning Electron Microscope (SEM). The crystal structure was identified by X-Ray Diffraction (XRD).

2. RESULTS AND DISCUSSION

2.1 Hot-Press Density Vs. Pressure

In Fig.1, the density of PZN-PZT ceramics is shown as a function of HP pressure. There are almost two linear parts in low and high-pressure stages separated by the pressure around 20MPa. This may indicate different densification mechanisms. Density of $8.06g/cm^3$, which is 99.8% of the theoretical density, was obtained by holding the samples for 1hr. with 49Mpa at 950°C, which is much higher than conventional sintering samples (about 95%). High temperature annealing normally reduces density insignificantly (from $8.06g/cm^3$ to $8.04g/cm^3$), due to phase transformation, oxygen diffusion into the sample, and variation of unit cell lattice parameter.

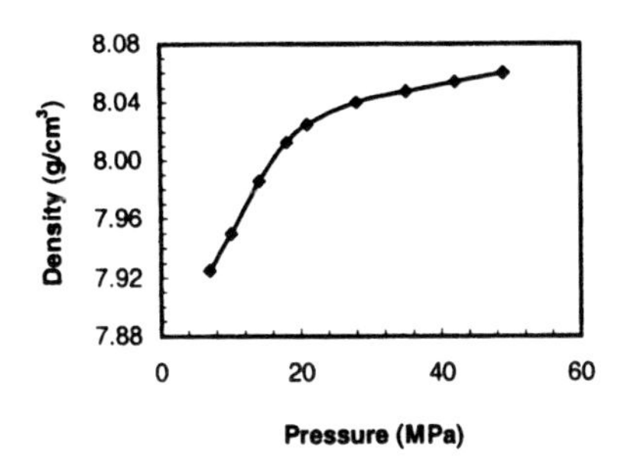

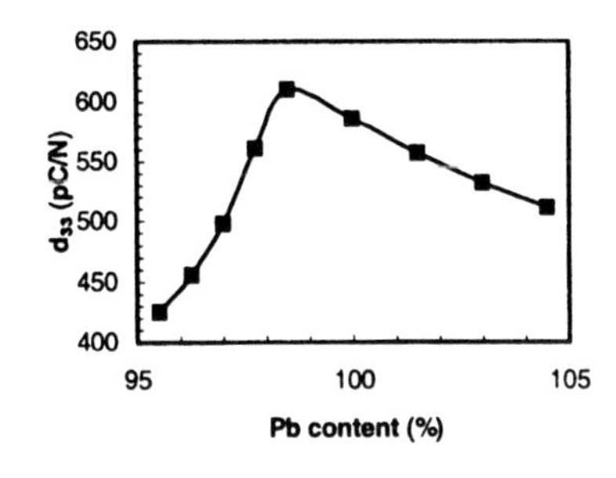

Fig. 1 Density of PZN-PZT vs. HP pressure

Fig.2 d_{33} vs. content of lead

2.2 The Effect of Pb Content on d_{33}

To study the effect of lead we changed the Pb_3O_4 content from 95.5% to 104.5%. The influence of Pb content on piezoelectric coefficient (d_{33}) at the same conditions (hot pressed: 950°C/42MPa/1hr.; annealing: 1240°C/5hrs.) is shown in Fig.2. The maximum d_{33} value is obtained when Pb content is 98.5%. This result coincides with that of the traditional sintering method. When the Pb content deviates from 98.5%, the piezoelectric coefficient decreases, however the decrease is more rapid in the lead deficiency side.

The effect of Pb content on piezoelectric properties is attributed to the formation of Pb^{2+} vacancies. In the perovskite structure, A-site vacancies are not only favorable for decreasing the *c/a* ratio, but also reducing crystal cell and enhancing bonding energy. When the amount of A-site vacancies is small, decreasing the *c/a* ratio will play a leading role, which makes piezoelectric ceramics more easily poled and piezoelectric properties increased. When the amount of A-site vacancies is excessive, enhancing bonding energy will dominate and space charges, which is against the motion of domain walls, thereby reducing the piezoelectric properties. The amount of Pb^{2+} vacancies continually increases with decreasing Pb content. When the amount of Pb^{2+} vacancies is small, piezoelectric properties continually increase up to 98.5%, 1.5% less than stoichiometry Pb value. Excessive Pb^{2+} vacancies will result in the decreasing of piezoelectric properties.

2.3 The Effect of Annealing Temperature on d_{33} and Microstructures

Fig.3 shows the d_{33} value as a function of annealing temperature (hot pressed: 950°C/42MPa/1hr., annealing time 5hrs.). The d_{33} value increases with increasing annealing temperature up to 1250°C, and then decreases rapidly. Fig.4 shows the d_{33} value as a function of annealing time (hot pressed: 950°C/42MPa/1hr.; annealing temperature: 1250°C). The d_{33} value reaches the maximum with the annealing time of 7hrs., which is 715pC/N. Any deviation will result in a decrease of d_{33} value. 7hrs seems too long compared with the traditional sintering parameters: 1250°C/1.5hr. However considering the high densification of the substrate, more time is needed for the maturing of the crystal structure and diffusion.

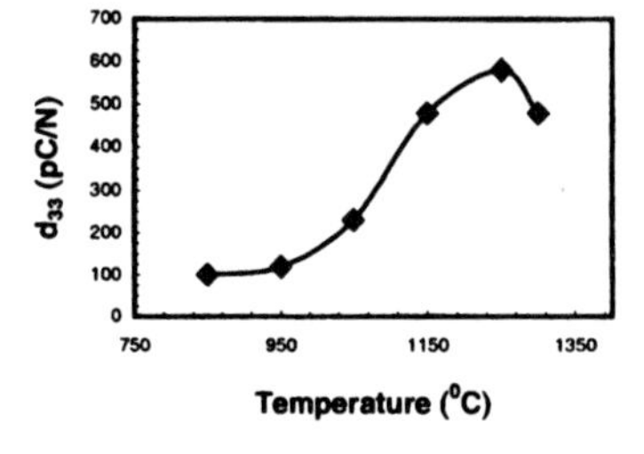

Fig. 3 d_{33} vs. annealing temperature

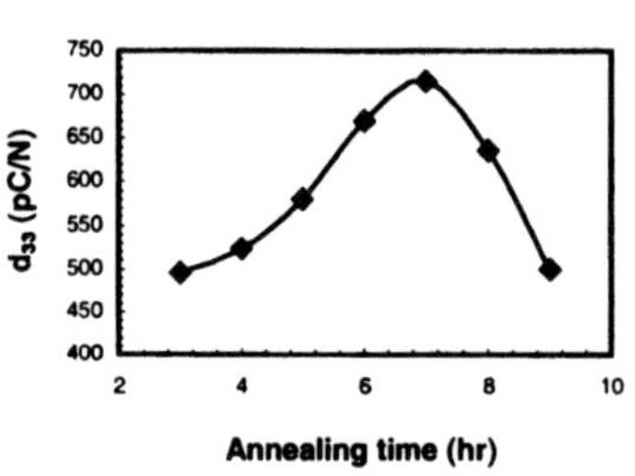

Fig. 4 d_{33} vs. annealing time

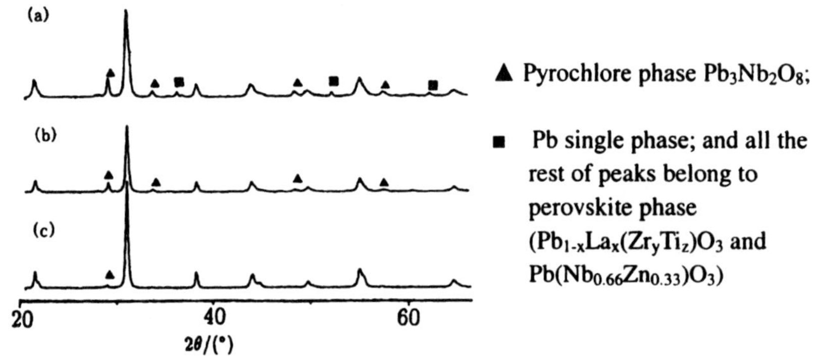

▲ Pyrochlore phase $Pb_3Nb_2O_8$;

■ Pb single phase; and all the rest of peaks belong to perovskite phase $(Pb_{1-x}La_x(Zr_yTi_z)O_3$ and $Pb(Nb_{0.66}Zn_{0.33})O_3)$

Fig.5 XRD pattern of hot-press samples
(a) The sample without annealing (b) Sample annealed at 850°C (c) Sample annealed at 1250°C.

Fig.5 shows the XRD pattern of the samples annealed at various temperatures. Pyrochlore phase ($Pb_3Nb_2O_8$), pure Pb phase, and perovskite phase $(Pb_{1-x}La_x(Zr_yTi_z)O_3$ and $Pb(Nb_{0.66}Zn_{0.33})O_3)$ are observed for the samples without annealing. The counts (peak intensity) for the pyrochlore phase decrease with increasing annealing temperature and become reduced when annealed at 1250°C/6hrs.

Precipitation of pure Pb phase during hot-press is probably due to the lower chemical potential of Pb in pure Pb phase than in perovskite phase at high pressure and temperature. The formation of pure Pb phase and pyrochlore phase, which changed the composition in the substrate, have resulted in a decrease in the perovskite phase. In addition, pyrochlore phase is not ferroelectric. So the presence of pyrochlore phase lowers piezoelectric properties significantly. The more the pure Pb phase and pyrochlore phase content, the worse the piezoelectric properties will be. Increasing annealing temperature, pyrochlore phase and some of the pure Pb phase transferred into perovskite phase, which increases d_{33} significantly.

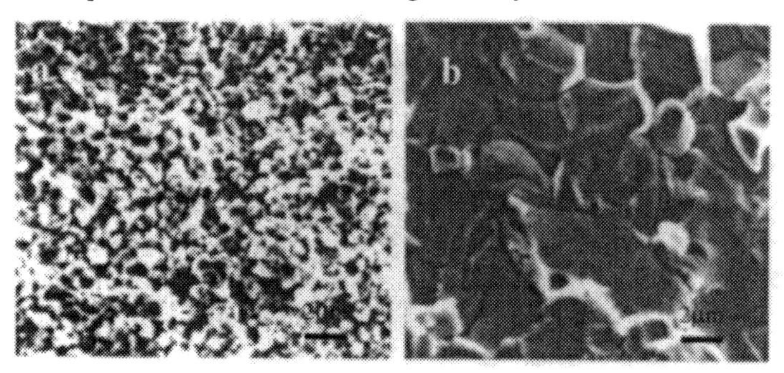

Fig.6 Microstructures of Hot-Pressed PZN-PZT Samples
(a) Annealed at 850°C for 3hrs. (b) Annealed at 1250°C for 3hrs.

Moreover, grain size has a significant influence on properties of piezoelectric ceramics. It has been reported that with an increase in grain size, the value of K_P, d_{33} and d_{31} increase while a dielectric constant decreases[6]. Fig.6 shows the coarse microstructure of PZN-PZT ceramics with rising annealing temperature. The average grain sizes increase from 0.5~1μm to 3μm when the annealing temperature rise from 850°C to 1250°C. Prolonging annealing time at the same temperature will further increase grain size, which increases piezoelectric properties but lowering the toughness and fatigue properties.

3. CONCLUSIONS

1) Under experimental conditions the density of HP PZN-PZT increases almost linearly under 20MPa, after which increases slowly. A relative density of 99.8% has been obtained with the pressure of 49PMa.
2) The highest d_{33} value has been obtained with Pb content of 98.5%, 1.5% less than the stoichiometry composition.
3) When Annealing the HP PZN-PZT sample at the normal sintering temperature (1250°C) the d_{33} value reaches its maximum at 7hrs., which is about three times longer than that of the normal sintering time.

4) With Pb content of 98.5%, annealing temperature at 1250°C and annealing time 7 hours, the piezoelectric properties of 0.3PZN-0.7PZT have reached at d_{33}=715 pC/N, K_P=0.65, tgδ=2.9%, ε_r=3457, Q_m=50.1.

4. REFERENCE

[1]S.E. Park and T.R. Shrout, Ultrahigh Strain and Piezoelectric Behavior in Relaxor Based Ferroelectric Single Crystals, J. Appl. Phys., 82[4]: 1804～1811 (1997)

[2]L. Biaorong, Technology Principle of Electric Ceramics, Edited by HUA ZHONG Technical College, 1986,7

[3]Z.W. Zhang and R. Raj, Influence of Grain Size on Ferroelastic Toughening and Piezoelectric Behavior of Lead Zirconate Titanate, J.Am Ceram. Soc., 78[12]: 3363-68 (1995)

[4]Q. Jiang, E.C. Subbarao and L.E. Cross, Grain Size Dependence of Electric Fatigue Behavior of Hot Pressed PLZT Ferroelectric Ceramics, Acta Metall. Mater., 42[11]: 3687-3694 (1994)

[5]S.J. Yoon, J.H. Moon, and H.J. Kim, Piezoelectric and Mechanical Properties of $Pb(Zr_{0.52}Ti_{0.48})O_3$-$Pb(Y_{1/3}W_{2/3})O_3$(PZT-PYW) Ceramics, Journal of Materials Science 32(1997), 779-782

[6]M. Kiyohara and K.I. Katoh, Grain Size Dependence of Piezoelectricity and Actuator Characteristics For Piezoelectric Ceramics, Journal of The Ceramic Society of Japan, 102:548-553 (1994)

RARE EARTH METAL DOPING EFFECTS ON THE PIEZOELECTRIC PROPERTIES OF $Pb(Zr,Ti)O_3$-$Pb(Mn,Sb)O_3$ CERAMICS

Yongkang Gao and Kenji Uchino
International Center for Actuators and Transducers,
Materials Research Institute, The Pennsylvania State University,
University Park, PA 16802, USA
and
Dwight Viehland
Naval Seasystems Command, Newport, RI 02841, USA

ABSTRACT

Improved piezoelectric materials with higher vibrational velocities are needed to meet the demands of advanced high power electromechanical applications. The Ce, Eu and Yb doping effects on the piezoelectric properties of $Pb(Zr,Ti)O_3$-$Pb(Mn,Sb)O_3$ ceramics have been investigated in this paper. Compared to the base $Pb(Zr,Ti)O_3$-$Pb(Mn,Sb)O_3$ ceramic, under high level driving condition, the maximum vibration velocity has been significantly increased by Ce, Eu or Yb doping. A new high power material with 2 at.% Yb doping has been developed, and its maximum vibration velocity was as high as 1.0 m/s under an electric field of 10 kV/m (rms value), which is about 1.6 times higher than that of the base material.

Keywords: high power, piezoelectric, maximum vibration velocity, doping, $Pb(Zr,Ti)O_3$-$Pb(Mn,Sb)O_3$, rare earth metals.

INTRODUCTION

The high-power characteristics of piezoelectric materials have recently been investigated for device applications in ultrasonic motors, piezoelectric actuators and piezoelectric transformers[1,2]. To achieve the requirements of these high-power applications, higher vibrational levels and velocities (v_0) at lower AC electric fields are required. Currently, the practical maximum vibrational velocity is restricted by heat generation, because above a certain vibrational level increasing hysteretic effects result in thermal instability. Consequently, approaches to enhance the maximum vibration velocity are an important issue, which to date has proven difficult to achieve. Heat generation under drive can be represented as a function of vibration velocity and other material constants.

The vibrational velocity v_o is proportional to the product of the mechanical quality factor Q_m and the electromechanical coupling factor k, i.e., $v_0 \propto Q_m k_{31}$ for a rectangular plate working under d_{31} mode.[3-5] Consequently, for high-power applications, a piezoelectric material is needed which has simultaneously both high Q_m and k_{31} values. Previous investigations of $Pb(Zr,Ti)O_3$ (PZT) have reported that v_0 is increased by lower valent substituents (for example, Fe on the B-site), whereas it is decreased by higher valent ones (for example, Nb on the B-site)[6]. However, practically, the development of a material with significantly higher values of v_0 has been in vain, as previous investigations have shown that either Q_m or k_{31} can be enhanced only at the expense of the other. Generally, substituents can be categorized into three classifications[7-9]: lower valent (effective acceptors), higher valent (effective donors), and isovalent. substituents with lower valence introduce "hard" piezoelectric characteristics, while higher valent substituent induce "soft" ones. “Hard” piezoelectrics have higher Q_m values, but lower k_{31} values. On the other hand, “soft” piezoelectrics have lower Q_m values, but higher k_{31} values.

Hagimura et al. have reported that RE elements smaller than 0.94 Å increase the coercive field (E_c) and field-induced strain (ε) of PZT ceramics[10]. No reports have been published concerning the effects of rare earth species on the piezoelectric vibration velocity. In this paper, Ce, Eu and Yb have been chosen as dopants, which has bigger (1.03 Å), close to (0.95 Å) and smaller (0.85 Å) ionic size than that critical radius, respectively. The ionic radii of Ce^{3+},

Eu^{3+} and Yb^{3+} are smaller than that of Pb^{2+}, but bigger than any of the various B-site species. Thus, no preferred site-occupancy for Ce^{3+}, Eu^{3+} and Yb^{3+} would seemingly exist, rather substitution on both site might be expected.

Pseudo-ternary crystalline solutions of $Pb(Zr,Ti)O_3$-$Pb(Mn,Sb)O_3$ have been reported to have significantly higher electromechanical coupling factors, higher mechanical quality factors and higher maximum vibration velocities than PZT.[11,12] In this system, a maximum vibration velocity (defined by an rms value which raises the temperature to 20°C above room temperature) of 0.6 m/s has been found.

The work presented in this paper focuses on optimizing the vibration velocity of $Pb(Zr,Ti)O_3$-$Pb(Mn,Sb)O_3$. The results demonstrate a new promising material for high power application with a maximum vibration velocity of 1.0 m/s.

EXPERIMENTAL

We have chosen a base composition of $0.90Pb(Zr_{0.52}Ti_{0.48})O_3$-$0.10Pb(Mn_{1/3}Sb_{2/3})O_3$, following a previous publication[12]. The following RE^{3+} modified compositions were synthesized by conventional mixed oxide processing: (i) $0.90Pb(Zr_{0.52}Ti_{0.48})O_3$-$0.10Pb(Mn_{1/3}Sb_{2/3})O_3$-x at.% Eu (x = 0, 2, 4, 6), (ii) $0.90Pb(Zr_{0.52}Ti_{0.48})O_3$-$0.10Pb(Mn_{1/3}Sb_{2/3})O_3$-y at.% Yb (y = 0, 2, 4, 6) (iii) $0.90Pb(Zr_{0.52}Ti_{0.48})O_3$-$0.10Pb(Mn_{1/3}Sb_{2/3})O_3$-z at.% Ce (z = 0, 4, 6). The sintered samples were cut into rectangular plates of dimensions 42mm × 7mm × 1mm. Gold sputtering was used to deposit electrodes on both surfaces. Then, the specimens were poled under 2.5 kV/mm for 15 min at 130°C in silicon oil. All electrical measurements were carried out about 36 h after poling.

The electromechanical properties were determined by two methods: (i) by impedance spectra under a low-level constant-voltage drive using an HP4194 impedance analyzer, and (ii) by impedance spectra under various vibration velocities using a constant current drive method[4]. The vibration velocity v_0 was measured using laser doppler vibrometer (LDV) models OFV-3001 and OFV-511 (Polytec PI). Temperature rise was determined by a thermocouple, which was put at the center of the vibrating sample (i.e., the nodal point).

RESULTS AND DISCUSSION

Low field behavior

Figures 1(a) shows Q_m as a function of x (atomic ratio) for Yb substituent. Yb substituent increased Q_m significantly, from ~1150 to ~1350 with increasing x, showing a maximum near x=0.04 and then decreased slightly upon further increment of x. Figure 1(b) shows the dielectric constant as a function of x (atomic ratio) for Yb subsituent in PZT-PMS. A significant decrease in the dielectric constant K was observed with increasing Yb concentrations.

Previous investigations have demonstrated that PZT-PMS has "hard" piezoelectric characteristics[11,12)]. The data presented in Figs. 1(a) and 1(b) demonstrates that Yb substituent increases the degree of "hard" characteristics, as mechanical quality factor was increased and dielectric constant was decreased. "Hardening" of piezoelectric behavior is believed due to the interaction between domain wall and acceptor impurity-oxygen vacancy dipoles[13)]. Accordingly, an internal bias field would pin ferroelectric domains, decreasing their contribution to the total permittivity and loss factors.

Figure 1(c) shows k_{31} as a function of Yb concentration. Upon increment of the Yb concentration between 0 and 4 at.%, k_{31} increased by 39%. Correspondingly, Fig. 1(d) shows that the absolute value of d_{31} increased with Yb concentrations. The absolute value of d_{31} increased from 44 to 53 pC/N upon Yb substitution. This result implies that Yb doping has "softening" side. The investigation of Ce and Eu substituents shows a similar combinatory "hardening" and "softening" effects on piezoelectric properties PZT-PMS.

Another intriguing behavior of Yb doping is shown in 1(e). With 2 at.% Yb doping, the elastic compliance of the $Pb(Zr,Ti)O_3$-$Pb(Mn,Sb)O_3$ ceramics has been decreased first, i.e., the elastic constant stiffened, then, it went up to 9.2×10^{-12} m^2/N, with further increasing doping concentration.

In general, substituents that increase the piezoelectric constant and electromechanical coupling coefficient are said to induce "soft" piezoelectric characteristics[7-9)]. Accordingly, the

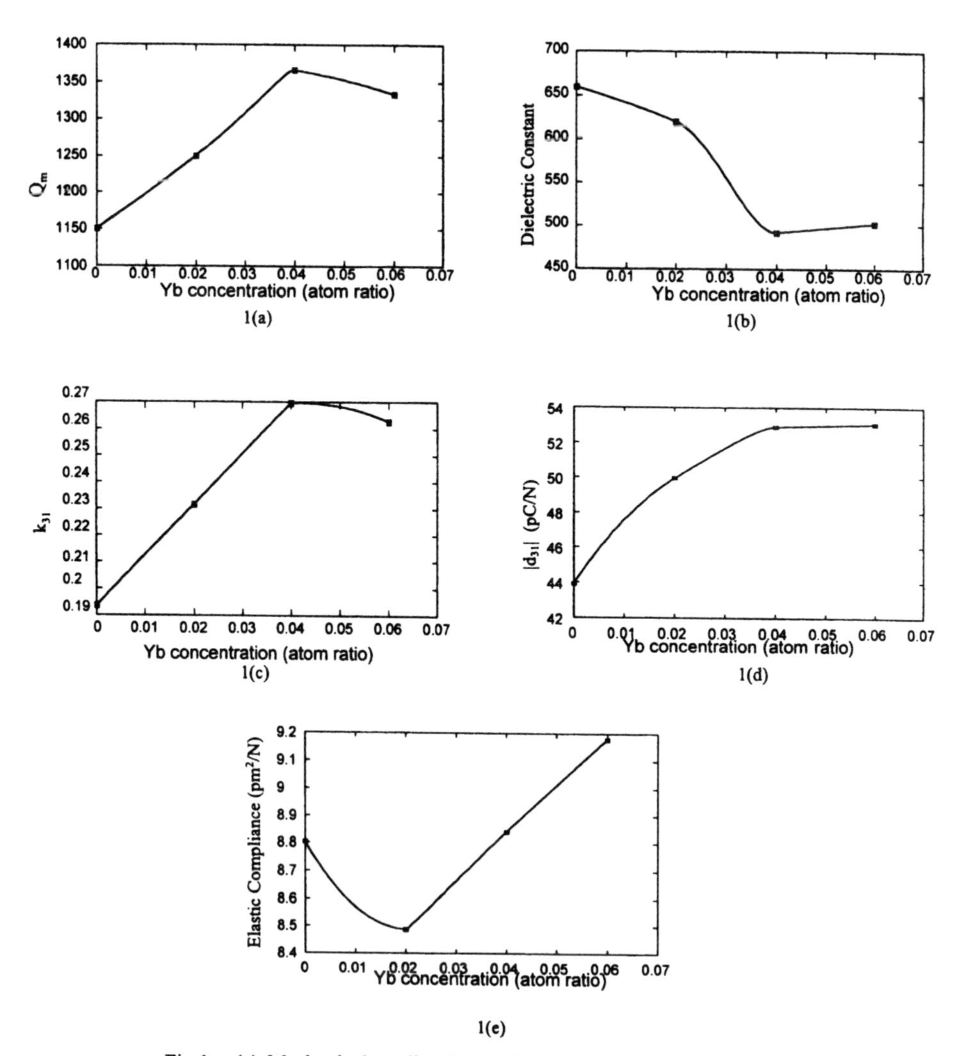

Fig.1 (a) Mechanical quality factor Qm (b) Dielectric constant K (c) Coupling factor k_{31} (d) Piezoelectric constant d_{31} (e) Elastic compliance of 0.90PZT-0.10PMS with 0,2,4,6 at.% Yb.

increase of d_{31} and k_{31} upon Ce, Eu and Yb substitution indicate that these species induce some degree of "soft" piezoelectric characteristics. The experiments indicate an unique effect of RE^{3+} substituents in PZT-PMS, as all of Ce^{+3}, Eu^{+3} and Yb^{+3} introduce combinatory "hard" and "soft" piezoelectric characteristics. Here, one possible explanation to understand these particular characteristics is proposed as following. "Hard" piezoelectric characteristics are believed to be due to pinning of domains by dipolar fields of defect complexes[13,14,15)]. Defect complexes form because of enhanced transport due to charge compensation by oxygen vacancies[13,14,15)]. "Soft" piezoelectric characteristics are believed to be due to defects of the random-field type[13,14,15)], as originally proposed by Imry and Ma[16)]. Combinatory "hard" and "soft" characteristics would require that defects are non-randomly distributed as extended structures (i.e., "hard") and at the same time exist as randomly quenched-in point defects (i.e., "soft"). These are contradictory situations, which at least in simple systems can not be simultaneously satisfied.

The results of this section demonstrate the importance of defect engineering of domain stability. However, to more precisely identify the structure of the defect complexes responsible for property optimizations, local structural and chemical probes are necessary.

High field properties

To have the highest vibration velocity possible, it is essential to have a combination of both high Q_m and high k_{31}. However, to achieve simultaneous high values of Q_m and k_{31} is against the conventional concepts of "hard" and "soft" piezoelectric characteristics. But, the results in the previous section demonstrate the existence of such combinatory characteristics in RE^{3+} modified PZT-PMS. Consequently, the high field properties of these materials were investigated.

Figure 2 summarizes the results of v_o ($\Delta T=20^oC$) for all Ce, Eu and Yb modified PZT-PMS as a function of x. From this figure, significant enhancements in v_o ($\Delta T=20^oC$) can be seen upon substitution with either element. With 2 at.% Yb doping, the rms vibration velocity has been increased from 0.60m/s (x=0 at.%) to

1.0m/s. It is also clear that Yb produces the highest values and at significantly lower concentrations. At higher concentrations, v_o ($\Delta T=20^oC$) decreased for all of Ce, Eu and Yb substituted materials.

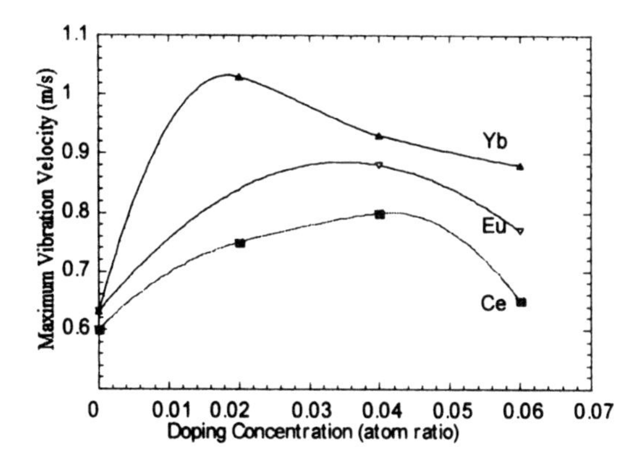

Fig.2 The maximum vibration velocity of PZT-PMS-Yb , PZT-PMS-Eu and PZT-PMS-Ce

SUMMARY

The effects of Ce, Eu and Yb substitutents on the piezoelectric properties of $Pb(Zr,Ti)O_3$-$Pb(Mn, Sb)O_3$ ceramics have been investigated. The following findings can be stated.

1) Combinatory "hard" and "soft" characteristics are found upon Eu and Yb substitution in PZT-PMS. Eu or Yb result in significant increases in Q_m, d_{31}, and k_{31}.

2) Under high drive level conditions, v_o has been significantly increased by Ce, Eu or Yb substitution, relative to that of the base PZT-PMS ceramic. The v_o seems to increase with decreasing the RE ionic size.

ACKNOWLEDGEMENT

This work was supported by the office of Naval Research through Contact No. N00014-99-J-0754.

REFERENCE

1) K. Uchino, *Piezoelectric Actuators and Ultrasonic Motors*, Kluwer Academic Publishers, Norwell, (1996).

2) K. Uchino, *Ferroelectric Devices*, Marcel Dekker, New York, (2000).

3) K. Uchino, H. Negishi and T. Hirose, Proc. Int. Conf. FMA-7, Jpn. J. Appl. Phys. 28 (1989) Suppl. 28-2, p.47.

4) K. Uchino, J. Zheng, A. Joshi, Y. H. Chen, S. Yoshikawa, S. Hirose, S. Takahashi and J. W. C. DE Vries, J. Electroceram. 2 (1998) 33.

5) Y. H. Chen, S. Hirose and K. Uchino, Jpn. J. Appl. Phys. 39 (2000) 4843.

6) S. Takahashi and S. Hirose, Jpn. J. Appl. Phys. 31 (1992) 2422.

7) D. Berlincourt and H. Krueger, J. Underwater Acoust. 15 (1965) 266.

8) D. Berlincourt and H. Jaffee, *Physical Acoustics*, ed. W. Cady (Academic Press, New York, 1964) vol. 1.

9) B. Jaffe, W. Cook and H. Jaffe, *Piezoelectric Ceramics* (Academic Press, London and New York, 1971).

10) A. Hagimura, M. Nakajima, K. Miyata and K. Uchino, Proc. Int. Symp. Applied Ferroelectrics 90 (1991) 185.

11) T. Ohno, N. Tsubouchi, M. Takahashi, Y. Matsuo and M. Akashi, Tech. Rep. IEICE J. (1972) US71-37.

12) S. Takahashi, Y. Sasaki, S. Hirose and K. Uchino, Mater. Res. Soc. Symp. Proc. 360 (1995) 305.

13) Q. Tan, Z. Xu and D. Viehland, J. Mater. Res. 14 (1999) 465.

14) Q. Tan, J.F. Li and D. Viehland, Philos. Mag. B 76 (1997) 59.

15) Q. Tan, PhD Dissertation, University of Illinois, Urbana, IL, USA, (1998).

16) Y. Imry and S. Ma, Phys. Rev. Lett. 35 (1975) 1399.

STUDIES ON DIELECTRIC BEHAVIOR OF $Ni_{0.8}Zn_{0.2}Fe_2O_4$ PROCESSED THROUGH NOVEL TECHNIQUES

R.V.Mangalaraja,S.Ananthakumar,P.Manohar and F.D.Gnanam
Centre for Ceramic Technology
Anna University
Chennai - 600 025, India

ABSTRACT

Dielectric behavior is one of the important properties of ferrites and depends on the processing conditions. Ni-Zn ferrite having composition of $Ni_{0.8}Zn_{0.2}Fe_2O_4$ was processed through various novel techniques such as flash combustion, citrate-gel decomposition and microwave processing techniques. The precursors were calcined at 900°C. XRD analysis confirmed the formation of ferrite phases. The fabricated samples were sintered in the temperature range 1150 –1350°C. Their dielectric properties such as dielectric constant and dielectric loss as the function of frequency range from 10 KHz to 13 MHz were studied. The densification and microstructural features were also studied. The properties were compared with ferrite prepared from the conventional solid-state route for the same composition. It is observed that the dielectric constant, ε'and dielectric loss, tanδ obtained for the ferrites prepared through these novel techniques possess lower value than that of the ferrites prepared by conventional ceramic method. The low value of dielectric constant and dielectric loss makes these ferrite samples suited for use at high frequency applications.

INTRODUCTION

Ni-Zn ferrites are important electronic ceramics which are used in electronic devices, suited for high frequency applications in the telecommunication fields and Ni-Zn ferrite have been commercially used in radio frequency circuits, high-quality filters, rod antennas, transformer cores, read/write heads for high speed digital tape and operating devices, which cannot be replaced by any other magnetic element because ferrites are more stable, relatively inexpensive, easily manufactured, good magnetic properties, low dielectric loss and high electrical resistivity. Several papers describe ferrite as a very structure sensitive material; its properties severely depend on differences in the manufacturing process. It is not easy to produce ferrites with excellent properties and high homogeneity. Therefore, extremely high quality control is necessary for their manufacture.[1-4] Nickel-Zinc ferrites are usually prepared by the conventional ceramic method. The wet chemical methods of preparation seem to offer a better alternative, overcoming the drawbacks of the conventional ceramic method.

In the present work, the wet chemical methods such as flash combustion[5,6] citrate-gel decomposition[7-9] and microwave technique[10] were used for the preparation of Ni-Zn ferrites and their dielectric properties were studied. Dielectric behavior is one of the most important properties of ferrites which is depend on the preparation conditions, sintering time and temperature, chemical composition , type and quantity of additives. The variations in dielectric constant with respect to various processing conditions are presented. The dielectric behavior have been studied for Ni-Zn ferrites prepared by conventional ceramic method.[11-15] The authors aimed to study the dielectric behavior of Ni-Zn ferrite system prepared using novel techniques and they were compared with the conventional method.

EXPERIMENTAL DETAILS

Ni-Zn ferrites of the composition $Ni_{0.8}Zn_{0.2}Fe_2O_4$ were prepared in the form of discs of thickness t and cross-sectional area A using novel techniques as mentioned earlier. The methods of preparation of ferrites were explained elsewhere.[5-10] The XRD analysis taken on Shimadzu Corporation Japan, Model XD-D1, Cu-kα X-Ray diffractometer, shows (Fig.1) the studied Ni-Zn ferrite samples have the single phase cubic spinel structure.[6,9]

The dielectric studies of silver paste coated on polished pellets were calculated from the capacitance measurements made on a Hewlett Packard 4192-A Impendence analyzer in the frequency range 10 KHz to 13 MHz. The values of dielectric constant(ε)' and dielectric loss(tanδ) were measured simultaneously as the function of frequency. The microstructure photographs of the fractured surfaces of the sintered samples were recorded on a Leica Stereo Scan 440 Scanning Electron Microscope(SEM).

RESULTS AND DISCUSSION

The variations in dielectric constant and dielectric loss as the function of frequency of Ni-Zn ferrites prepared by flash combustion, citrate gel-decomposition, microwave processing and conventional ceramic method are shown in Fig 2-9. Dielectric dispersion in ferrites can be explained on the basis of space change polarization which is a result of the presence of higher conductivity phases (grains) in the insulating matrix (grain boundaries) of a dielectric causing localized accumulation of change under he influence of an electric field. Since an assembly of space charge carriers in a dielectric requires finite time to line up their axes parallel to an alternating electric field, if the frequency of the field reversal increases, a point will be reached when the space charge carriers can not keep up with the field and the alternation of their direction lags behind that of the field. This results in the reduction in the dielectric constant of the material.[14,17] Fig.2-5 shows the variation of dielectric constant with frequency, observed in the range 10 KHz to 13 MHz for the samples prepared by various techniques and sintered in the temperature range 1150° C – 1350° C. Both dielectric constant and dielectric loss are gradually decrease as the frequency increases from 10 KHz to 5 MHz for the ferrites prepared through novel techniques. Dielectric constants of the ferrites prepared by novel techniques are observed to be quite low, between 40-60 for flash combustion, 50-70 for citrate gel-decomposition and 50-85 for microwave processing techniques. The samples prepared by conventional ceramic

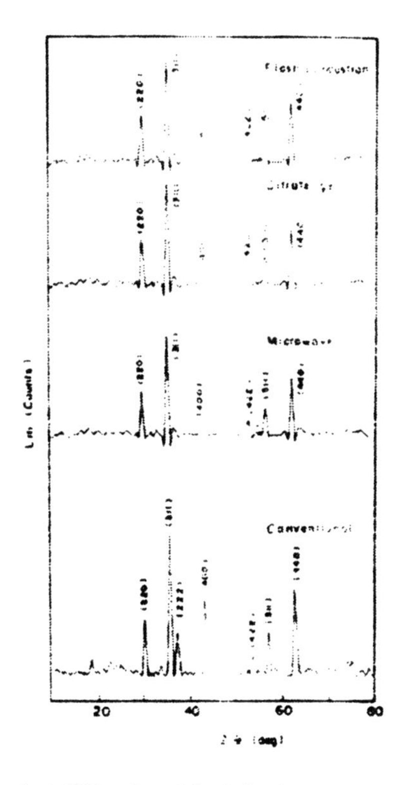

Fig 1 XRD pattern of Ni-Zn ferrite samples processed by various techniques and calcined at 900°C

method shows dielectric constant of 200-600 for sample sintered at 1150° C and 6000-25,000 for samples sintered above 1250°C. The samples sintered at 1350° C shows comparatively high values of dielectric constants for all the processing techniques. Very high values of dielectric constant is obtained in the case of Ni-Zn ferrites prepared by the conventional method.

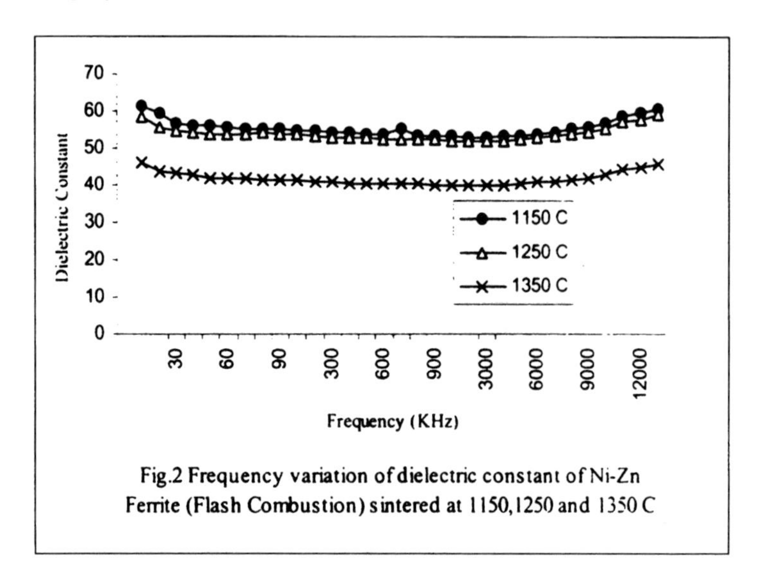

Fig.2 Frequency variation of dielectric constant of Ni-Zn Ferrite (Flash Combustion) sintered at 1150,1250 and 1350 C

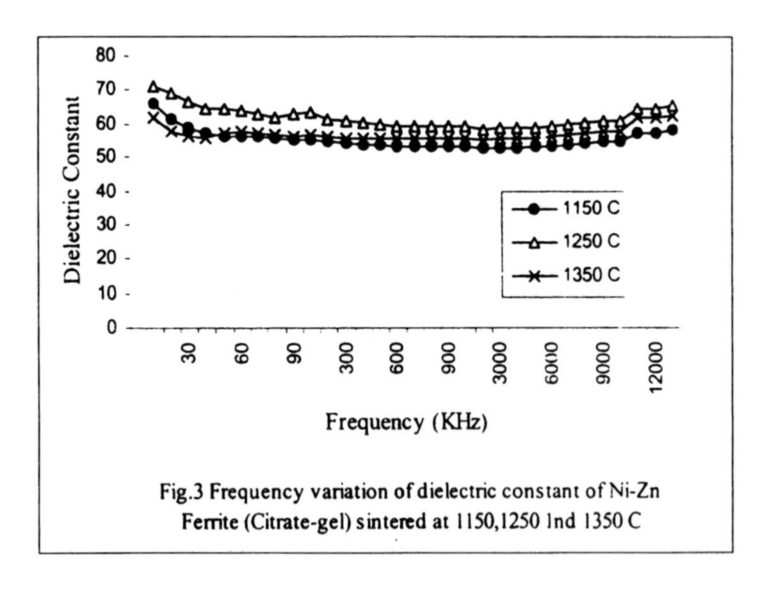

Fig.3 Frequency variation of dielectric constant of Ni-Zn Ferrite (Citrate-gel) sintered at 1150,1250 Ind 1350 C

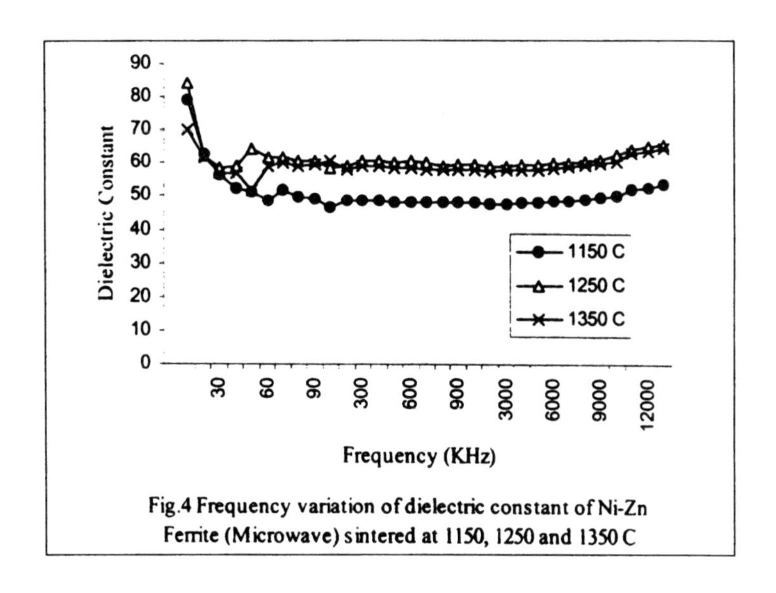

Fig.4 Frequency variation of dielectric constant of Ni-Zn Ferrite (Microwave) sintered at 1150, 1250 and 1350 C

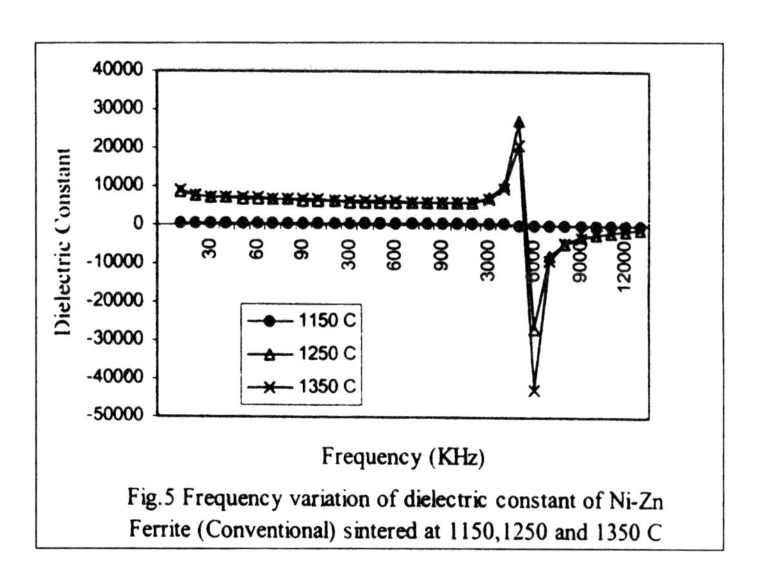

Fig.5 Frequency variation of dielectric constant of Ni-Zn Ferrite (Conventional) sintered at 1150,1250 and 1350 C

The ferrites prepared by the conventional method are likely to have more Fe^{2+} and, therefore, a high dielectric constant.[16] Considerable increase in the grain size is seen in the Ni-Zn ferrite prepared by conventional route which also responsible for the increase in the dielectric constant.[21] Density also plays a role in the variation of dielectric constant. Table I shows the density obtained for the ferrite samples prepared by various techniques.

Table I. Density (gm/cc) of $Ni_{0.8}Zn_{0.2}Fe_2O_4$ processed through various techniques and sintered at different temperatures

Processing techniques	1150°C	1250°C	1350°C
Flash Combustion	4.05	4.25	4.46
Citrate-gel Decmposition	4.15	4.46	4.59
Microwave Process	4.31	4.72	4.99
Conventional Process	4.44	4.82	5.10

It can be seen from the dielectric loss curves shown in Fig. 6-9 that the losses are frequency dependent. The dielectric loss in each sample is found to decrease with the increase in frequency. The dielectric loss arises due to the lag of the polarization behind the alternating electric field and is caused by the impurities and imperfections in the crystal lattice. It is observed that the values of dielectric loss are significantly lower compared to those reported for the ferrite samples prepared by conventional method.[14,15] The occurrence of peaks in $\tan\delta$ versus frequency curves can be explained qualitatively. There is a strong correlation between the conduction mechanism and dielectric behavior. The conduction in these ferrites is considered as due to the hopping of electrons between divalent and trivalent iron ions over the octahedral sites. A peak in the dielectric loss tangent is observed when this hopping frequency is approximately equal to the frequency of the external applied electric field.[11,14] The dielectric constant and dielectric loss are due to the combined effect of crystal structure perfection, microstructure, chemical homogeneity and the amount of Fe^{2+} in the sample.

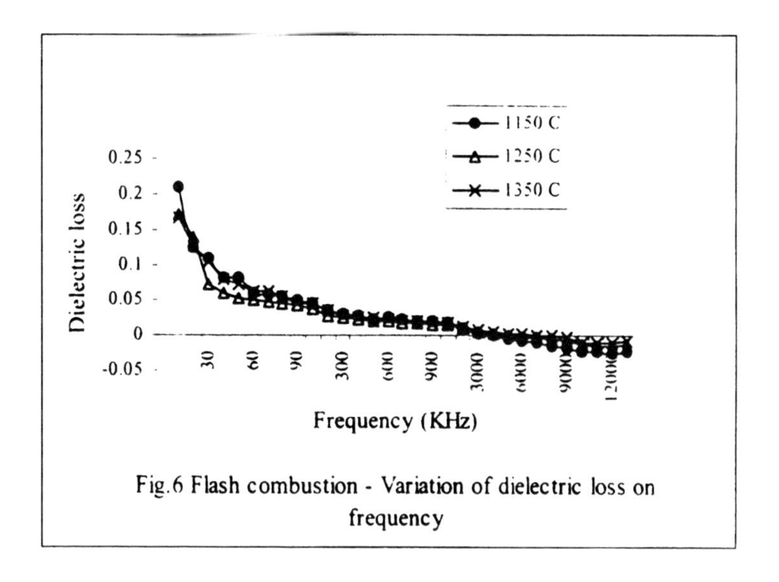

Fig.6 Flash combustion - Variation of dielectric loss on frequency

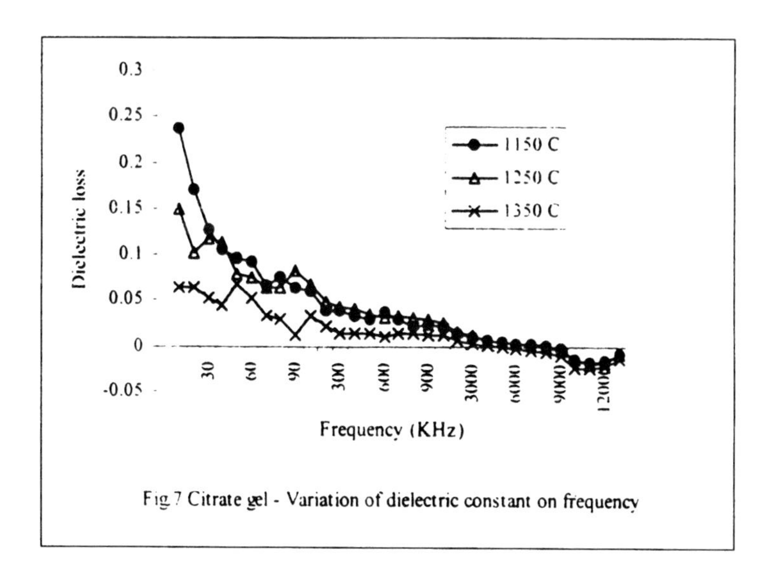

Fig.7 Citrate gel - Variation of dielectric constant on frequency

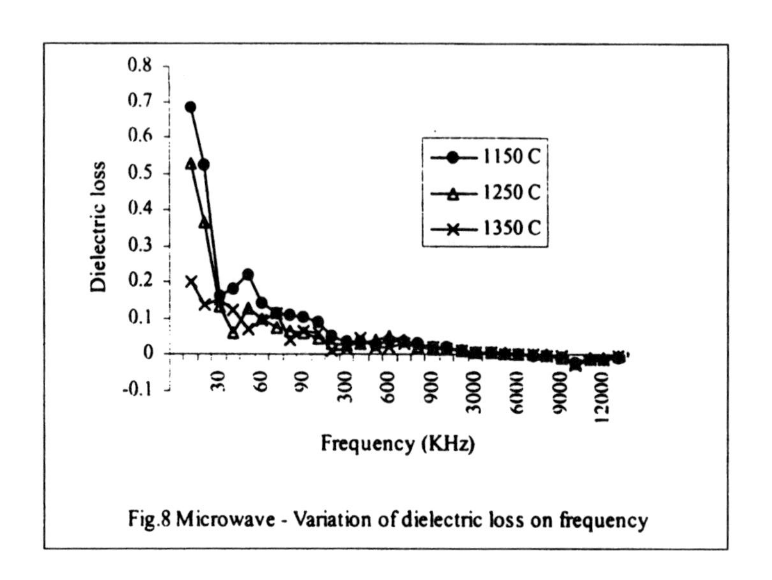

Fig.8 Microwave - Variation of dielectric loss on frequency

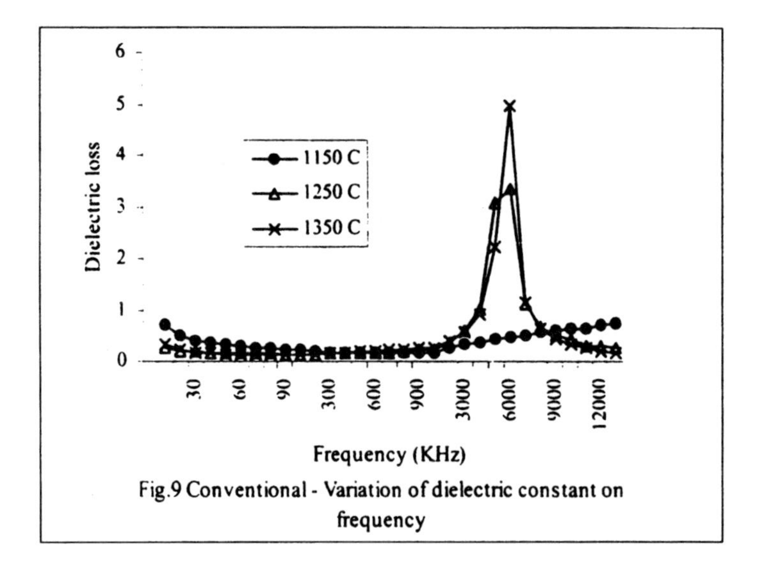

Fig.9 Conventional - Variation of dielectric constant on frequency

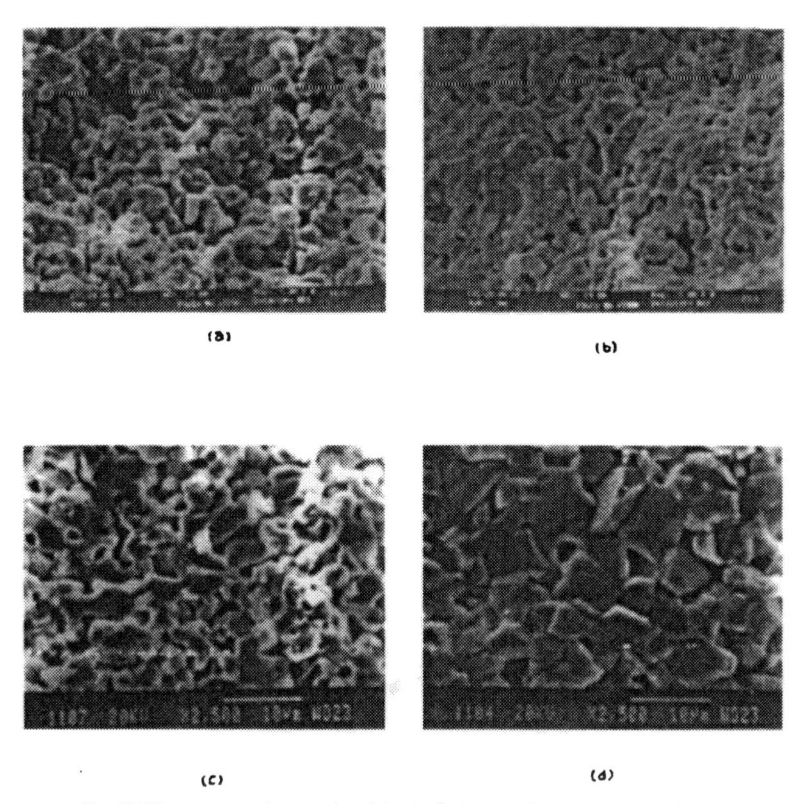

Fig. 10 Microstructure photographs of Ni-Zn ferrite samples processed at various techniques (a) Flash combustion (b) Citrate-gel decomposition (c) Microwave processing and (d) Conventional route and sintered at 1250°C

SEM photographs of the ferrite samples sintered at various temperatures and prepared by various techniques were taken to understand the microstructure. The photographs of these samples are shown in Fig.10 and it can be seen from the photograph, the microstructure consists of small grains of size 2-3 μm for sample prepared by various novel techniques and it is 6-10 μm for sample prepared by conventional method. Small grains are preferred in ferrites as oxidation advances faster in smaller grains and the probability of Fe^{2-} formation in smaller grains is low resulting in lower dielectric constants.

CONCLUSIONS

The results of the present study can be summarized as follows.

1. The dielectric constant and dielectric loss of Ni-Zn ferrite samples prepared by various novel techniques such as flash combustion, Citrate-gel decomposition and microwave processing were studied as the function of frequency and the results compared with conventional method. These values are much lower than those for the samples prepared by conventional ceramic method.
2. The sintered density measurements were calculated and compared. The microstructure analysis was also studied for various ferrite samples.
3. The present investigation suggests that the non-conventional techniques adopted in this work are convenient for the preparation ferrite used in the high frequency applications.
4. High quality ferrites can be prepared by these non-conventional techniques gives better microstructure and chemical homogeneity.

REFERENCES

[1]M.Sugimoto, "The Past, Present and Future of Ferrites," *Journal of American Ceramic Society*, **82**[2] 269-80 (1999).

[2]D.L.Fresh, "Methods of Preparation and Crystal Chemistry of Ferrites," Proceedings of IRE, **44**, 1303-41(1956).

[3]D.W.Johnson Jr. and B.B.Ghate. "Scientific Approach to Processing of Ferrites," pp.27-38, *Advances in Ferrites*, Vol.15, Fourth International Conference on Ferrites. Part I Edited by F.F.Y.Wang, American Ceramic Society, Columbus. OH.(1986).

[4]E.Roess, "Modern Technology for Modern Ferrite," pp.129-36 in Proceedings of Fifth International Conference on Ferrites. *Advances in Ferrites*. Edited by C.M.Srivastava and M.J.Patni. Oxford and IBM Pulishing Co. Pvt.Ltd., New Delhi, Bombay, and Calcutta, India, (1989).

[5]N.Balagopal, K.G.K.Warrier and A.D.Damodaran, " Alumina-Ceria Composite Powders Through Flash Combustion Technique," *Journal of Materials Science Letters*,**10**,1116-18 (1991).

[6]R.V.Mangalaraja, S.Ananthakumar, P.Manohar and F.D.Gnanam, "Synthesis and Characterization of Ni-Zn Ferrite Derived Through Flash Combustion Technique," Proceedings of the 8th International Conference on Ferrites, Kyoto,Japan (2000).

[7]P.A.Lessing, "Mixed Cation Oxide Powders via Polymeric Precursors, *American Ceramic Society Bulletin*, **68**,1002-7 (1989).

[8]A.Verma, T.C.Geol, R.G.Mendiratta and M.I.Alam, "Dielectric properties of NiZn Ferrites Prepared by the Citrate Precursor Method," *Materials Science & Engineering*, **B60**,156-162(1999).

[9]R.V.Mangalaraja, S.Ananthakumar, P.Manohar and F.D.Gnanam, "Powder Characteristics,Sintering Behavior and Microstructure of Ni-Zn Ferrite Derived Through Citrate-Gel Decomposition technique," Proceedings of the 8th International Conference on Ferrites, Kyoto,Japan (2000).

[10]Willard H.Sutton, "Microwave Processing Ceramic Materials," *American Ceramic Society Bulletin*, **68** (2),376-86 (1989).

[11]V.R.K.Murthy and J.Sobhandri, "Electrical and Dielectric Properties of Nickel-Zinc ferrites," Proceedings of the seminor on Electronic and Special Ceramic Materials, Hyderabad,India, 139-45 (1976).

[12]Kozo Iwauchi, "Dielectric Properties of Ferrites," *Journal of Applied Physics*, **10**,1520 (1971).

[13]L.G.Vanuiter, "Dielectric Properties and Conductivity of Nickel - Zinc ferrites," Proceedings of IRE, **44**,1294-319 (1956).

[14]B.Parvatheeswara Rao and K.H.Rao, " Effect of Sintering Conditions on Resistivity and Dielectric Properties of Ni-Zn Ferrites," *Journal Materials Science*, **32**,6049-6054 (1997).

[15]A.M.Abdeen, "Dielectric behavior in Ni-Zn Ferrites," *Journal Magnetism and Magnetic Materials*, **192**, 121-129 (1999).

[16]A.Verma, T.C.Geol, R.G.Mendiratta and M.I.Alam, "Dielectric Properties of NiZn Ferrites Prepared by the Citrate Precursor Method," *Materials Science & Engineering*, **B60**,156-162 (1999).

[17].Manas Chanda, "Science of Engineering Materials," Vol.3, The Macmillan Company of India Ltd., New Delhi, 103 (1980).

[18]J.Smit and H.P.J.Wijn, "Ferrites," Philips Technical Library, Eindhoven, The Netherlands, pp .234 (1959).

[19]A.Goldman, "Modern Ferrite Technology," Van Nostrand Reinhold, New York, pp.71 (1990).

[20]A.Verma, T.C.Geol, R.C.Mendiratta and R.G.Gupta, " High-Resistivity Nickel-Zinc Ferrites by the Citrate Precursor Method," *Journal of Magnetism and Magnetic Materials*,**192**, 271-276 (1999).

[21]Ram Narayan, R.B.Tripathi and B.K.Das, "Advances in Ferrites," Vol.1, pp.267, Edited by C.M.Srivastava and M.J.Patni, Oxford & IBH Publishing Co., New Delhi, India (1989).

HIGH BREAKDOWN STRENGTH AND HIGH DIELECTRIC CONSTANT CAPACITORS IN THE STRONTIUM ZIRCONATE AND STRONTIUM TITANATE SOLID SOLUTION SYSTEM

Stephen J. Lombardo and
Rajesh V. Shende
Department of Chemical Engineering,
University of Missouri
Columbia, Missouri 65211, USA

Daniel S. Krueger
Honeywell
Federal Manufacturing
& Technologies, LLC
Kansas City, Missouri 64141, USA

ABSTRACT

The development of high-voltage breakdown, V, for $SrTiO_3$ (V=40 $V/\mu m$) and $SrZrO_3$ (V=35 $V/\mu m$) dielectrics has been previously reported. Although these materials exhibit high breakdown strength, the moderate dielectric constant of these materials limits their charge storage capability. In this work, we first describe the strategies used to achieve the high-voltage breakdown and then address methods of compositional modification to increase the dielectric constant. We report on the relationship between the dielectric properties and the processing and composition of the substrates.

INTRODUCTION

The perovskite materials strontium zirconate ($SrZrO_3$), strontium titanate ($SrTiO_3$), and mixtures thereof have found widespread use in a number of applications [1-3]. For charge storage applications, however, the dielectric constants, κ, of the pure $SrTiO_3$ and $SrZrO_3$ materials are 380 and 60, respectively, and are low as compared to other commonly used perovskite materials. In addition, monolithic capacitors formed by tape casting are typically prone to a number of processing related defects that can lower the breakdown voltage, V. The effect of κ and V on the volumetric energy density [4], u, of a capacitor is given by

$$u = \frac{1}{2}\kappa\varepsilon_o\left(\frac{V}{d}\right)^2 \qquad (1)$$

where ε_o is the permittivity of free space and d is the thickness of the dielectric. The energy density can thus be enhanced if either the breakdown voltage or the dielectric constant or both can be increased.

Although the theoretical estimate for the upper limit for the breakdown voltage is 1600 $V/\mu m$, ceramic oxide dielectrics prepared by powder processing routes typically have values of 8-20 $V/\mu m$ [5]. In contrast to the breakdown behavior of ceramic dielectrics, polymeric materials exhibit breakdown voltages of the range of 40-200 $V/\mu m$. The lower values obtained for ceramic oxides is generally attributed to the presence of various defects, such as multiple phases and porosity, that arise in the synthesis and processing of polycrystalline materials.

The perovskites $SrZrO_3$ and $SrTiO_3$ have a number of attributes, however, which may make them especially suitable for high-voltage applications. Because neither material is ferroelectric, they are not prone to experience significant electromechanical deformation that may contribute to voltage breakdown. For $SrZrO_3$, the cations (Sr^{2+}, Zr^{4+}) also do not exhibit multiple oxidation states. Finally, under typical sintering conditions, no species in the $SrTiO_3$-$SrZrO_3$ system volatilizes to any great extent.

In the present work, the Pechini method [6,7,8] was used to synthesize single phase, high purity, and highly crystalline powders. To achieve high sintered density, slurry formulations were developed which lead to high green density and the optimum sintering temperature was established. Pure $SrZrO_3$ and $SrTiO_3$ substrates were prepared and the breakdown voltage characteristics were determined. The dielectric constants of the base substrates were then increased by doping these materials with Y_2O_3 as per the procedure of Burn and Neirman [9].

EXPERIMENTAL

Powders of $SrZrO_3$ and $SrTiO_3$ were prepared from the reagent grade precursors zirconium acetate and strontium nitrate (99+), both obtained from Aldrich Chemical (Milwaukee, WI), and from Tyzor GBA/GBO-titanium acetyl acetonate obtained from Dupont (Wilmington, DE). The details on the synthesis procedure have been reported elsewhere [10].

The synthesized powders were used in various slurry formulations to obtain high solids loadings. The processing related details can be found elsewhere [10]. The cast substrates were sintered in air at temperatures between 1300-1560°C for one hour in a box furnace. After sintering, the density was determined from the sample weight and dimensions and by the Archimedes technique. The green density was established by the former method only.

Doping of the substrates was accomplished by wet milling the $SrTiO_3$ powder with Y_2O_3 (0.8 mol%) in ethanol for six hours with poly (vinyl butyral) as binder at 1% by weight. The solvent was evaporated and the dried powders were pressed into disc specimens of 14.9 mm diameter and 1.0 mm height. These samples were then sintered at 1500°C for 15 hours.

X-ray diffraction using CuKα radiation was performed on the calcined powder with a Bruker D5005 θ/2θ Bragg-Brentano diffractometer equipped with a curved, graphite-crystal diffracted beam monochromator and a NaI scintillation detector. An HP 4263B LCR meter was used for determination of the dielectric constant, κ, and dissipation factor, *tan δ*, from frequencies of 10^2-10^5 Hz. For high-voltage breakdown testing, circular monolithic films of 100-150 μm by 0.95 cm diameter were used. The electrodes were gold sputtered (~10 nm on each side) and testing was conducted in a fluorocarbon-based fluid.

RESULTS AND DISCUSSION

Prior to performing the powder synthesis, the starting precursors were assayed by thermogravimetry with a Perkin Elmer TGA 7 analyzer. The yield as oxide for each of the precursors is listed in Table I. The x-ray pattern for $SrZrO_3$ (see Fig. 1) synthesized by this method exhibits nominal phase purity and is consistent with the x-ray patterns reported by others [11,12].

Table I. Thermogravimetric analysis of the precursors used for the synthesis of $SrZrO_3$ and $SrTiO_3$ powders

Precursors	Assay as oxides (w%)
Strontium nitrate	48.4
Zirconium acetate	18.4
Titanium acetyl acetonate	14.9

The x-ray pattern for $SrTiO_3$ synthesized using the precursor stoichiometry as determined by thermogravimetry is shown in Fig. 1. Although the diffraction peaks were shifted to somewhat higher 2θ angles as compared to published values, the pattern shows sharp diffraction peaks and is consistent with the results of other investigators [13,14]. There was no evidence of peaks corresponding to secondary phases such as Ti_4O_7 at 28-32° 2θ and $Sr_3Ti_2O_7$ at 31-32° 2θ as reported by other investigators [14,15].

The optimum sintering temperature was established based on the density and is 1520°C and 1400°C for $SrZrO_3$ and $SrTiO_3$, respectively. For both the substrates, the sintered density was >99% of the theoretical. The microstructure of the as-cast surfaces of the two materials sintered at the optimum conditions is shown in Fig. 2. For both compositions, a very dense microstructure is observed with an average grain size of 2-4 μm.

The dependence of the relative dielectric constant of $SrZrO_3$ as a function of frequency and sintering temperature is displayed in Fig. 3, where κ is seen to decrease with increasing frequency and with increasing sintering temperature. For samples fired at 1520°C, the percent decrease in κ from 10^2 to 10^5 Hz is 28% versus a 50% decrease for samples fired at 1300°C. A similar dependence of κ on frequency and temperature has been reported for $SrTiO_3$ [13].

The individual breakdown strengths for tape cast substrates of both compositions are reported in Table II. For $SrZrO_3$, the average breakdown strength is 40 V/μm with a standard deviation of 14 V/μm; for $SrTiO_3$, the corresponding values are 35 and 10 V/μm. Individual values as high as 57 V/μm were observed on multiple substrates of $SrZrO_3$ while the lowest value was 24 V/μm. For $SrTiO_3$, the high and low values were 44 and 26 V/μm. For comparison, a value of 25 V/μm was reported [16] for thin films of $SrTiO_3$ fabricated by pulsed excimer laser. Scanning electron microscopy was used to search for extrinsic flaw origins such as voids in the samples, but as is often the case for breakdown origins, it was not possible to identify the strength-limiting defects.

Table II. Substrate thickness and breakdown voltages for tape cast $SrZrO_3$ and $SrTiO_3$ capacitors

Sample	Substrate Thickness (μm)	Breakdown Strength (V/μm)
$SrZrO_3$		
1	140	57
2	153	45
3	137	57
4	168	24
5	188	37
6	163	25
7	171	36
Average	**160**	**40±14**
$SrTiO_3$		
1	117	26
2	125	26
3	125	44
4	97	42
Average	**116**	**35±10**

For the undoped $SrTiO_3$, the values of κ and $tan\ \delta$ measured at 25°C and 10^3 Hz were 300 and 0.01, respectively, whereas for undoped $SrZrO_3$, the corresponding values were 55 and 0.013, respectively. On doping $SrTiO_3$ with Y_2O_3, the dielectric constant was found to increase. As seen in Fig. 4, κ for the doped material is enhanced by a factor of 30-50 over the pure composition. The higher values of κ arising from doping have been ascribed to size and structure effects of semiconducting grains within an insulating matrix [9].

Although doping of these substrates with Y_2O_3 leads to an enhancement in the dielectric constant, it remains to be established what effect the doping will have on the breakdown strength. In the next phase of this work, Y_2O_3-doped substrates of $SrTiO_3$ (~100 μm) will be fabricated either by using the same approach described in the present work or by using the internal boundary layer (secondary diffusion) approach [17].

CONCLUSIONS

The Pechini method has been used to synthesize nominally phase-pure $SrZrO_3$ and $SrTiO_3$ powders. To minimize the occurrence of voids, well-dispersed slurries of both powders were developed which lead to high solids loading and high green densities. Sintered densities greater than 99% were achieved for both materials. The breakdown strengths of the pure substrates fabricated with this approach were 40 and 35 V/μm for $SrZrO_3$ and $SrTiO_3$, respectively, and correspond to energy densities of 0.4 and 2 J/cm^3 for 100 μm substrates. Doping of the $SrTiO_3$ material with Y_2O_3 lead to a factor of 30-50 increase in the dielectric constant.

ACKNOWLEDGEMENT

This project was funded by Honeywell, FM&T, LLC, which is operated for the United States Department of Energy, National Nuclear Security Agency, under contract No. DE-AC04-01AL66850.

REFERENCES

[1]H.M. Christen, L.A. Knauss, and K.S. Harshavardhan, "Field-dependent Dielectric Permittivity of Paraelectric Superlattices Structures," *Materials Science and Engineering*, **B 56** [2-3] 200-03 (1998).

[2]M.S. Chu, "High Voltage Dielectrics (SRT 103), " US Patent No. 4,388,415 June 14, 1983.

[3]Y. Yajima, H. Suzuki, T. Yogo, and H. Iwahara, "Protonic Conduction in $SrZrO_3$-Based Oxides," *Solid State Ionics*, **51** [1-4] 101-07 (1992).

[4]H. Fröhlich, "*Theory of Dielectrics; Dielectric Constant and Dielectric Loss*," Clarendon Press, Oxford, 1958.

[5]W.D. Kingery, H.K. Bowen, and D.R. Uhlmann, "*Introduction to Ceramics*," John Wiley & Sons, New York, 1976.
[6]S.J. Lombardo, R.V. Shende, D.S. Viswanath, G.A. Rossetti Jr., D.S. Krueger and A. Gordon, "Synthesis, Processing and Dielectric Properties of Compositions in the Strontium Titanate: Strontium Zirconate Solid Solution System," to appear in *Proceedings of The American Ceramic Society 102nd Annual Meeting,* St. Louis, 2000 (American Ceramic Society, Westerville, OH).
[7]M.P. Pechini, "Methods of Preparing Lead and Alkaline Earth Titanates and Niobates and Coating Method Using the Same to Form a Capacitor," U.S. Pat. No. 3,330,697, July 11, 1967.
[8]N.G. Eror and H.U. Anderson, "Polymeric Precursor Synthesis of Ceramic Materials," pp. 571-77 in Materials Research Society Symposium Proceedings, Vol. 73, *Better Ceramics through Chemistry* (Palo Alto, California, 1986). Edited by J.C. Brinker, D.E. Clark and D.R. Urich, Publishers Choice Book Mfg. Co., Pennsylvania, 1986.
[9]I.Burns and S. Neirman, "Dielectric Properties of donor-doped polycrystalline $SrTiO_3$, " *Journal of Materials Science*, **17** [12], 3510-24 (1982).
[10]R.V. Shende, D.S. Krueger G.A. Rossetti Jr., and S.J. Lombardo, "Strontium Zirconate and Strontium Titanate Ceramics for High Voltage Applications: Synthesis, Processing and Dielectric Properties," to appear in *Journal of American Ceramic Society.*
[11]T.R.N. Kutty, R. Vivekanandan, and S. Philip, "Precipitation of Ultrafine Powders of Zirconia Polymorphs and their Conversion to $MZrO_3$ (M=Ba, Sr, Ca) by the Hydrothermal Method," *Journal of Materials Science*, **25** [8] 3649-58 (1990).
[12]J.S. Smith, R.T. Dolloff and K.S. Mazdiyasni, "Preparation and Characterization of Alkoxy-Derived $SrZrO_3$ and $SrTiO_3$," *Journal of American Ceramic Society*, **53** [2] 91-95 (1970).
[13]C.F. Kao and W.D. Yang, "Preparation and Electrical Properties of Fine Strontium Titanate Powder from Titanium Alkoxide in a Strong Alkaline Solution," *Materials Science and Engineering,* **B 38** [1-2] 127-37 (1996).
[14]P.A. Lessing, "Mixed-Cation Oxide Powders via Polymeric Precursors," *Ceramic Bulletin*, **68** [5] 1002-07 (1989).
[15]G. Pfaff, "Peroxide Route to Synthesize Strontium Titanate Powders of Different Composition," *Journal of the European Ceramic Society*, **9** [2] 121-25 (1992).
[16]G.M. Rao and S.B. Krupanidhi, "Study of Electrical Properties of Pulsed Excimer Laser Deposited Strontium Titanate Films," *Journal of Applied Physics*, **75** [5] 2604-11 (1994).

[17]M. Fujimoto and W.D. Kingery, "Microstructure of $SrTiO_3$ Internal Boundary Layer Capacitors During and After Processing and Resultant Electrical Properties, " *Journal of American Ceramic Society*, **68** [4] 169-73 (1985).

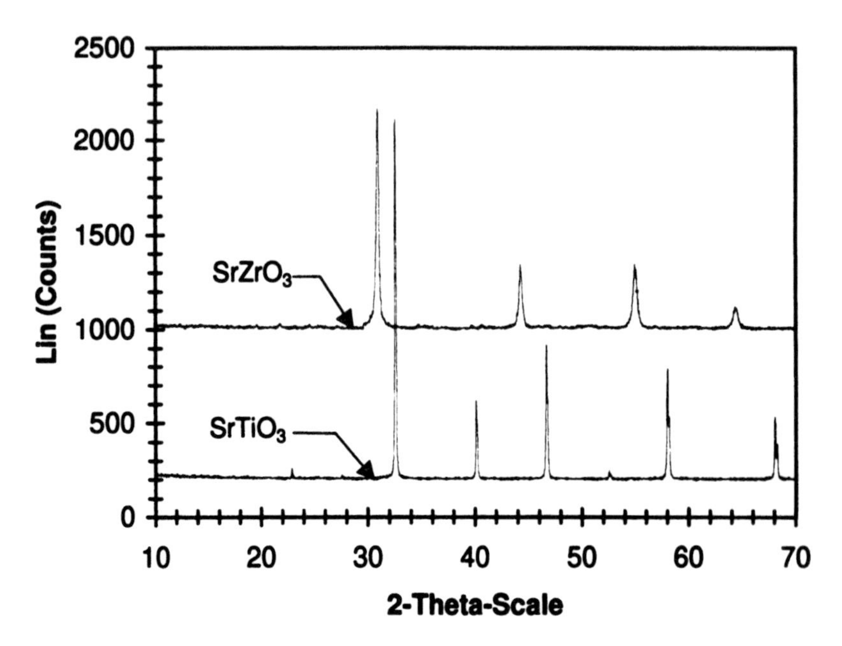

Figure 1. X-ray diffraction patterns of $SrZrO_3$ and $SrTiO_3$ synthesized by the Pechini method.

Figure 2. SEM microstructure of as-cast surfaces of $SrZrO_3$ (left) and $SrTiO_3$ (right).

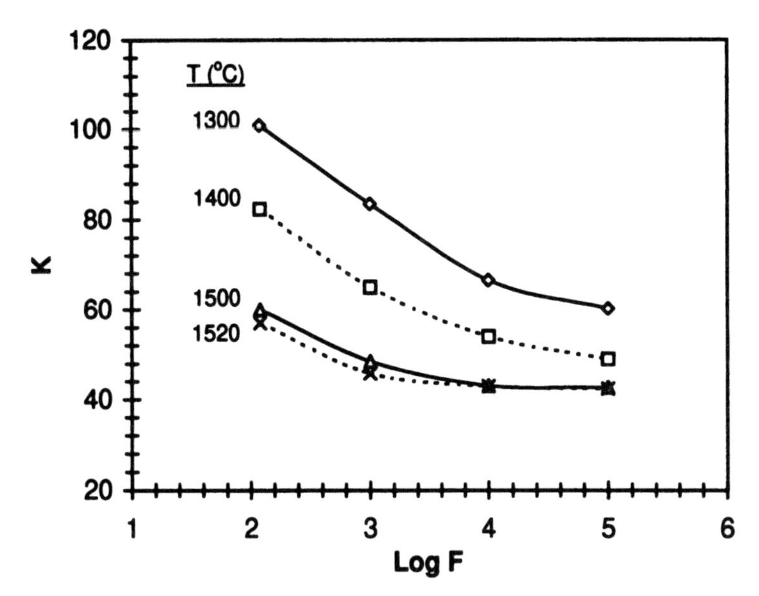

Figure 3. Dielectric constant for $SrZrO_3$ as a function of frequency and sintering temperature.

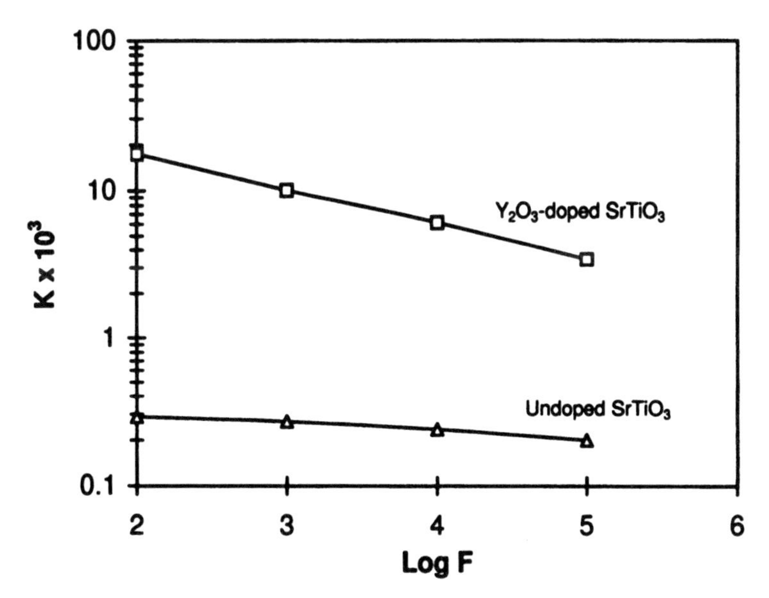

Figure 4. Dielectric constant as a function of frequency for pure $SrTiO_3$ and Y_2O_3-doped $SrTiO_3$.

PREPARATION AND CHARACTERIZATION OF $Sr_{0.5}Ba_{0.5}Nb_2O_6$ CERAMIC FIBERS THROUGH SOL-GEL PROCESSING

Masahiro Toyoda* and Kyougo Shirono
Fukui National College of Technology, Geshi, Sabae, 916-8507 Japan
*Corresponding author. e-mail address: toyoda22@fukui-nct.ac.jp (M. Toyoda)

ABSTRACT

The sol-gel processing was applied to the fabrication of $SrBaNb_2O_6$ ceramics fibers. The $Sr_{0.5}Ba_{0.5}Nb_2O_6$ ceramic fibers were prepared by using Ba methoxide, Sr metal and Nb pentaethoxide as starting materials, and then refluxed and distilled with stirring to form complex alkoxide. The hydrolysis and polycondensation of its obtained alkoxide gave polymerized products, and as a result the viscosity of the solution increased, its suggested that linear polymer products were obtained. The condition to get viscous solution was decided 0.20 μl distilled water for 0.24 mol/l complex alkoxide solutions. The $Sr_{0.5}Ba_{0.5}Nb_2$ Gel fibers were drawn from its viscous solution. Its gel fibers were crystallized into tetragonal tungsten bronze phase at 900 °C. Its heat treated fibers were 10 centimeters long and from 30 to 200 μm in diameter, crack free and dense. The dielectric constant and dielectric loss of its fiber fired at 1100 °C were 370 and 3.0 %, respectively.

INTRODUCTION

$Sr_xBa_{1-x}Nb_2O_6$ (SBN) has a tetragonal tungsten bronze structure and they have large pyroelectric coefficients and piezoelectric and electro optic properties. SBN has been receiving great attention for applications in pyroelectric sensors, SAW filters and electro-optic devices.[1-4] In particular, $Sr_{0.48}Ba_{0.52}Nb_2O_6$ has an attractive effect on the piezoelectric effect, linear optoelectric and high refraction index and then it is expected to be applied for a piezo type infrared sensor[5] and an elastic surface wave filter.[6] Its ceramics is also very important material because of its lead-free composition.

Control of the Sr/Ba ratio is one of the key factors to optimize the properties of $SrBaNb_2O_6$ ceramics. However, the control of composition is usually difficult in CVD and sputtering processing. Chemical processing such as sol-gel has some advantages such as feasible control of chemical stoichiometry and low temperature processing.[7-13] These sol-gel processing was adopted to prepare $Sr_xBa_{1-x}Nb_2O_6$ thin films ceramics due to these advantages.[14] Recently, the demand for fiber shapes processing has increased because of the integrated device and improved functionality of sensor. The sol-gel processing has also advantages such as fabricating of fibrous shape of ceramics.[15-19]

In this paper, an attempt is made to decide the quantity of addition of water and to fabricate the $Sr_{0.5}Ba_{0.5}Nb_2O_6$ ceramics fibers through sol-gel processing and to measure the electric properties.

EXPERIMENTAL PROCEDURE

Figs. 1 and 2 show the experimental procedure of preparation of complex alkoxide solution. Ba methoxide [Ba $(OCH_3)_2$], Sr metal and Nb pentaethoxide [$Nb(OC_2H_5)_5$] as raw materials and 2-ethoxyethanol as organic solvent were selected. The 2-ethoxyethanol was dried over molecular sieves and distilled before use. Ba methoxide and Sr metal were dissolved in the 2-ethoxyethanol at around boiling point, 125℃, with stirring for 48 h respectively. The Nb $(OC_2H_5)_5$ with excess 2-ethoxyethanol was refluxed and distilled to get the Nb-ethoxyethoxide separately. These obtained alkoxide solution were mixed with excess its organic solvent, and then reacted at 2-ethoxyethannol, 125℃, with stirring for 48 h. The obtained precursor solution was concentrated and removed by-products by vacuum distillation. The final solutions, which had a concentration 0.24 to 0.52 mol/l were yellowish transparent solution without any suspension of particles. All procedures were conducted in dry N_2 atmosphere, because the starting materials are extremely sensitive to moisture. For the investigation of the gelation, the 0.24 and 0.52 mol/l Sr-Ba-Nb complex alkoxide solution were weighed and combined with distilled water in a sample bottle (Table 1). The gelation reaction was designated by the molar ratio of water to alkoxide.

The spinnability of the sol-solutions was examined from the capability of fiberization of the sols by dipping a glass stick and pulling it up. The obtained fibers were heat-treated at 600 and 1200 °C. The observation of morphology on fibers was conducted by using SEM (Hitachi, S-4100). The crystalline phase of its fibers was identified by XRD (Rigaku, Rad-2B). The dielectric measurement was examined using Hewlett Packard impedance analyzer HP4284A.

Table 1 Hydrolysis condition

Concentration of complex alkoxide solution (mol / l)	Amount of alkoxide solution (ml)	Addition of H_2O or H_2O – 2-ethoxyethanol solution (μl)
0.240	3	5 (a)
	3	15 (a)
	3	25 (a)
	3	50 (a)
	3	75 (a)
	3	125 (a)
0.376	5	350 (b)
0.523	3	125 (b)
	3	150 (b)
	3	175 (b)
	3	230 (b)

(a) H_2O, (b) H_2O – 2-ethoxyethanol

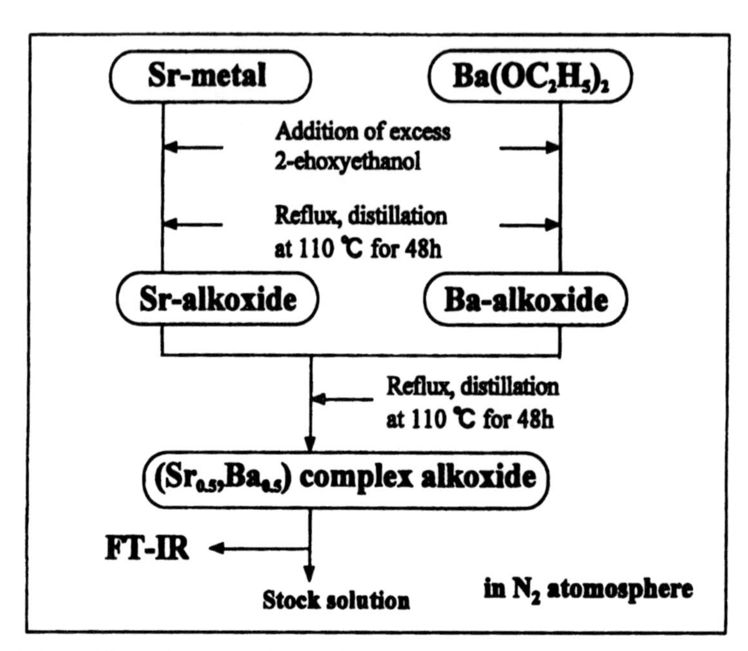

Fig. 1 Flow diagram of experimental procedure for preparation of Sr-Ba double alkoxide

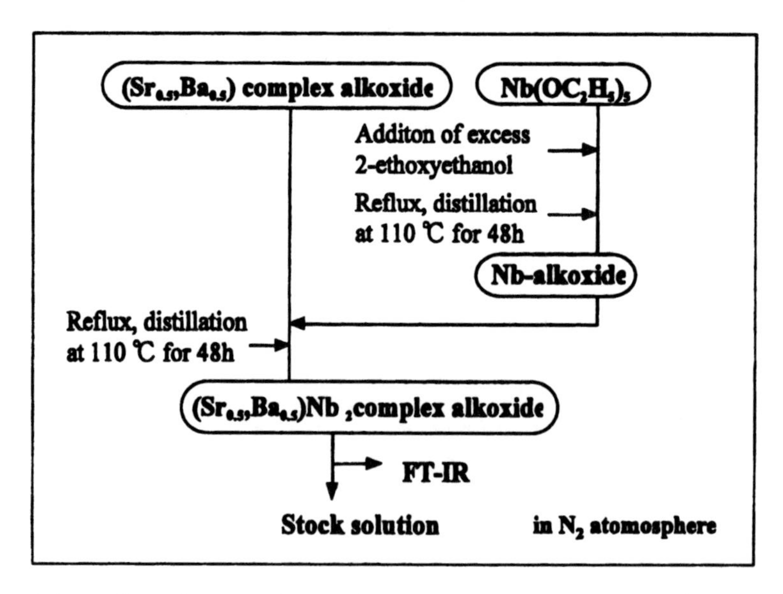

Fig. 2 Flow diagram of experimental procedure for preparation of Sr-Ba-Nb complex alkoxide

RESULTS and DISCUSSION

The viscosity of the Sr-Ba-Nb complex alkoxide solution with 0.20 μl distilled water increased with the passage of time (96 – 120 h) by hydrolysis polycondensation and viscoused its complex alkoxide solution without water increased with the passage time of 48 days. The gel fibers were spinnabled through viscous sol-solutions by dipping a glass stick and pulling. The gel fibers could be drawn after the viscosity reached about 1 kg/m·sec. Its drawn gel fibers were yellowish transparent and homogeneous. Fig. 3 shows morphology of ceramics fiber after heat-treated at 1100 °C [low magnification (a), middle magnification (b), and high magnification (c)]. The heat-treated fibers were 10 cm long and from 30 to 200 μm in diameter, crack free and densed. As seen, grains of its fibers are rod like and small fibril. Size of rod like grain was about 0.3 X 1.5μm. Fig. 4 indicates XRD patterns of the fiber. The Sr-Ba-Nb gel fibers heat-treated at 900 °C is proved to have the crystallized tetragonal tungsten bronze phase. Molar ratio of Sr-Ba-Nb complex alkoxide solution and composition of $Sr_{0.5}Ba_{0.5}Nb_2O_6$ ceramics fiber analyzed close to the expected stoichiometry (Sr/Ba = 0.475/0.525). To measure the ferroelectric properties, both ends of fiber, 2 cm long and 100μm in diameter, was fixed on glass substrate by using Ag paste. The dielectric constant and dielectric loss of its fiber fired at 1100 °C were 370 and 3.0 %, respectively at room temperature and 400 Hz.

CONCLUSIONS

1. $Sr_{0.5}Ba_{0.5}Nb_2O_6$ ceramics fiber was successfully prepared through sol-gel processing
2. The condition of gelation to obtain the gel fiber was decided 0.20 μl distilled water for 0.24 mol/l complex alkoxide solutions.
3. Sr-Ba-Nb complex alkoxide solution produce tungsten bronze phase was obtained heat-treatment at 900 °C.
4. Grains of $Sr_{0.5}Ba_{0.5}Nb_2O_6$ fibers were rod like shaped. Size of these grains was about 0.3 X 1.5 μm.
5. The dielectric constant and dielectric loss of $Sr_{0.5}Ba_{0.5}Nb_2O_6$ ceramic fiber fired at 1100 °C were 370 and 3.0 % at room temperature at 400 Hz, respectively.

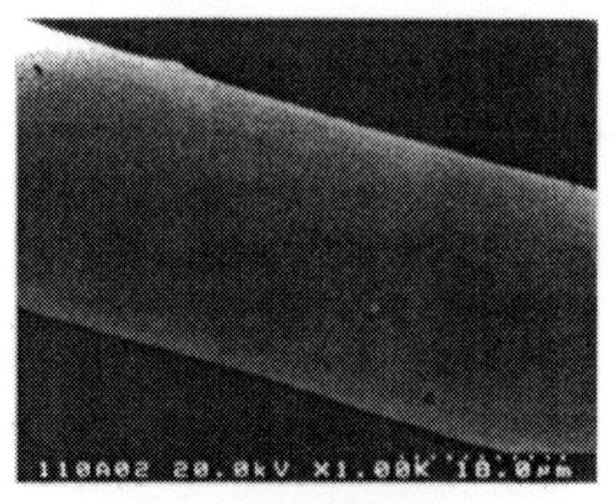

(a) low magnification,

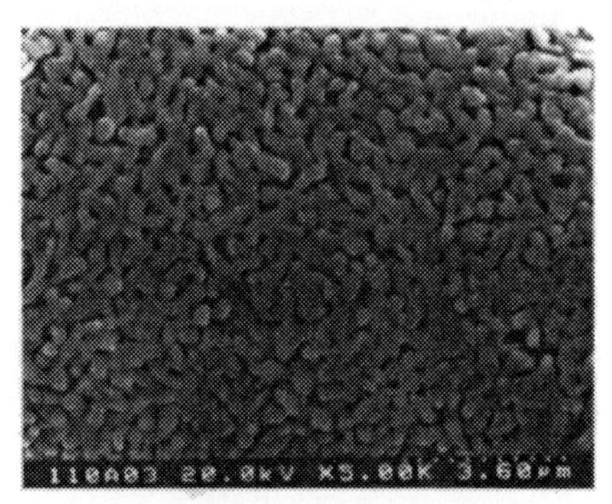

(b) middle magnification

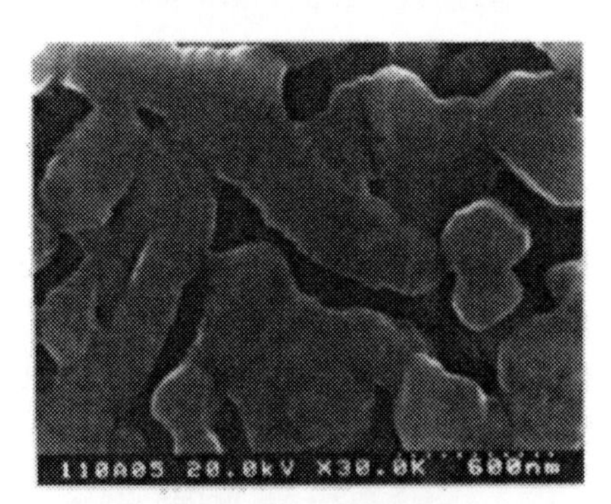

(c) high magnification

Fig. 3 SEM micrograph of $Sr_{0.5}Ba_{0.5}Nb_2O_6$ fiber heat-treated at 1100 °C

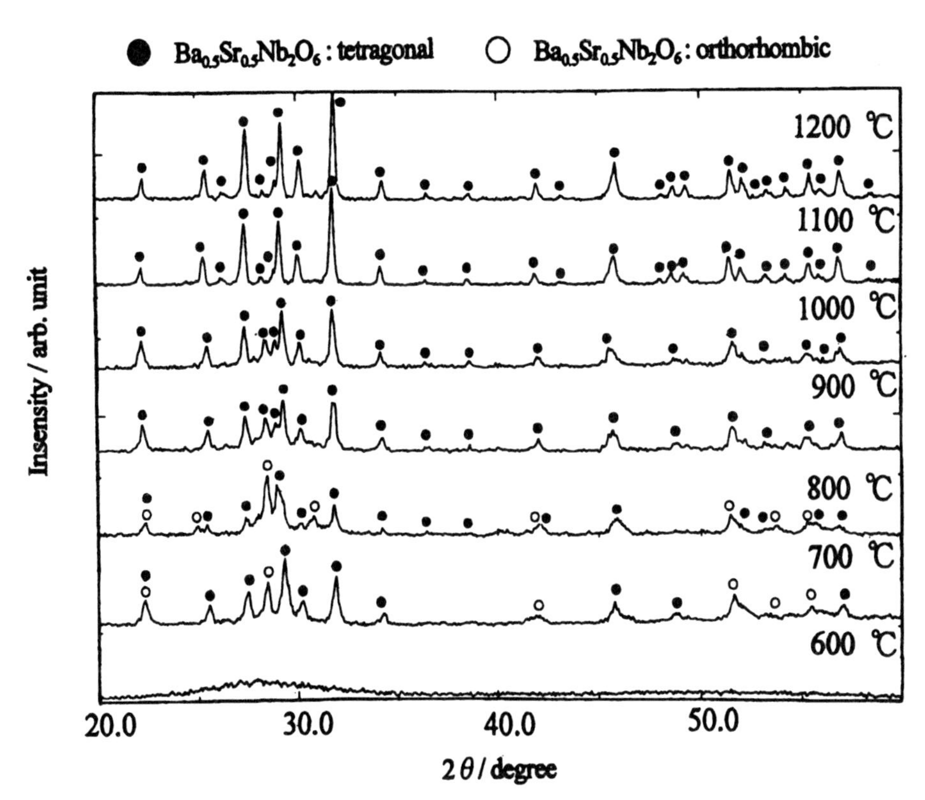

Fig. 4 XRD patterns of $Sr_{0.5}Ba_{0.5}Nb_2O_6$ ceramics fiber fired at 600 and 1200 °C

REFERENCES

[1] M. P. Trubeja, E. Ryba and D. K. Smith, "A Study of Positional Disorder in Strontium Bariumu Niobate," J. Mater. Sci., **31**, 1435-1443(1996).

[2] R. R. Neurgaonkar, W. F. Hall, J. R. Olover, W. W. Ho and W. K. Cory, Tungsten Bronze $Sr_{1-x}Ba_xNb_2O_6$: A Case History of Versatility." Ferroelectrics, **87**,167-179(1988)

[3] A. M. Glass, "Investigation of the Electrical Properties of $Sr_{1-x}Ba_xNb_2O_6$ with Special Reference to Pyroelectric Detection," J. App; Phy., **40**[12],4699-4713(1969)

[4] D. Ryta, B. A. Wechsler, R. N. Schwartz, C. C. Nelson, C. D. Brandle, A. J. Valentino and G. W. Berkstresser, "Temperature Dependence of Photorefractive Properties Strontium-Bariumu Niobate ($Sr_{0.6}Ba_{0.4}Nb_2O_6$)," J. Appl. Phys.,

66[5],1920-1924(1989)

[5] S. Nishiwaki, T. Yogo, K. Kikuta, K. Ogiso, A. Kawase and S. Hirano, "Synthesis of Strontium-Bariumu Niobate Thin Films Through Metal Alkoxide," J. Am Ceran, Soc., **79**[9], 2283-2288(1996)

[6] R. R. Neurgaonka, M. H. Kalisher, T. C. Cim, E. J. Staples and K. L. Keester, "Czochralski Single Crystal Growth of $Sr_{0.61}Ba_{0.39}Nb_2O_6$ for Surface Acoustic Wave Applications," Mater. Res. Bull., **15**,1235-1240(1980)

[7] W. Sakamoto, T. Yogo, A. Kawase and S. Hirano, "Chemical Processing of Potassium-Substituted Strontium Barium Niobate Thin Films through Metallo-Organics." J. Am. Cera. Soc., **81**[10], 2692-2698(1998)

[8] K. Nishio, N. Seki, J. Thongrueng, Y. Watanabe and T. Tsuchida, "Preparation and Properties of Highly Oriented $Sr_{0.3}Ba_{0.7}Nb_2O_6$ Thin films by a Sol-Gel Processing." J. Sol-Gel Sci. Tec., **16**, 37-45(1999)

[9]M. Toyoda, Y. Muhamm, Preparation and Characterization of (Ba,Ca)(Ti,Zr)O_3 Thin Films through Sol-Gel Processing J. Sol-Gel Sci. Tech., **16**, #1/2, pp7-12 (1999)

[10]M. Toyoda, J. Watanabe and T. Matsumiya, Evolution of Structure of the precursor during Sol-Gel Processing of Zinc Oxide and Low Temperature Formation of Thin Films J.Sol-Gel. Sci. Tech., **16**, #1/2, pp93-102 (1999)

[11]M. Toyoda, Y. Hamaji, K. Tomono, D. A. Payne, Ferroelectric Properties and Farigue Characteristics of $Bi_4Ti_3O_{12}$ Thin Films by Sol-Gel Processing. Jpn. J.Appl. Phys. **33**, pp.5543-5548 (1994)

[12]M. Toyoda, T. Hamaji, K. Tomono, D. A. Payne, Synthesis and Characterization of $Bi_4Ti_3O_{12}$ Thin Films by Sol-Gel Processing. Jpn. J. Appl. Phys. **32**, pp.4158-4162 (1993)

[13]M. Toyoda, D. A. Payne, Synthesis and Characterization of an Acetate Alkoxide Precursor for Sol-Gel derived $Bi_4Ti_3O_{12}$. Materials Letter. **18**, pp.84-88 (1993)

[14] W. Sakamoto, T. Yogo, A. Kawase and S. Hirano, "Chemical Processing of Potassium-Substituted Strontium Barium Niobate Thin Films through Metallo-Organics." J. Am. Cera. Soc., **81**[10], 2692-2698(1998)

[15] M. Toyoda, Y. Hamaji and K. Tomono, "Fabrication of $PbTiO_3$Ceramics Fibers by Sol-Gel Processing." J. Sol-Gel Sci. Tec., **9**, 71-84(1997)

[17]M. Toyoda, Y. Hamaji and K. Tomono, "Fabrication of $Pb(Ti,Zr)O_3$Ceramics Fibers by Sol-Gel Processing." Nippon Kagakukai-shi, **1995**; 150-156(1995)

[18]M. Toyoda, Y. Hamaji and K. Tomono, "Fabrication of $PbTiO_3$Ceramics Fibers by Sol-Gel Processing." Nippon Kagakukai-shi, **1994**; 1118-1126 (1994)

[19]M. Toyoda and K. Shirono, "Prepapration and Characterization of $(Sr_{0.48}Ba_{0.52})Nb_2O_6$ Ceramics Fibers Through Sol-Gel Processing" Ceramics Transaction, (accepted), in press.

CURRENT TOPICS IN THE FIELD OF MATERIALS TECHNOLOGY OF BME-MLCCS

Takeshi Nomura and Yukie Nakano
Materials Research Center, TDK Corporation
570-2,Matsugashita, Minami-Hatori, Narita-shi, Chiba, 286-8088 Japan
FAX : 81-476-1637, e-mail : tnomura@mb1.tdk.co.jp

ABSTRACT

The present paper focuses on the recent progress of materials technology for highly reliable BME (base metal electrode)-MLCCs. Recently, thinning dielectric layer and multiplication have been accelerated in order to achieve higher capacitance. With increase of the number of the dielectric layer, precise control of process parameters is required for obtaining fault-free MLCCs. In this paper, current topics in the field of materials technology of BME-MLCCs have been reviewed with special reference to the effect of the binder burn-out conditions on the mechanical and electrical properties. It is shown that the binder burn-out process is a key step for obtaining highly reliable MLCCs. Not only faults such as delaminations and cracks but also mechanical strength of MLCCs are strongly affected by the binder burn-out conditions. This is well explained by the residual carbon of MLCCs.

INTRODUCTION

In recent years, electronic components have been rapidly developed with the production of new electronic devices. Multilayer ceramic capacitors (MLCCs) are particularly important electronic components that are used in almost all areas of electronics. Recently, required are reduction of production cost, miniaturization, high capacitance, and high reliability, in order to increase the range of applications of capacitors. Among these requirements, low cost is of prime importance from the viewpoint of industrial product. The cost of the electrode materials occupies a

large part of the total cost. So far, palladium or silver-palladium has been the majority of the internal electrode material. In order to lower the production cost, much effort has been devoted to develop the base-metal electrode system.
Recent trend of electronic equipments or facilities is down sizing, which accelerates the miniaturization of chip components. Chip components show a rapid growth despite the gloomy economic circumstances. Among them, especially MLCCs market expands on a large scale overcoming the economic situation. The price of MLCCs has yet to be reduced because of the strong demand of low price. It is known that MLCCs is composed of alternately layers of dielectrics and internal electrode such as Pd or Ni. Since Pd is a precious metal, the cost is another major problem. Ni-electrode MLCCs are a promising way of reducing the cost. In the initial stage of the R&D, Ni-electrode MLCCs had a serious problem about reliability. However, since the authors had succeeded in achieving highly reliable MLCCs with Ni-electrodes, it shows a remarkable expansion in both quantity and extent of application. Multiplication and thinning of dielectrics and Ni-electrode layers are strongly requested for attaining larger capacitance. Recently, MLCCs with 400-600 layers of 2-3 μ m dielectrics in 3216 (mm) type have been developed. Multiplication and thinning of dielectrics, however, may cause some faults such as delaminations and cracks.
These faults cause inferior IR properties, so that these faults should be avoided. It is well known that delamination and crack are caused by the insufficient removal of binder during binder burn-out process in case of MLCCs with Pd-electrodes.

In the present paper, the effect of binder burn-out condition on the mechanically related failures of MLCCs with Ni-electrodes, in order to overcome the barrier for larger capacitances by multiplication and thinning of dielectrics.

MATERIALS FOR BASE METAL ELECTRODE

High capacitance MLCCs are expected to extend the application field in which electrolytic capacitors or plastic film capacitors are much used.
In such circumstances, MLCCs of base metal electrodes occupy the attention of ceramists, because of the new material-technology for the firing under low oxygen partial pressure without the reduction of $BaTiO_3$-based ceramics. An Ni internal electrode is easily oxidized during firing in air. Firing should be therefore carried out in a reducing atmosphere using a non-reducible dielectric material. When a non-reducible material is used and fired in a reducing atmosphere, the capacitors with internal Ni electrodes can provide initial characteristics similar to those of Pd-electrode MLCCs.
Accordingly, non-reducible dielectric materials have been developed by paying much effort. The techniques are as follows[1-3].

Control of the molar ratio $(Ba+Ca)O/(Ti+Zr)O_2$ larger than the unity (1)

$$\text{Substitution of } Ca^{2+} \text{ for } Ba^{2+} \quad (2)$$

$$\text{Addition of MnO} \quad (3)$$

The biggest theme of Ni-electrode MLCCs was about the lifetime of insulation resistance (IR) under highly accelerated life testing (HALT). Nowadays, Ni-electrode MLCCs show superior reliability[4-10]. Moreover, much effort is being devoted to develop more reliable MLCCs with larger capacitance.
The most effective ways for prolongation of lifetime have been reported as reoxidation after firing under reducing atmosphere, donor doping, control of grain boundary chemistry, and so on.
Donor addition is effective to prolong the lifetime of IR as well as reoxidation after firing. The dielectric material contains Mn in order to improve initial IR. Mn ion is considered to become an acceptor which induces oxygen vacancies during firing. The addition of donor to the dielectrics was expected to compensate the oxygen vacancies generated by acceptor doping.

$$MnO \rightarrow Mn_{Ti} + O_O + V_O \quad (4)$$

$$Y_2O_3 \rightarrow 2Y_{Ba} + 3O_O + V_{Ba} \quad (5)$$

$$V_O + V_{Ba} \rightarrow n \quad (6)$$

Table 1 summarizes the effect of dopants on the lifetime. Y_2O_3 produced the most significant prolongation of life.
Oxides of V, W, Dy, Ho, etc. are also effective to prolong the lifetime under HALT. Figure 1 shows TEM observations of MLCCs with and without addition of Y_2O_3. The addition of Y_2O_3 eliminated dislocation loops without the annealing for

additive	life time
no additive	1
Y_2O_3	15.5
V_2O_5	6.7
Nb_2O_5	0.3
MoO_3	2.8
Ta_2O_5	0
WO_3	1.8
La_2O_3	0
Ce_2O_3	0
Pr_6O_{11}	0
Nd_2O_3	0
Tb_2O_3	2.3
Dy_2O_3	7.4
Ho_2O_3	1.8
Er_2O_3	1.8
Tl_2O_3	4

Table 1 Effect of additives on the lifetime (Relative value)

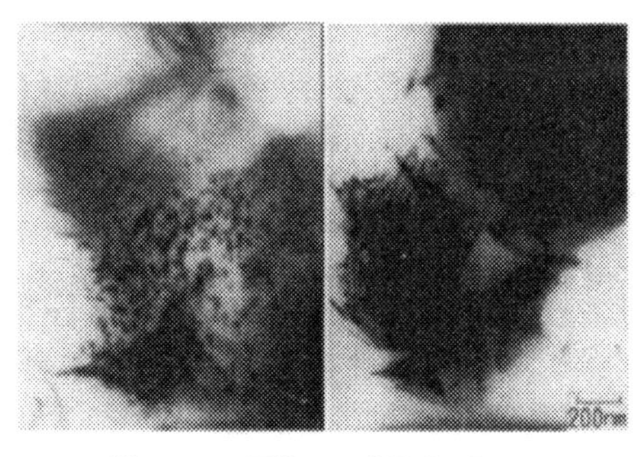

Figure 1 Effect of Y^2O^3 doping on the microstructure.

re-oxidation after firing. The addition of Y_2O_3 is considered to reduce the oxygen vacancies. The effect of the amount of Y_2O_3 addition on capacitance-temperature characteristics showed that the addition of Y_2O_3 lowered the Curie temperature. Consequently, Y_2O_3 was incorporated into the $BaTiO_3$ lattice.
The MTTF obtained here is larger than 10^3 times of that for conventional composition. This technique enabled high reliability as well as Pd-electrode MLCCs.

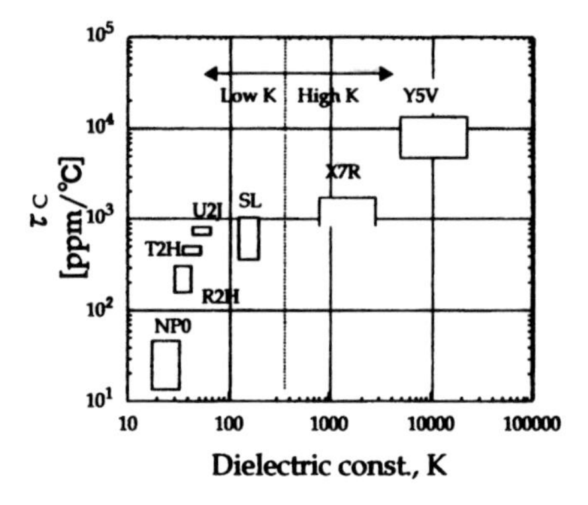

Fig. 2 Dielectric materials for BME MLCCs in the diagram of dielectric constant and temperature coefficient of dielectric constant.

Today, several kinds of temperature-characteristics materials have been developed as shown in Fig.2. Pd electrode MLCCs of temperature compensation use can be replaced by Ni electrode MLCCs. New application of Ni electrode MLCCs to the field of plastic film and electrolytic capacitors is expected to progress more rapidly.

SITE OCCUPATION OF RARE EARTH ELEMENTS

Recently, C.Randall, et al[11]. reported that Y was thought to occupy Ti-site of $BaTiO_3$ lattice so that it was thought to act as an acceptor. According to their speculation, ionic radius of Y is not large enough to occupy Ba-site. Change in XRD profiles of $BaTiO_3$ with Y doping is shown in Fig.3. In the case that Ba/Ti is greater than the unity, peaks of Ba_2TiO_4 can be observed and their intensity become smaller with increasing amount of Y-doping. This is thought that Y reacts with Ba_2TiO_4 to form perovskite structure. This means Y might occupy Ti-site in the case of Ba-rich $BaTiO_3$. In the case of stoichiometric $BaTiO_3$, a small amount of Y-doping results in the formation of BaO phase. This is because Y occupies Ba-site of $BaTiO_3$ and excess Ba precipites from $BaTiO_3$ lattice. When Ba/Ti is less than the unity, second phases,$Ba_6Ti_{17}O_{40}$ and $BaTiO_3$ present in the case of no Y-doping. A small amount of Y-doping accelerates the reaction between $Ba_6Ti_{17}O_{40}$ and $BaTiO_3$ and one phase $BaTiO_3$ is obtained. This indicates Y occupies Ba-site. A large amount of Y-doping, however, introduces precipitation of BaO. Summarizing above, Y prefer to occupy Ba-site in case of Ba/Ti$\leqq$1 but Ti-site in case of Ba/Ti>1. Non-reducible dielectrics require Ba/Ti>1, and it is 1.004 in this paper. MnO is also indispensable for non-reducible $BaTiO_3$. Mn occupies Ti-site, then Ba/Ti ratio become smaller than the nominal composition. Moreover, SiO_2 or Al_2O_3 is generally doped as a sintering aid. These lower Ba/Ti ratio because they forms Ba-rich second phase. Consequently, actual composition of non-reducible $BaTiO_3$ is Ba/Ti<1, so that Y is thought to occupy the Ba-site in $BaTiO_3$ lattice.

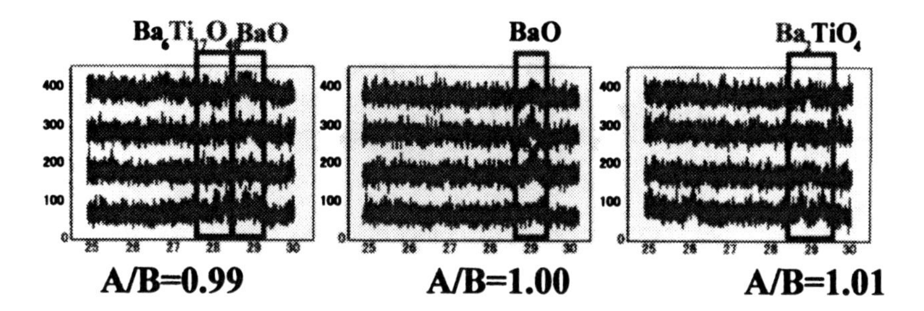

Fig. 3 Effect of Y_2O_3 doping on XRD profile of $BaTiO_3$.

Kishi, et al.[12] have reported about the effect of occupational sites of rare-earth element on the microstructure in $BaTiO_3$ with various Ba/Ti ratios. Their conclusions are as follows. The grain growth behavior strongly affected by the kind of rare-earth element as well as by the Ba/Ti ratio. Larger ion such as La and Sm occupy A-sites, while Yb ions occupy B-sites irrespective of the Ba/Ti ratio. On the other hand, in the case of Dy, Ho, and Er, which have intermediate ionic radius, the grain size changed gradually in a wide range of Ba/Ti ratio. Dy Ho, and Er are considered to occupy both sites, depending on the Ba/Ti ratio.

THINNING AND MULTIPLICATION

Nowadays, multilayer ceramic capacitor tends to be processed downsizing and attained high capacitance by miniaturization and high-performance of electronic equipments. A substitution to the ceramic capacitors from the other capacitors is expected, and the expansion of the use is also being expected. Capacitance of MLCCs can be expressed by the following equation. It depends on an effective area of a layer, the number of dielectric layers, dielectric constant, and dielectric thickness.

$$C = \varepsilon_0 \cdot \varepsilon_r \cdot n \cdot S/d \quad (7)$$

C: capacitance, ε_0:dielectric constant of vacuum, ε_r:specific dielectric constant, n: number of dielectric layers, S: an effective area, d: dielectric thickness

Therefore, dielectric thickness should be made thin in order to increase the capacitance and it is necessary to increase the number of those layers. It proceeds with multiplication and thinning eagerly. The change of layer thickness of dielectric of multilayer ceramic capacitor is shown in Fig. 4. Obtaining 2-3 μm must have been thought even impossibility 10 years ago. If dielectric grain size was taken into consideration, there were many researchers who thought that 5-6 μm was a limit at most. The progress of materials technology and process technology is remarkable in this period. Although this progress is being considered to reach a limit of thick films technology later, it still proceeds with the research vigorously toward multiplication and thinning technology at present.

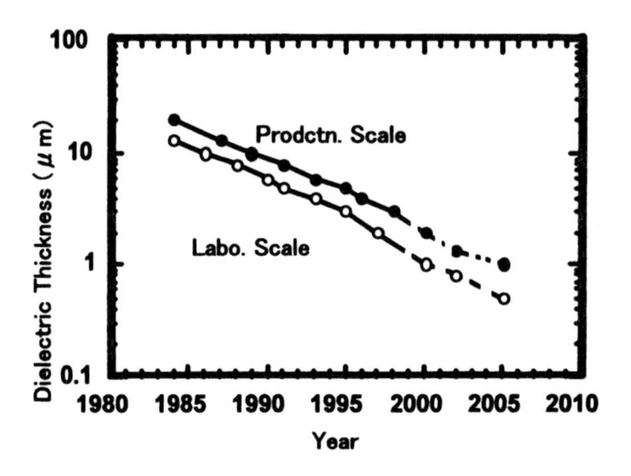

Fig. 4 Transition of dielectric thickness of MLCCs.

It is afraid of the problems that make insulation resistance, break down voltage and the reliability decrease even though the thin layer of dielectric increases performance. In addition, the phenomenon that a temperature characteristics changes by the thickness of dielectric is confirmed.

The effect which dielectric thickness exerts on the temperature dependence of capacitance is shown in Fig. 5, as an example. As for these problems numerous research is done from the point of chemical composition and microstructure. A difference in density around margin point and a part of an internal electrode of multilayer ceramic capacitor become significant with increase the number of dielectric layers and hence the distortions or structural defects are easy to bring about. The examples of crack caused by thinning and multiplication is shown in Fig. 6. Also, even if a structural defect is not caused, it is guessed that there do still exist residual stress. Therefore, the solution against the problems, which residual stress that exists in chip and control of structural defect while multiplication and thinning, is a key for mass production in the near future.

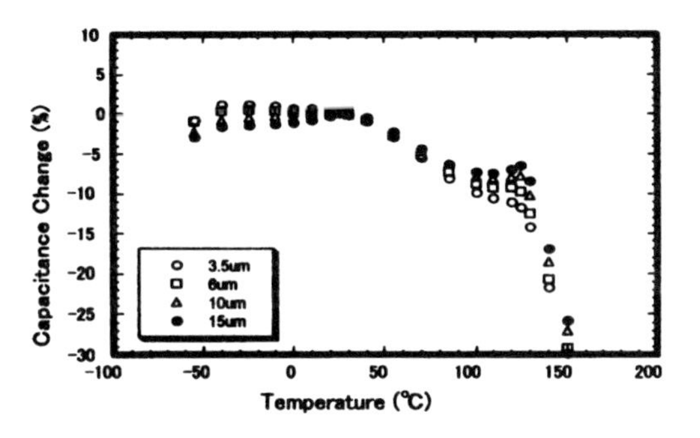

Fig. 5 Effect of dielectric thickness on the temperature dependence of dielectric constant.

Fig.6 Typical crack caused by thinning and multiplication.

Highly reliable MLCCs with Ni-electrodes were achieved by adopting donor doping, reoxidation treatment after firing, control of grain boundary chemistry, etc. Much more reliable dielectrics are, however, required for thinning dielectric layers in order to attain larger capacitance. There might be a limit for thinning dielectric thickness, much effort is now devoted to thin the thickness of dielectric layers as less than 2 μ m. So that the detailed understanding of IR-degradation mechanism is a key to achieve superior properties. Figure 7 shows the effect of thickness of dielectric layers on the meantime to failure (life time) under HALT for the dielectric composed of $BaTiO_3$-MgO-MnO-Y_2O_3-V_2O_5-$Ba_{0.4}Ca_{0.6}SiO_3$. Lifetime strongly depends on the thickness of dielectric thickness, nevertheless the same electric field. Moreover, the slope is larger in lower electric field than that in higher electric field. The dependence of the lifetime on the dielectric thickness is considered due to the distance of oxygen vacancies migration. Electric field dependence of the slope might be explained by the distance of

oxygen vacancies migration alone. This is because the slope is larger in the case of the lower electric field. Mobility of oxygen vacancies can be considered to be proportional to the electric field strength. Then, the slope is proportional to the reciprocal rate of oxygen vacancies migration. In the case of ultra-thin dielectric layers, the effect of Ni electrodes, however, should be taken into consideration, because Ni can be oxidized and diffuses into dielectrics at the annealing step.
Figure 8 shows the effect of electric field on the lifetime under HALT. Logarithm of MTTF is a linear function of logarithm of electric field. The slope which means the voltage acceleration factor, however, depends on the dielectric thickness.

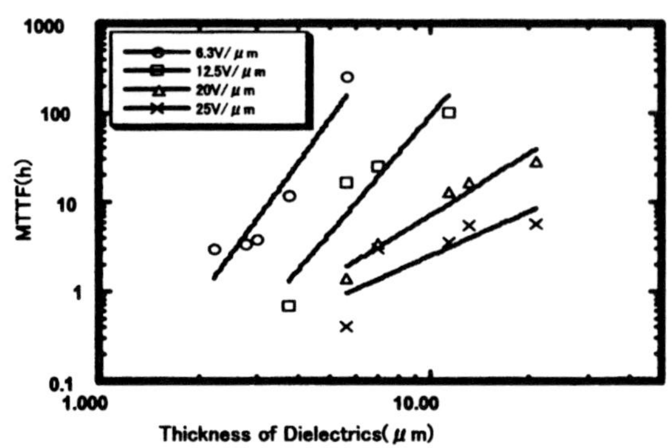

Fig. 7 Variation of lifetime under HALT with the thickness of dielectric layers.

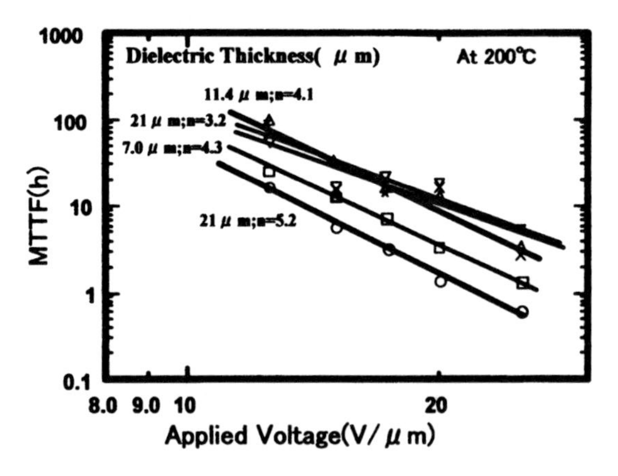

Fig. 8 Effect of applied field on the lifetime under HALT.

Here, the effect of Ni electrodes should also be taken into consideration, or another intrinsic problem might be the cause of the dielectric thickness dependence of voltage acceleration factor. The variation of voltage acceleration factor with thickness of dielectric layers is shown in Fig. 9. The voltage acceleration factor is not constant nevertheless the chemical compositions of dielectrics are the same. It abruptly increases under 10 μ m of dielectric thickness. The authors are now studying about this phenomenon in detail.

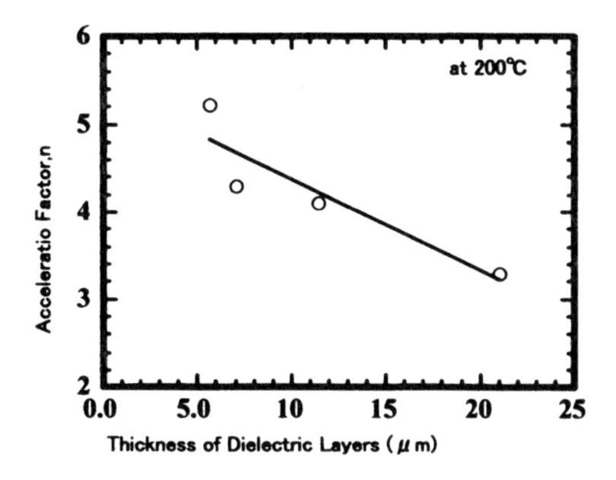

Fig. 9 Variation of voltage acceleration factor with dielectric thickness.

BINDER BURN-OUT PROCESS

Binders such as ethylcellulose and acrylic resin generally burn at around 250℃ in air. Rapid burn out of binders may cause the delamination because a large amount of exhaust gas generates in an instance. Accordingly, binder should be slowly burned out without any combustion. Even if the binder burn-out is slowly performed, it is not perfect for Ni-electrode MLCCs, because Ni easily oxidize in air and it generates O_2 gas during firing under reducing atmosphere. Moreover, the discrepancy of shrinkage between dielectrics and Ni electrode will be stressed by the multiplication. This is remarkable for the case of fine powders as dielectrics and nickel. Typical photographs of delaminetion and crack are shown in Figs.10 and 11.

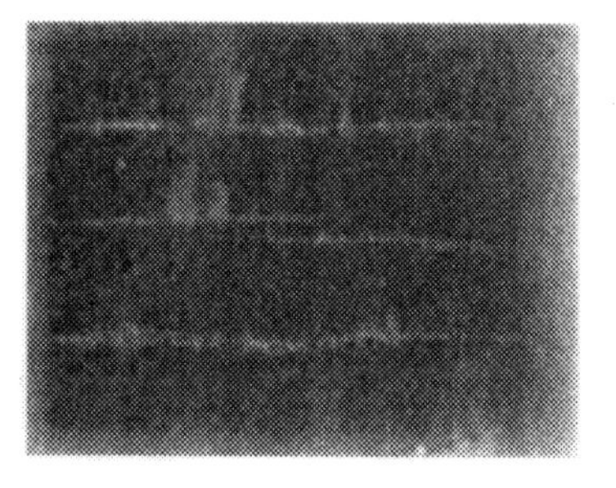

Fig.10 Typical delamination observed in large capacitance Ni electrode MLCC.

The main causes of these faults are considered as the expansion by the oxidation of Ni or the gas generation caused by the reduction of NiO. Figure 12 shows the effect of temperature for binder burn-out process on the residual carbon of the de-bindered chips. Higher temperatures than 300 ℃ in air is effective to remove the binder of chips. Nickel, however, easily oxidize so that high temperature in air can not be adopted. On the contrary, even if in a reducing atmosphere, binder can be effectively removed in the case of higher temperatures than 600 ℃. The effect of temperature for binder burn-out process on the fraction oxidized of Ni is shown in Fig. 13. Even if it is at lower temperatures than 200 ℃, few percent of Ni oxidized in air. Of course, this can be strongly affected by the particle size and crystallinity of Ni oxidized in air. Of course, this can be strongly affected by the particle size and crystallinity of Ni powder. This is not preferable for suppressing the expansion during de-binder process and degassing during firing of electrode layers.

Fig.11 Typical cracks observe in large capacitance Ni electrode MLCC.

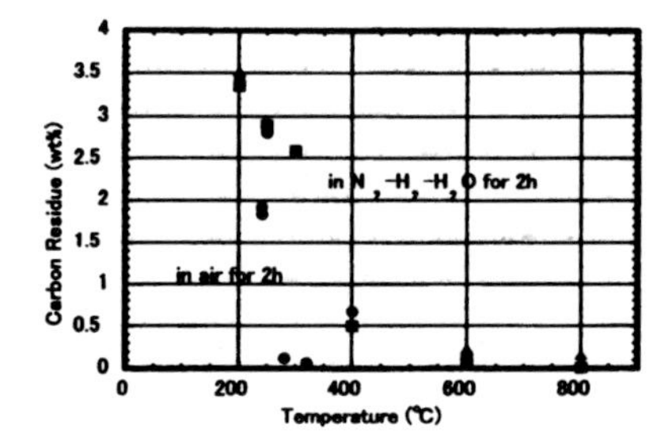

Fig. 12 Effect of binder burn-out temperature on the carbon residue.

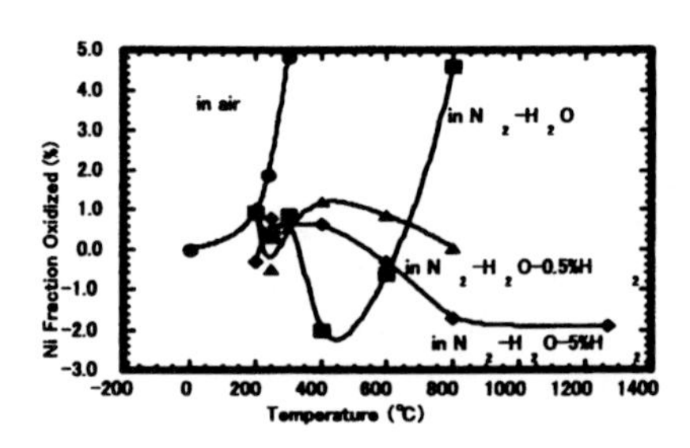

Fig. 13 Effect of binder burn-out temperature on the Ni fraction oxidized.

Figure 14 shows the relationship between residual carbon content and fraction oxidized of Ni. Binder burn-out in N_2-H_2-H_2O is so effective way to remove the binder without any oxidation of Ni, so that it is perfect to restrain the delaminations or cracks as is shown in Fig. 15. As a conclusion, binder burn-out in air, which is widely used in an industry scale, is not adequate for multiplication and thinning of dielectrics and Ni electrodes. Contrarily, binder burn-out under reducing atmosphere such as H_2-H_2O-N_2 effectively reduces residual carbon without any oxidation of Ni electrodes.

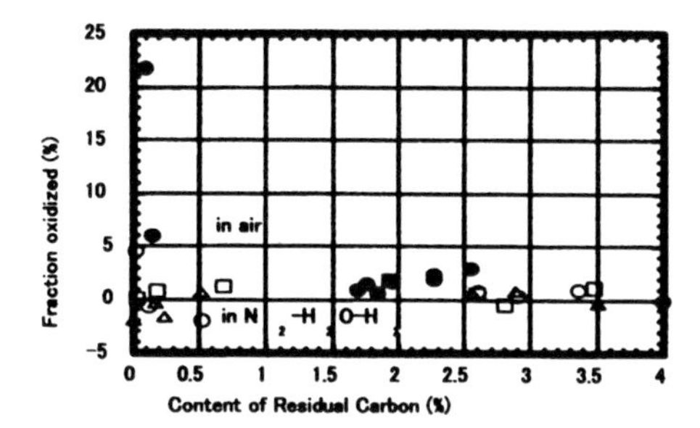

Fig. 14 Relationship between carbon residue and the Ni fraction oxidized.

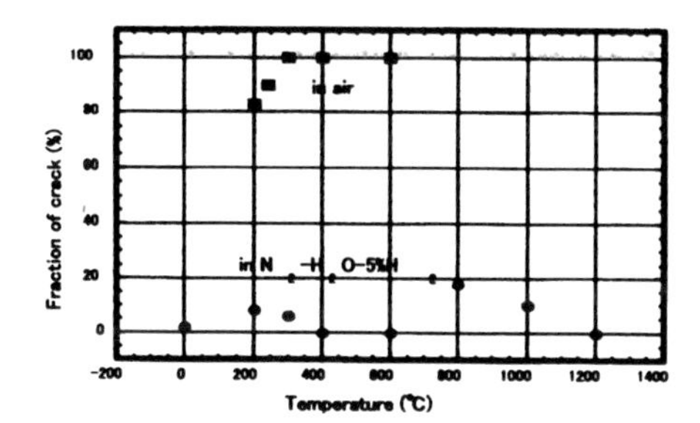

Fig. 15 Effect of binder burn-out temperature on the fraction of cracks.

In general, MLCCs are automatically mounted and soldered on printed circuit board, then higher mechanical and thermal strength is required. From this point of view, delaminations and cracks are so serious because they are the decisive factors for strength.

The authors have been paid much effort to ensure higher mechanical and thermal strength, and discovered that the residual carbon content is an important factor for deciding mechanical and thermal strength. Figure 16 shows the effect of residual carbon after firing on the bending strength of MLCCs with Ni electrodes. Lower content of carbon residue is preferable for achieving higher mechanical strength. In any case, delamination or crack is not detected. Residual carbon content is affected not only by the conditions of binder burn-out process but also by the firing conditions.

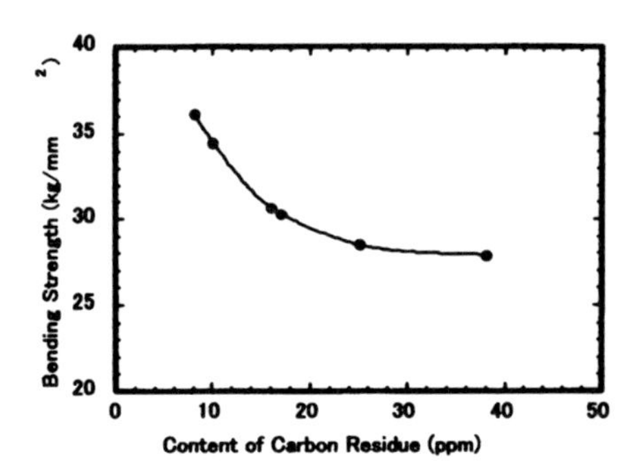

Fig. 16 Effect of carbon residue after firing on the bending strength of MLCCs.

Unfortunately, the state of carbon residue is not clarified up to the present. Residual carbon is not detected by TEM observation, because its content is very low level by gas analysis.

CONCLUSION

In this paper, recent progress in the field of BME capacitors has been reviewed. Binder burn-out process has been studied from the view point of residual carbon and oxidation of Ni. This work can be summarized as follows:

(1) Binder burn-out in air is not adequate for multiplication and thinning of dielectrics and Ni electrodes, because Ni could be oxidized.

(2) Reducing atmosphere such as N_2-H_2-H_2O is so effective to remove the binder without any oxidation of Ni electrodes.

(3) Residual carbon after firing is an important factor for the mechanical strength of MLCCs.

REFERENCES

[1]J.M.Herbert, Trans. "High-Permittivity Ceramics Sintered in Hydrogen," *Brit.Ceram*, Soc., **62** 645-658 (1963).

[2]I.Burn, and G.H.Maher, "High Resistivity $BaTiO_3$ Ceramics Sintered in CO-Co_2 Atmospheres," *J.Mater. Sci.*, **10** 633-640 (1975).

[3]Y.Sakabe, K.Minai and K.Wakino, "High-Dielectric Constant Ceramics for

Base Metal Monolithic Capacitors," *Jpn. J.Appl.Phys.*, **20**[4]147-150 (1981).
[4]T.Nomura, A.Sato and Y.Nakano, "Effect of Microstructure on the Stability of the Electrical Properties of $BaTiO_3$-Based Dielectric Matierials," *J. Soc. Mater.Eng.Resour*, **5** 44-53(1992).
[5]T.Nomura, S.Sumita, Y.Nakano and K.Nishiyama, "A STUDY ON THE DEGRADATION OF Ni ELECTRODE CERAMIC CHIP CAPACITORS," *Proc.5th US-Jpn. Seminar Dielectric Piezoelectric Ceramics*, 29-32(1990).
[6]S.Sumita, M.Ikeda, Y.Nakano, K.Nishiyama and T.Nomura, "Degradation of Multilayer Ceramic Capacitaors with Nickel Electrodes," *J.Amer.Ceram. Soc.,* **74** 27392746(1991).
[7]T.Nomura and Y.Nakano, "High Capacitance Multilayer Ceramic Capacitors with Ni Electrode," *Denshi Tokyo*, **31** 168-172(1993).
[8]Y.Nakano, A.Sato, J.Hitomi and T.Nomura, Ceram.Trans., "Microstructure and Related Phenomena of Multilayer Ceramic Capacitors with Ni-Electrode," *Ceramic Transactions*, **32** 119-128(1993).
[9]J.Yamamatsu, N.Kawano, T.Arashi, A.Sato, Y.Nakano and T.Nomura, "Reliability of multiplayer ceramic capacitors with nickel electrodes," *Journal of Power Sources*, **60** 199-203 (1996).
[10]T.Nomura, N.Kawano, J.Yamamatsu, T.Arashi, Y.Nakano, and A.Sato, "Aging Behavior of Ni-Electrode Multilayer Ceramic Capacitors with X7R Characteristics," *Jpn. J. Appl. Phys.*, 34-9B 5389-5395(1995).
[11] C.A.Randall and T.Shrout, "Hypothesis on Rare Earth Doping of BaTiO3 Ceramic Capacitors," *Proc. of the Eighth US-Japan Seminar on Dielectric and Piezoelectric Ceramics*, 44-47(1997).
[12] H.Kishi, N.Kohzu, Y.Mizuno, and Y.Iguchi, "Occupational Sites of Rare-Earth Elements in $BaTiO_3$," *Proc. of the 9th US-Japan Seminar on Dielectric and Piezoelectric Ceramics*, 311-314 (1999).

FORMATION OF TITANIUM DIOXIDE MICROPATTERN BY DIRECT SYNTHESIS FROM AQUEOUS SOLUTION AND TRANSCRIPTION OF RESIST PATTERN

Naoshi Ozawa, Hiroki Yabe and Takeshi Yao
Department of Fundamental Energy Science
Graduate School of Energy Science
Kyoto University
Yoshida-honmachi, Sakyo-ku
Kyoto 606-8501 Japan

ABSTRACT

Submicron-scale resist pattern was printed on the surface of CrN-coated silicon single crystal wafer substrate by photolithography. The substrates was soaked in a mix solution of $(NH_4)_2TiF_6$ and B_2O_3 under ordinary temperature and ordinary pressure. Formation of Titanium dioxide thin film was confirmed by thin film X-ray diffraction measurement. Then the substrate was soaked in acetone with ultrasonic vibration. The resist material was dissolved off with Titanium dioxide thin film formed just on. Micropattern of Titanium dioxide with minimum size of 0.3 μm transcribing the resist pattern was formed on the substrate. This method is promising for producing micropattern of electronic devices and so on.

INTRODUCTION

Ceramics have many functional uses and the thin film formation is very advantageous for the advanced and efficient use of the functions, light and compact system construction, and so on. For preparing thin films of ceramics, chemical vapor deposition, sputtering, vacuum evaporation and sol-gel methods are used. However, these methods have some disadvantages in actual use. For chemical vapor deposition, sputtering, and vacuum evaporation, expensive vacuum equipments are required and the film areas are restricted. For the sol-gel method, film geometries are limited, and moreover, heat-treatment, which is injurious because of the possibility of a change in the shape and/or size of the film and of crack formation, is required to form ceramic film.

Methods for forming films from an aqueous solution are expected to be advantageous because no vacuum, no high temperature and no expensive apparatus will be required, and substrates, even those with wide areas and/or complicated shapes, are available. It is believed to be very important to develop reactions for synthesizing ceramics from aqueous solutions. Recently, the authors have developed the method for synthesizing functional ceramics such as zirconium dioxide, lanthanum transition metal oxide, titanium dioxide, and so on, from aqueous solutions at ordinary temperature and ordinary pressure by using the hydrolysis reaction of fluoro-complexes[1-4].

Today, micro- and nano-size control of materials is paid a great attention as the future technology. The downsizing of electronic devices such as dynamic random access memory (DRAM), ferroelectric random access memory (FeRAM), metal oxide semiconductor (MOS) transistor and so on, is actively developed in order to increase the accumulation. Because the electric capacity of capacitor or insulator made by using silicon dioxide or silicon nitride is not sufficient for the downsizing, the usage of ceramics with high dielectric constant is investigated eagerly. Micro-patterning of ceramic thin film is very important. Furthermore, micro-patterning will be applicable to the formation of optical devices, bit patterns of magnetic memory materials, and the production of the parts of micro machines.

In this study, the combination of titanium dioxide thin film synthesis reaction from aqueous solution and transcription of the resist pattern was made to fabricate submicron-scale micropatterns of titanium dioxide thin films. This method is very advantageous, because the technique to form minute resist pattern by photolithography is highly established in manufacturing of semiconductor devices. Titanium dioxide has higher dielectric constant than that of silicon dioxide or silicon nitride and is promising for capacitor material of DRAM and gate insulator material of MOS transistors.

EXPERIMENT

Formation of titanium dioxide thin film: Thin film of Titanium dioxide was formed by hydrolysis of titanium fluoro complex[5]. It is considered that the chemical equilibrium between hexafluorotitanate ion and titanium dioxide holds as in reaction (1),

$$TiF_6^{2-} + 2H_2O \rightleftharpoons TiO_2 + 6F^- + 4H^+ \qquad (1)$$

and that, when borate ion is added, fluoride ion is consumed by reaction (2),

$$BO_3^{3-} + 4F^- + 6H^+ \rightarrow BF_4^- + 3H_2O \qquad (2)$$

then the chemical equilibrium in reaction (1) is shifted from left to right to increase fluoride ions, resulting in the formation of titanium dioxide.

CrN-coated silicon single crystal wafer was used as the substrate. Resist pattern was formed on the substrate by photolithography. Scanning electron micrographs of the finest part of the resist pattern are shown in Fig.1. Both the minimum width of the line pattern and the minimum diameter of the dot pattern are 0.3 μm.

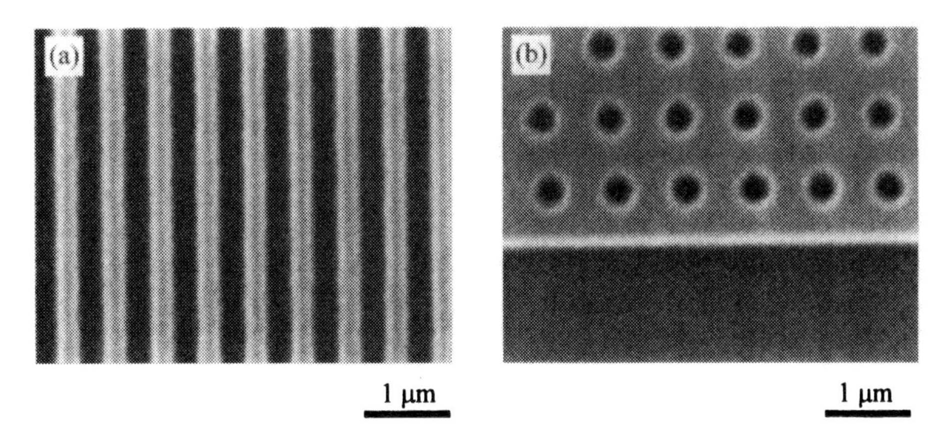

Fig. 1 SEM micrographs of the finest part of the tresist pattern.

$(NH_4)_2TiF_6$ was dissolved in distilled water and B_2O_3 was added. The composition of the mixed solution was 25 mmol·dm^{-3} $(NH_4)_2TiF_6$ and 240 mmol·dm^{-3} B_2O_3. The substrate with resist pattern on was soaked in the solution at 30°C for 24 h. After the soaking, the substrate was washed with distilled water and dried at room temperature. No heat treatment was conducted.

The surfaces of the substrates were analyzed by thin film X-ray diffraction (TF-XRD; RINT 2500, Rigaku Co., Japan), scanning electron microscopy (SEM; ESEM-2700, Nikon Co., Japan), energy-dispersive X-ray analysis (EDX; DX-4, Philips Analytical, Netherlands), atomic force microscopy (AFM; Nanoscope III, Digital Instruments Inc., USA).

Transcription of resist pattern 1 -soak in acetone: After the formation of titanium dioxide thin film, substrates were soaked in acetone with ultrasonic vibration for 5 h. The resist material was dissolved off with the titanium dioxide thin film formed just on.

Transcription of resist pattern 2 -UV exposure and soak in acetone: After the formation of titanium dioxide thin film, the substrates were exposed to UV rays for 6 h. The wave length of UV was 253.7 nm and the irradiance was about 1.8 $mW \cdot cm^{-2}$. Then the substrates were soaked in acetone with the same condition described above.

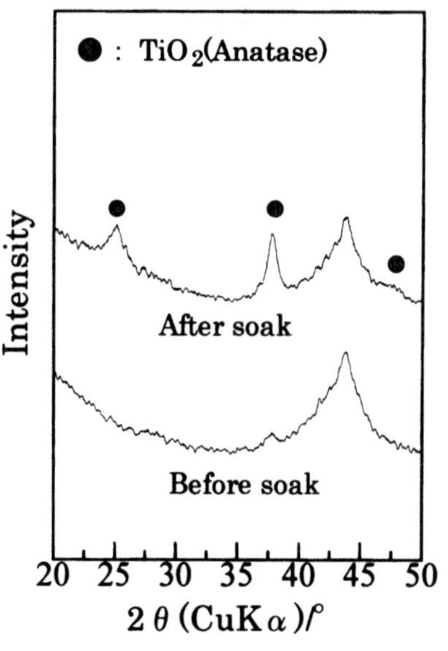

Fig. 2 TF-XRD patterns of the substrate before and after the soak in $(NH_4)_2TiF_6$ and B_2O_3 mixed solution.

RESULTS AND DISCUSSION

Figure 2 shows the TF-XRD patterns of the substrate soaked in $(NH_4)_2TiF_6$ and B_2O_3 mixed solution for 24 h. Characteristic peaks attributed for anatase type titanium dioxide were observed. The intensity of the 004 peak at about 38° was the strongest. This indicates the orientation of (001) plane.

Figure 3 shows SEM micrographs of the surfaces of the substrate soaked in $(NH_4)_2TiF_6$ and B_2O_3 mixed solution for 24 h and then soaked in acetone for 5 h with ultrasonic vibration. It is seen from Fig.3 that the resist material was dissolved off with

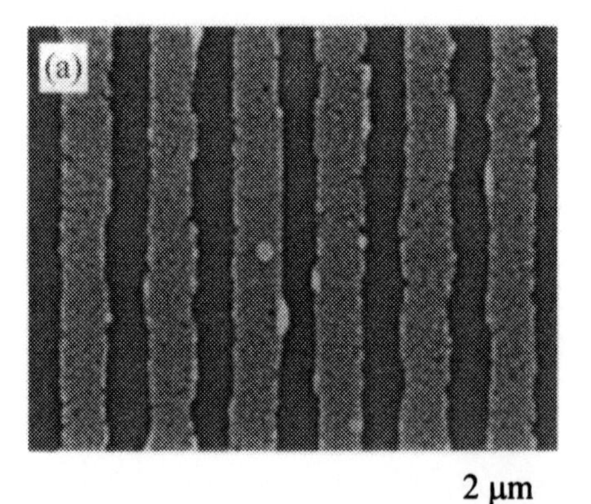

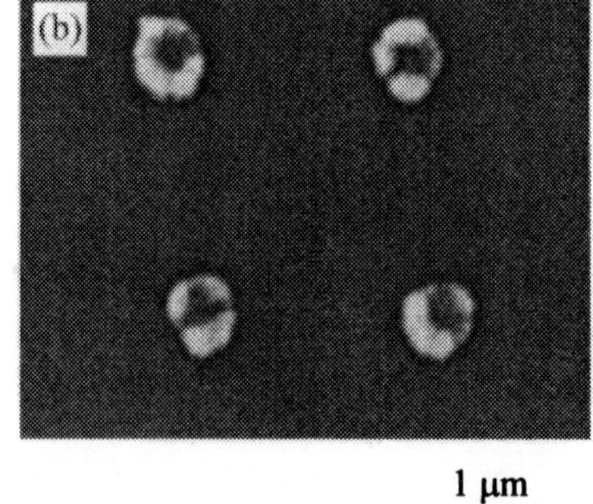

Fig. 3 SEM micrograph of the substrate soaked in $(NH_4)_2TiF_6$ and B_2O_3 mixed solution and then soaked in acetone with ultrasonic vibration for 5 h. (a) Line pattern. (b) Dot pattern.

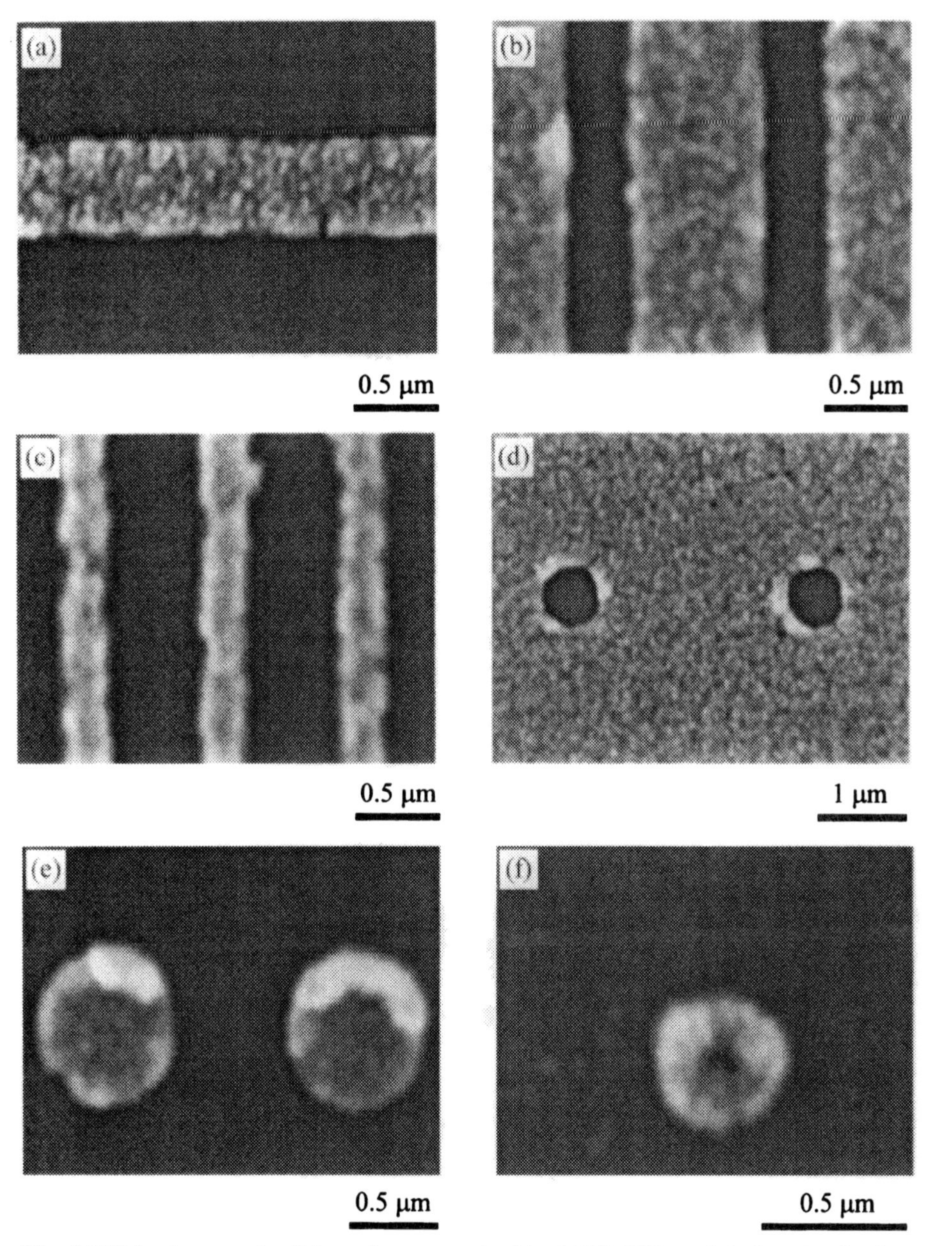

Fig. 4 SEM micrograph of the substrate soaked in $(NH_4)_2TiF_6$ and B_2O_3 mixed solution, exposed to UV rays for 6 h and then soaked in acetone with ultrasonic vibration for 5 h.

titanium dioxide thin film just on and the micropattern of titanium dioxide thin film transcribing the resist pattern was formed. The minimum line width is about 1 μm and the minimum diameter of the dot pattern is about 0.5 μm.

Figure 4 shows SEM micrographs of the surfaces of the substrate soaked in $(NH_4)_2TiF_6$ and B_2O_3 mixed solution for 24 h, exposed to UV rays for 6 h and then soaked in acetone for 5 h with ultrasonic vibration. The submicron-scale minute patterns transcribing the resist pattern were formed. It is shown from Fig.4(a) to (c) that line pattern with minute line width about 0.3 μm, which was the finest part of the resist pattern used in this study, was formed. It is seen from Fig.4(d) to (f) that dot pattern with minimum diameter about 0.3 μm, which was also the finest part of the resist pattern used in this study, was formed. It is considered that the titanium dioxide thin film acted as a photocatalyst by UV irradiation, the resist material was decomposed partially, and the contact between titanium dioxide thin film and resist material was weakened, then the damage of the titanium dioxide thin film during the dissolving of the resist material was reduced.

Figure 5 shows the results of EDX line analysis. It is seen from the Fig.5 that the intensity of Ti-Kα increases at the part of the micropattern. These result support that the submicron-scale macro patterns were formed by titanium dioxide.

Figure 6 shows the SEM micrograph of the submicron-scale micropattern. The view angle is 60°. It can be seen from Fig.6 that the resion of titanium dioxide rises sharply and that the edge of the pattern was clearly outlined.

Figure 7 shows the result of AFM observation of the surface of the substrate soaked in $(NH_4)_2TiF_6$ and B_2O_3 mixed solution for 24 h, exposed to UV rays for 6 h and then soaked in acetone with ultrasoncic vibration for 5 h . The hollow about 1 μm in width was formed by the dissolution of the resist material. The flatness of the bottom of the hollow means that the surface of silicon wafer was exposed. The edge is clear. It is seen that the titanium dioxide thin film has very uniform thickness. From the step of the micropattern, the thickness of the thin

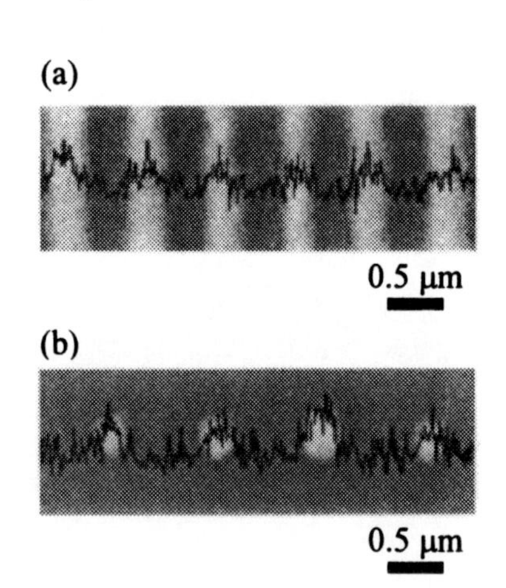

Fig. 5 EDX analysis of the micropatterns. (a) Line pattern. (b) Dot pattern.

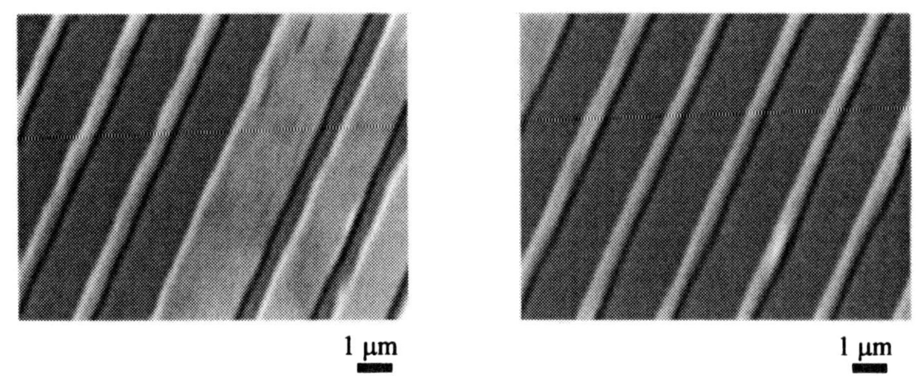

Fig. 6 SEM micrograph of the submicron-scale titanium oxide micropattern. View angle is 60°.

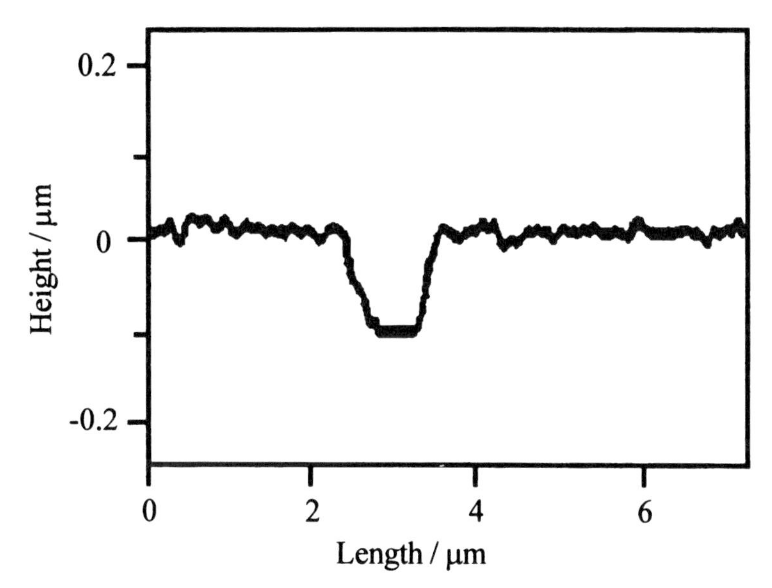

Fig.7 AFM observation of the substrate soaked in $(NH_4)_2TiF_6$ and B_2O_3 mixed solution, exposed to UV rays for 6 h and then soaked in acetone with ultrasonic vibration for 5 h.

film measured about 130 nm. It is indicated that the resist material dissolved off without damaging the titanium dioxide thin film.

CONCLUSION

The combination of titanium dioxide thin film synthesis reaction from aqueous solution and transcription of the resist pattern was made to fabricate submicron-scale micropatterns of titanium dioxide thin films.

Minute line and dot pattern of titanium dioxide with minimum size of 0.3 μm transcribing the resist pattern was well formed on the substrate. The fabrication of submicron-scale micropattern was very important for the application to electronic devices. This method is promising for producing capacitor and gate insulator, etc. of electronic devices of the next generation. This method is also considered applicable for the production of optical devices, bit patterns of magnetic memory devices, and the production of the parts of micro machines and so on.

REFERENCES

[1]T. Yao, T. Inui and A. Ariyoshi, *J. Am. Ceram. Soc.*, **79** [12], 3329-30 (1996).

[2]T. Yao, A. Ariyoshi and T. Inui, *J. Am. Ceram. Soc.*, **80** [9], 2441-44 (1997)

[3]T. Yao, *J. Mater. Res.*, **13** [5], 1091-1098 (1998).

[4]T. Yao, Y. Uchimoto and H. Yabe, *Mat. Res. Soc. Symp. Proc.,* **623**, 423-426 (2000).

[5]T. Yao, Y. Uchimoto, M. Inobe, K. Satoh and S. Omi, *Abstract of the 1999 Joint International Meeting of the Electrochemical Society and Electrochemical Society of Japan*, No.1399 (1999).

STUDY OF SURFACE DONOR-ACCEPTOR ACTIVE CENTERS DISTRIBUTIONS DURING CERAMICS BALL MILLING

M.M. Sychov, O.A. Cheremisina, V.G. Korsakov,
S.V. Mjakin, V.V. Popov and N.V. Zakharova
St. Petersburg Institute of Technology,
26 Moscovsky prospect, 198013 St. Petersburg, Russia

L.B. Svatovskaya
St. Petersburg Institute of Railway Transportation
9 Moscovsky prospect, 198013 St. Petersburg, Russia

ABSTRACT

Donor-acceptor properties of dispersed materials determine the energy of interface interactions thus being responsible for structuring processes and consequently for the performances of ceramic and composite materials. We report study of distributions of surface donor-acceptor adsorption centers and acidity function of barium titanate and $YwiiO_3$-based material VS-1 and their changes after surfactant adsorption and in the process of ball milling in aqueous surfactant solution. Distribution of adsorption centers was studied by adsorption of calorimetric indicators. Total quantity of active surface centers was also controlled. Approach is proved to be useful for choice of the surfactant, which have groups reactive to specific type of surface adsorption centers, and for control of dispersing process kinetics and efficiency.

INTRODUCTION

Dispersed ferrielectric materials are used in the fabrication of sensors, high-k polymer composites and capacitors[1]. Important step in this technology is obtaining of well-dispersed and stable water suspensions of ceramic material. Interface interactions play critical role at this stage controlling adsorption processes, polymer-solid interactions, structuring in the system and its properties[2]. Therefore information about the quantity and distribution of surface active centers is crucial for chemical characterization and certification of materials and during study of interface interactions[3]. Particularly knowledge of distributions of surface donor-acceptor centers may be used for evaluation of interactions between surfactant and powder material during obtaining of water suspensions of ceramic materials.

We studied changes of distributions of surface donor-acceptor centers and acidity function of barium titanate and YwiiO3-based material VS-1 as well as changes of their surface properties after surfactant adsorption and in the process of ball milling in aqueous surfactant solution.

EXPERIMENT

Subjects of study were dispersed $Ywiie_3$ and $Ywiie_3$-based ceramic material VS-1 – mixture of BaTiO3 and percentage amounts of $CaZrO_3$, Nd_2e_5 Nb_2e_5, ZnO, $MnCO_3$ (Kulon Ltd.). Dispersing process was carried out in the ball mill in the aqueous solution of dispersant with rotation speed 48 rpm. Oxyethylidene-diphosphonat OEDF-3A was used as dispersant because it provided maximum content of solid phase (75...85 %wt.) for both $Ywiie_3$ and VS-1 in suspensions at dispersant content 0,4...0,6 %wt. Optimum composition of suspensions was found from rheologycal studies[4].

The surface properties were characterized by absorption of acid-base calorimetric indicators with $àa_w$ values of the transition between acidic and basic forms in the range of pK_a -5…13. These indicators are undergoing selective adsorption from standard aqueous solutions onto the corresponding surface centers of dispersed solids[5]. The content of various adsorption centers on the surface was determined by measurements of the optical density at the wavelength of maximum light absorption of the acidic or basic form of a certain indicator for the following solutions:

(a) the initial aqueous solution of the indicator of a certain concentration (D_0);
(b) the same solution containing analyzed powder (D_1) where the optical density is affected by both the indicator adsorption and pH change of the medium caused by the water-surface interaction;
(c) solution of the indicator in the water decanted after the contact with a sample of the same weight as in the measurement (2) where the optical density is only affected by medium pH change due to water-surface interaction allowing elimination of this effect at the subsequent subtractive calculation (D_2).

These measurements allow the evaluation of quantity of active centers with a certain pK_a value on the studied surface according to the following equation:

$$q_{pKa} = \left| \frac{|D_0 - D_1|}{m_1} \pm \frac{|D_0 - D_2|}{m_2} \right| C_{ind} V_{ind} / D_0, \qquad (1)$$

where C_{ind} is the concentration of the indicator solution, mole/L; V_{ind} - volume of the indicator solution used in the analysis, L. The m_1 and m_2 – weights of the samples for measuring D_1 and D_2 correspondingly, g. Positive sign corresponds the case when D_1 and D_2 are oppositely changed related to D_0, ($D_1<D_0$ and $D_2>D_0$, i.e. the changes in optical density caused by adsorption and water-surface interaction are opposite and the decrease of optical density due to the indicator adsorption is greater than the increase due to the water-surface interaction). Negative sign corresponds to one-sided optical density change caused by both adsorption and water-surface interaction ($D_1<D_0$ and $D_2<D_0$, or $D_1>D_0$ and $D_2>D_0$, i.e. either water-surface interaction results in the decrease of the optical density or it results in the growth of this value but the decrease caused by adsorption is not sufficient to compensate it).

The integral surface acidity was evaluated as:

$$H_0 = \frac{\Sigma(pK_a \cdot q_{pKa})}{\Sigma pK_a} \qquad (2)$$

Particle size distribution was studied by laser beam scattering using Mastersizer μ Ver.2.12 from Malvern Instruments Ltd.

RESULTS AND DISCUSSION

Surface Properties of $Ywiie_3$ and VS-1 before and After Adsorption

As one can see from the fig. 1, on the surface of powder $Ywiie_3$ there are three main types of centers. Centers with $àa_w$= -4.4 supposedly formed by two-electron orbital of oxygen atom bonded to barium, capable to form coordination bond with free orbitals of absorbed indicator molecule. Brensted acid centers ($àa_w$= 2.0-2.5) are ≡ii–ed radical. Brensted base centers ($àa_w$= 10-11) were identified as –Yw–ed radicals. Acidity function of the surface of $Ywiie_3$ is d_0 = 1.9.

Accordingly to fig. 1 surface of $Ywiie_3$ contains maximum amount of centers with $àa_w$ = -4.4 and $àa_w$ = 2.5. Therefore for its absorption modification it is necessary to use dispersants containing electron-acceptor functional groups (e.g. -OH) capable to interact with electron-donor surface groups of dispersed solids thus reducing their surface activity. Indeed, after the treatment of barium titanate in water solution of OEDF-3A (fig. 1, graph 2) decrease of quantity of surface centers with $àa_w$ -4.4 and 2.5 took place. Possible mechanism of interactions is formation of hydrogen bonds between surface adsorption centers of $Ywiie_3$ and hydroxyl groups of surfactant (after hydrolysis of ammonium sol).

Content of centers with àa$_w$ 10-11 did not change after the treatment since dispersant molecules do not contain groups reactive to Ywiie$_3$ surface centers.

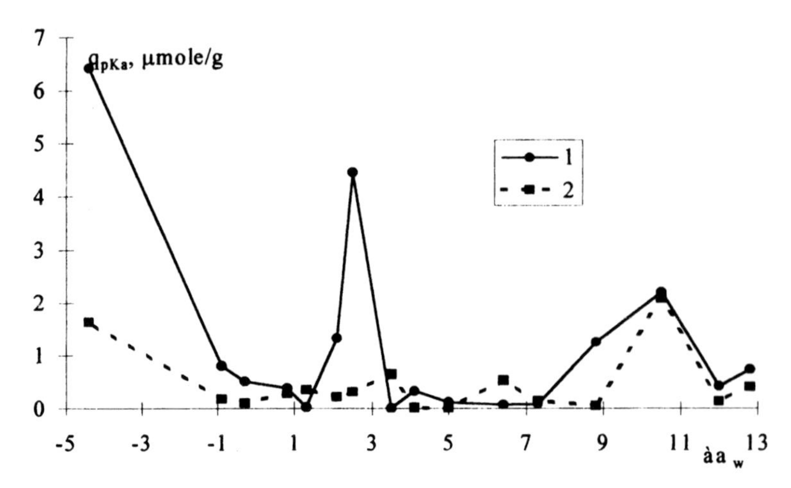

Figure 1. Distribution of adsorption centers on the surface of Ywiie$_3$:
1 - initial; 2 - after treatment with surfactant solution

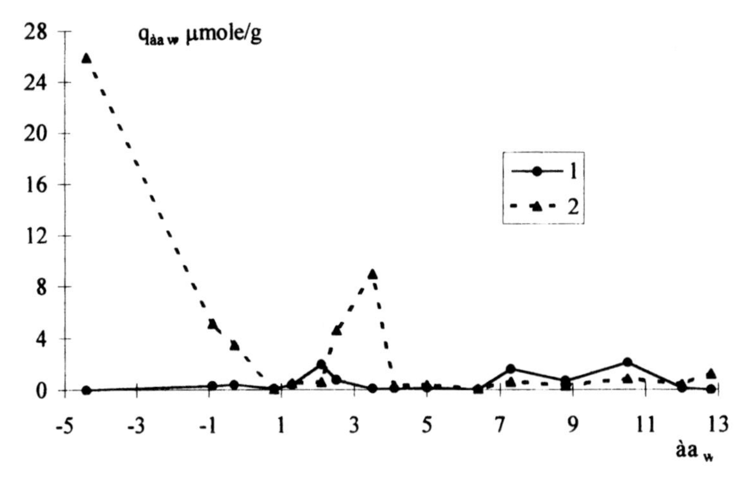

Figure 1. Distribution of adsorption centers on the surface of VS-1:
1 - initial; 2 - after treatment with surfactant solution

For the ceramic material VS-1 (fig. 2, graph 1) distribution is different in spite of the fact that it is 90% Ywiie$_3$ – total content of active centers (especially acid ones) is lower. It may be supposed that absence of the centers with àa$_w$ = -4.4 is caused by donor-acceptor interactions of unshared electron pair with free 4d-orbitals of Nb (or 4f of Nd) that leads to the centers blocking. Brensted centers with àa$_w$ 7.3 may be caused by Nb–ed (Nd–ed) groups. Due to higher relative content of Brensted base centers acidity function of the surface of VS-1 is higher than that of barium titanate and have the value 5.6.

Distribution of adsorption centers on the surface of VS-1 after the treatment with water solution of dispersant is shown on the fig. 2, graph 2. One can see that treatment leads to the appearance of significant amount of centers with pK_a -4.4 and 2-3 which are almost absent on the initial surface and distribution becomes very alike that of barium titanate. In the same time significant decrease of mean particle diameter take place - from 2.3 to 0.8 μm.

Since no mechanical treatment was made, this facts supports assumption that absence of these centers on the surface of initial VS-1 is caused by their blacking by center-center interactions between surface of $BatiO_3$ comprising the base of VS-1 and admixture oxides. As a result of treatment with surfactant solution particles aggregates decompose and unblocking of centers take place. Besides centers characteristic for barium titanate, treated VS-1 surface contains centers with pK_a 3.5 and 7.5 due to presence of admixture materials.

Changes of Surface Properties of Barium Titanate During Milling

Change of content and distribution of adsorption centers in the process of ball milling is caused by two main factors: dispersant adsorption/desorption (interactions with surface centers) and mechanic-chemical processes leading to the formation of new surface and sequentially new centers. As one can see on the fig. 3, after the treatment of $BaTiO_3$ in the water solution of dispersant total amount of surface centers Σq_{pKa} decreases three time due to adsorption of dispersant. During following ball milling Σq_{pKa} increases due to disintegration of aggregates and particles and formation of new surface. Maximum content of surface centers is after one hour of milling. Then it is decreasing due to adsorption of dispersant. Most change of mean particle size (from 2.6 to 1.6 μm) takes place during first 5-6 hours and therefore following decrease in surface activity must be caused by mechanically-induced desorption of surfactant. During last twelve hours of milling mean particle size changes from 1.6 to 1 μm.

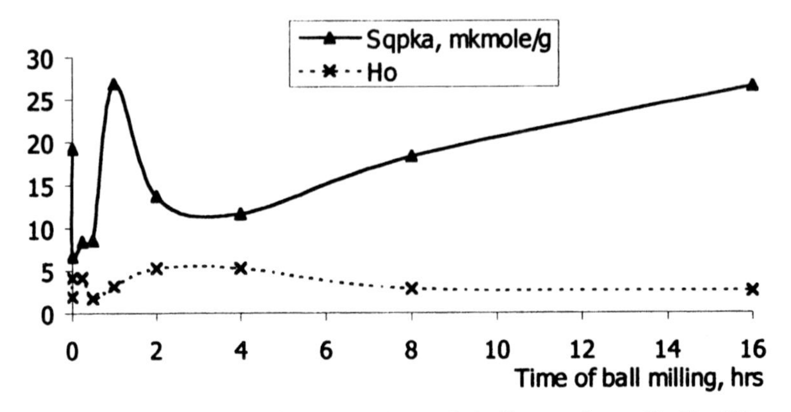

Figure 3. Dependence of Σq_{pKa} and H_o of Ywiie$_3$ on time of ball milling

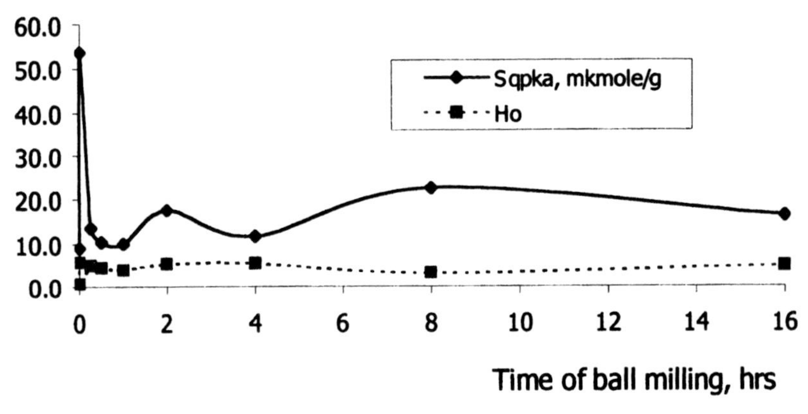

Figure 4. Dependence of Σq_{pKa} and H_o of VS-1 on time of ball milling

Interesting to note that changes of acidity function H_o are opposite to the changes of Σq_{pKa}. After the treatment of $BaTiO_3$ in the water solution of dispersant H_o changes from 1.9 to 4.1 due to discussed fact that Brensted base centers content did not change while content of centers with pK_a -4.4 and 2.5 increased drastically. Following decrease of H_o to the value 1.7 which close to that of initial surface is caused by increase of content of Brensted acid centers of freshly formed surface (the value of Σq_{pKa} is also close to that of initial surface). Further changes follow this scheme.

Changes of Surface Properties of Ceramic Material VS-1 During Milling

Treatment of VS-1 in water solution of dispersant caused significant increase of total amount of surface centers due to above discussed unblocking of centers. Unlike $BaTiO_3$ most changc of mean particle size of VS-1 take place without mechanical action and during following ball milling almost no change of particle size take place. Therefore milling led to the decrease of the total amount of adsorption centers during first half of hour due to adsorption of dispersant on the freshly formed surface. Further changes of Σq_{pKa} during milling caused by mechanically-induced adsorption/desorption processes.

Changes of H_0 of VS-1 surface as in case of barium titanate are opposed changes of Σq_{pKa}. Since during treatment in surfactant water solution basically centers with pK_a -4.4 and 2-4 were formed, increase of Σq_{pKa} led to was followed by decrease of the value of acidity function from 5.6 to 0.6. Adsorption of dispersant during following milling caused backward change of H_0 to the values around 5. That is close to the H_0 value for $BaTiO_3$ (5.2) at the time corresponding the lowest total content of absorption centers on surface of barium titanate (3-4 hours of milling), when mechanically-induced desorption of surfactant did not start yet.

Usage of Obtained Results

Obtained results allow to make a conclusion that optimum ball milling time for the studied sample of barium titanate sample is 4 hours, that provides properties of material required for the successful mixing with water dispersions of polymers for the following fabrication of polymer composites. Optimum time for the VS-1 powder is one hour.

Obtained suspensions nicely mixed with butadiene acrylonitrile latex that allowed to obtain uniform composite films of 20-40 mm thickness with the value of dielectric constant 35-45. These films were used for the fabrication of flexible electroluminescent capacitors[6].

CONCLUSIONS

Study of distributions of surface adsorption centers of powder barium titanate and $BaTiO_3$-based ceramic mixture VS-1 allowed to make a choice of needed dispersant type. It was found that surface of VS-1 contains significantly lower amount of centers that was explained by VS-1 blocking of surface centers take place due to center-center interactions of barium titanate and admixture oxides. After treatment in surfactant solution deblocking take place.

It was found that during treatment and milling of studied ceramic materials in surfactant water solutions the value of acidity function decreased when the amount of total surface adsorption centers increased and visa versa. That is due to the fact that during adsorption/desorption processes and disintegration of materials basically centers with pK_a values -4.4 and 2-3 appeared and disappeared.

Thus distribution of adsorption centers on the surface of powder materials allows to study particularities of interface interactions and proved to be useful tool for the choice of the surfactant, which have groups reactive to specific type of surface adsorption centers, and for control of dispersing process kinetics and efficiency.

Work resulted in optimization of technology of obtaining of water suspensions of dispersed ceramic materials for following preparation of their mixtures with a butadiene acrylonitrile latex and fabrication of uniform dielectric films.

ACKNOWLEDGMENTS

Authors gratefully acknowledge support provided by TRACE Photonics Inc., IL.

REFERENCES

[1] Richard E. Mistler, Eric R. Twiname. Tape Casting. Theory and Practice. American Ceramic Society, Westerville, Ohio, 2000, 298p.

[2] P.I. Ermilov, L.A. Tsvetkova, E.A. Indeykin, "Adsorption-Dispersion Equilibrium in Paint Systems", *Lakokrasochnye Materialy*, [6] 24—26 (1994).

[3] O.A. Raevsky, "Recognition Problem in Chemistry", *Teoreticheskaya I Experimentalnay Khimiya*, **22** [4] 450—463 (1986).

[4] O.A. Cheremisina, M.M. Sychov, V.V. Popov, V.G. Korsakov, "Regulation of Properties of Functional Latex Composites for Electroluminescent Sources of Light", *Jurnal Prikladnoy Khimii*, Dep. VINITI ©3960-Y99 1-20 (1999).

[5] L.F. Ikonnikova, T.S. Minakova, A.P. Nechiporenko, " Usage of Indicator Method for Study of Surface Acidity of Zinc Sulfide", *Jurnal Prikladnoy Khimii* [8] 1709-1715 (1990).

[6] M.M. Sychov, O.A. Cheremisina, N.V. Zakharova, V.V. Popov, N.V. Sirotinkin, V.G. Korsakov, A.I. Kuznetsov, "Flexible Electroluminescent indicator Based on Latex Functional Composite", *Proceedings of International Aerospace Conference APEP-2000*, Saratov, 470-474 (2000).

MODELING OF NON-LINEAR PHENOMENA DURING DEFORMATION OF INTERPARTICLE NECKS BY DIFFUSION-CONTROLLED CREEP

A. Maximenko and O. Van Der Biest
KU Leuven, Department MTM
Kasteelpark Arenberg 44
3001 Heverlee, Belgium

E.A. Olevsky
San Diego State University
5500 Campanile Dr.
San Diego, CA 92182-1323, USA

ABSTRACT

The deformation of a neck between two spherical particles of the same size by coupled grain-boundary and surface diffusion has been simulated. When a force is applied, the rate of approach of the particles and the rates of neck evolution are usually treated as linear functions of external stress in the neck area. In general, this assumption has some limitations.

During modeling of compressive hydrostatic loading, a non-linearity of the neck growth rate was detected for the case of active grain-boundary diffusion and high level of stresses. The non-linearity appeared only at the moment of transition from one stage of densification to another. It was confirmed that under compression shrinkage rate is a linear function of external loading throughout all stages of densification.

During tensile loading, non-linear effects were detected for the moment close to the beginning of neck reduction when external loading compensated compressive influence of sintering stress.

INTRODUCTION

Solid state sintering of ceramics is a result of the matter transport by different diffusion mechanisms activated by high temperature and high surface energy of aggregates of fine ceramic particles. Grain-boundary and surface diffusion dominate sintering if particles are small enough. In the present paper only these two mechanisms are considered.

If external loading is applied, the diffusional creep with the same transport mechanisms is observed. It is generally assumed that the densification rate of a powder compact and growth rates of necks between particles during creep are linear functions of the magnitude of external loading [1]. Linearity has been confirmed in many experimental observations, but from the theoretical point of view its existence seems surprising because of a strong non-linearity of the surface diffusion equations. Furthermore, it is known that the rate of the pore growth along the grain boundary during diffusional creep is a non-linear function of external loading for the crack-like non-equilibrium pores [2]. It is reasonable to assume that the similar non-linear behavior can be observed during pressure-assisted sintering or HIP. This is the subject of the present investigation.

STATEMENT OF THE PROBLEM

Deformation of a neck between spherical particles of the same size by the coupled grain-boundary and surface diffusion has been simulated numerically. The schematics of particle necks are given in Fig.1. The system is assumed axially symmetric with respect to the center-to-center line OO_1 given in Fig.1. During creep, plane circular necks of radius x between particles of radius R are developed. The dihedral angle at the neck edge is denoted as ψ. In the calculations ψ was taken equal to 0.8π. A co-ordination number of powder packing is approximately incorporated into the model through the packing angle φ. Cubic packing of powder particles is modeled as if there is a band of contacts all around the equator of the particle [1]. For the cubic packing, φ was taken equal to $\pi/4$. This approximation is convenient for retention of axial symmetry of the problem. A symmetry boundary condition is superimposed at the element of the particle surface corresponding to the packing angle φ. In numerical calculations it is sufficient to consider only a part of a grain boundary and a free surface within the angle φ (See Fig.1.) because all necks between particles are assumed identical. For the axisymmetrical statement of the problem, the picture in Fig.1 can be approximately treated as a section of a powder packing through centers of particles.

If effect of particle elasticity on grain boundary diffusion is neglected, the matter flux at the edge of the axisymmetric neck is [1]:

$$J_r = \frac{4D_g\Omega\gamma_s}{RxkT}\left(RK(x) + 2\frac{R}{x}\sin\left(\frac{\psi}{2}\right) - \bar{\sigma}\right)$$

where J_r is the volume of the material passing out of the grain boundary, x

is the neck radius, D_g is the diffusivity in the grain-boundary, Ω is the atomic volume, k is Boltzmann's constant, T is the absolute temperature, γ_s is the specific surface energy, K(x) is the sum of the principal curvatures at the edge of the neck, ψ is the dihedral angle, σ is the average external stress on the neck and $\bar{\sigma}$ stands for the dimensionless combination $\sigma R / \gamma_s$,. Curvature is defined to be positive if the center of the curvature is outside of the particle.

The shrinkage rate $\dot{w}$ can be found through J_r as:

$$\dot{w} = -\delta_g \frac{J_r}{x}$$

where δ_g is the thickness of the grain-boundary and $2\dot{w}$ is the rate of approach of one particle towards the other (see Fig.1).

The diffusion flux J_s along the free surfaces of particles is driven by chemical potential gradients depending on the surface curvature:

$$J_s = -\frac{\delta_s D_s \Omega \gamma_s}{kT} \frac{dK}{ds}$$

where δ_s is the thickness of the surface layer in which diffusion takes place and D_s is the surface diffusivity. The surface flux, which results in the deposition of material with the displacement rate ν normal to the surface, is equal to

$$\nu = \frac{1}{r} \frac{d(rJ_s)}{dr}$$

where r is the distance from the centerline of the particles (see Fig.1). The condition of the flux continuity at the edge of the neck can be given in the form:

$$J_r = 2J_s$$

It provides the boundary condition for the problem of the matter redistribution by the surface diffusion. In the numerical calculations time of the process was normalized by the characteristic time τ_g where

$$\tau_g = \frac{kTR^4}{\delta_g D_g \Omega \gamma_s}$$

If porosity of a powder compact is small and neck radii are large with $x/R > 0.3$, the axisymmetrical approach is not valid anymore but it becomes very close to the plain (2-D) one [3]. In other words, for the final densification, the modeling corresponds to the diffusional creep of packing of long wires and Fig.1 shows a cross-section of the packing by the plane normal to the wires.

The idea of the numerical approach for the modeling of the neck development during sintering was close to the one proposed in Ref. [3]. The numerical method was based on the method of lines where all derivatives along the surface of the particle were replaced by their finite-difference approximations. In that way the partial differential equations were converted to the set of ordinary differential equations in time space attributed to the coordinates of the particle surface in the nodal points of the finite difference mesh. The implicit numerical scheme with Newton iterative procedure for the solution of non-linear equations was used for the approximation of the ordinary differential equations. As a result of the modeling evolutions of the shrinkage rate $\dot{w}$, neck growth rate $\dot{x}$ and shape of the particles during different regimes of pressure assisted sintering were obtained. Different sintering conditions were taken into account by variation of the external stress σ and the relative rate ξ of grain-boundary diffusion compared to surface diffusion where ξ was defined as :

$$\xi = \frac{\delta_g D_g}{\delta_s D_s}$$

NON-LINEAR EFFECTS DURING DIFFUSION CREEP UNDER COMPRESSIVE LOADING.

Non-linear response to an external loading during diffusional creep can be observed only if loading favors considerable deviation of pores from their equilibrium shapes. It is possible in two cases: if the surface diffusion is comparatively slow or in the case of a high external loading. Traditionally it is assumed that diffusion along free surfaces of particles is much faster than diffusion along the grain boundaries. However, direct experimental measurements have proved that parameter ξ is very sensitive to the temperature and for high temperatures about $0.8T_m$ it can reach values ≈ 0.1 and even higher for many metal and ceramic materials [2,4]. In the numerical examples, parameter $\xi = 0.2$ was taken. The powder compact was subjected to hydrostatic loading with constant external stress σ throughout the densification.

The results of the modeling demonstrate that process of densification during pressure assisted sintering can be divided up into three stages in the way similar to the free

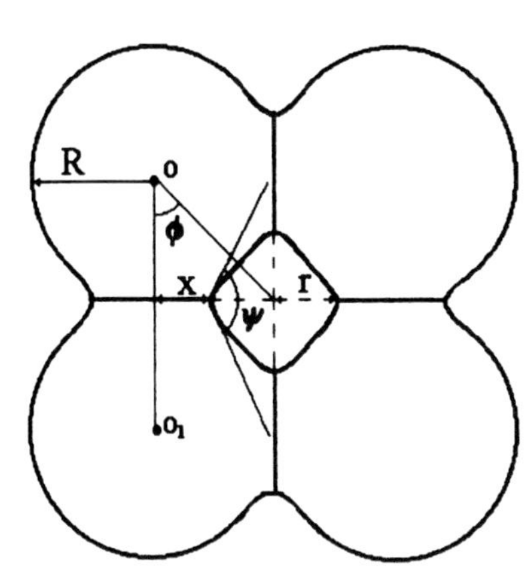

Fig.1 Schematics of particle necks

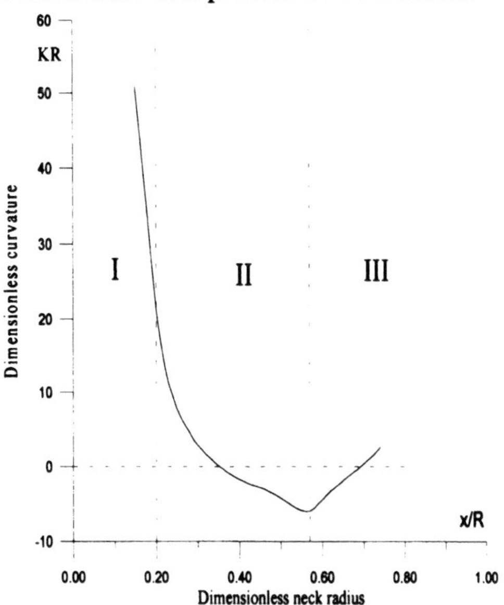

Fig. 2 Stages of densification

sintering. During the first and the third stages high curvature of pores dominates the densification. At the first stage, the curvature at edges of the small necks between particles is high and in the third stage the uniform curvature of small pores is high. During the first stage the curvature rapidly decreases and during the third stage it increases with a decrease of porosity. Both stages can be observed in Fig.2 where the dimensionless curvature KR at the edge of the neck is given as a function of the dimensionless neck radius for the densification with $\overline{\sigma} = -100$. These two stages are separated by the second stage where the external loading dominates. During this stage, the pores can be far from their equilibrium shapes. As it is clear from Fig.2, high levels of loading even change the sign of the curvature at the edge of the neck due to active extrusion of the material out of the grain boundary. Examples of non-equilibrium shapes of pores with different radii r/R are given in Fig.3 for $\overline{\sigma} = -300$. During transition from the second to the third stages, pores regain their equilibrium shapes. A higher level of loading induces later beginning of the third stage and higher rate of return to equilibrium.

The calculations predict that the shrinkage rate $\dot{w}$ is a linear function of external loading σ throughout all stages of densification. For the neck growth rate $\dot{x}/R$, considerable deviation from linearity was detected at the moment of transition from the second to the third stages (Fig.4). Position of the peaks at the curves in Fig.4 correlates well with the beginning of the third stage for the densification with $\overline{\sigma} = -100$ which is used as a reference process in Fig.4. As was mentioned above a higher compressive stress means transition to equilibrium pores at a smaller level of porosity. Comparison of neck growth rates at the same neck radius but from different stages of densification violates linearity of the process.

BEHAVIOUR OF NECK BETWEEN PARTICLES UNDER TENSILE LOADING.

In the case of neck deformation under tensile loading with a fixed positive $\overline{\sigma}$ it is also convenient to set apart three different stages. During the first stage initial small necks between particles grow due to high surface curvature in its vicinity. Neck growth rate decreases with the increase of its radius and decrease of curvature. During the second stage external loading dominates the process and neck radius decreases under tensile stress. If level of tensile stress is not very high the final separation of particles does not take place, at some moment neck reduction is ceased and all process is stopped. The system

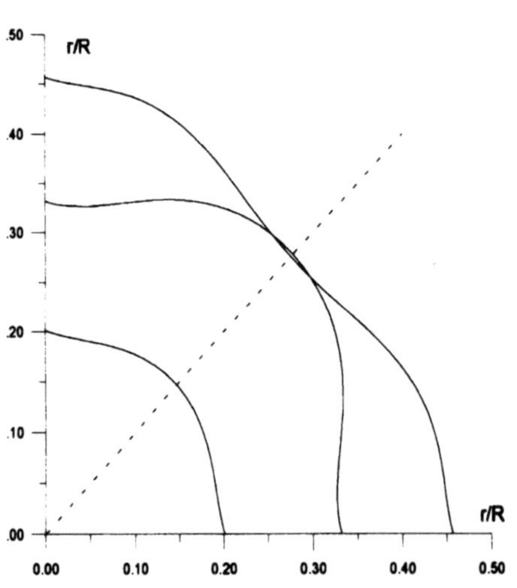

Fig. 3 Non-equilibrium pore shapes

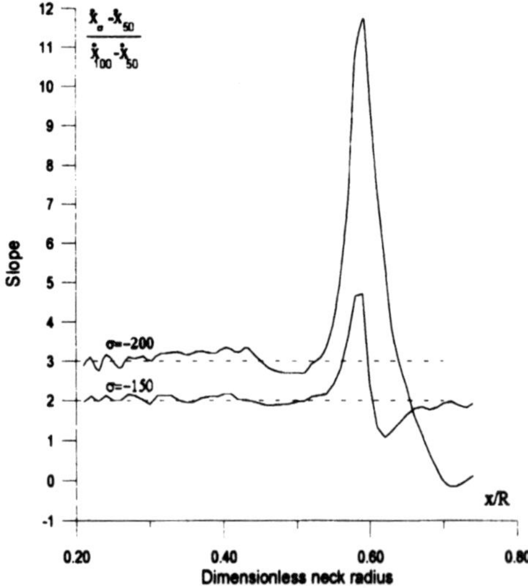

Fig. 4 Slope of the line dx/dt = f(σ) as a function of dimensionless neck radius

reaches equilibrium. The equilibrium neck radius as a function of the external stress is given in Fig. 5. With an increase of the tensile stress the equilibrium disappears and a very fast and unstable fracture of small necks is observed. The similar process of the development of crack-like cavities in the neck area under high tensile loading was described in [5]. The process of the unstable fracture of necks can be treated as a third stage of the diffusional creep.

The non-linear behavior of the shrinkage rate and neck growth rate as functions of external loading are observed only during transition from the first to the second stage of creep. The neck growth rates as functions of $\overline{\sigma}$ for the neck radius $x / R = 0.2$ and the parameters $\xi = 0.2$, $\xi = 2$ are given in Fig. 6. Numerical results predict that a tensile stress can prevent densification only if $\overline{\sigma} > 2.7$. This value is indicated in Fig.6.

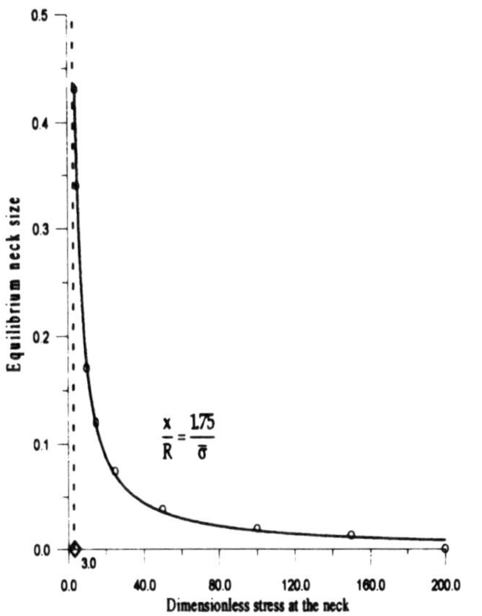

Fig. 5 Radius of equilibrium neck as function of external loading.

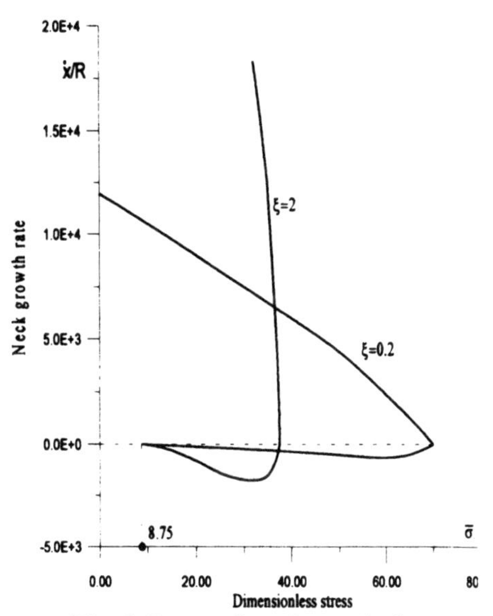

Fig.6 Rate of neck evolution a as a function of $\overline{\sigma}$ for x/R=0.2.

5. CONCLUSIONS

In general, during diffusional creep under hydrostatic loading the shrinkage rate of powder compacts and the average rate of the neck growth between particles are linear functions of the level of external loading.

Exceptions from this rule reveal themselves only at the moments of transitions from one stage of the diffusional creep to another. Under compression the shrinkage rate demonstrate almost ideal linear response to the level of loading in accordance with experimental observations. More detailed experiments are still needed for better understanding of rheological behavior of powder compacts under tension and for the analysis of the neck growth rate as a juncture of external loading.

ACKNOWLEDGEMENT

This work has been partially supported by the NSF Division of Manufacturing and Industrial Innovation, Grant DMI-9985427.

REFERENCES

1. D. Bouvard, R.M. McMeeking, Deformation of interparticle necks by diffusion controlled creep, J. Am. Ceram.Soc., **79** [3] 666-72 (1996)
2. T.-J. Chuang, K.I. Kagawa, J.R. Rice, L.B. Sills, Non-equilibrium models for diffusive cavitation of grain interfaces, Acta Metall., **27** 265-284 (1979)
3. J.M. Dynys, R.L. Coble, W.S. Coblenz, R.M. Cannon, Mechanisms of atom transport duting initial sintering of Al_2O_3, pp.391-404 in Proc. Fifth Int. Conf. Sintering and Related Phenomena, Edited by G.C. Kuczynski, Plenum Press, New York, 1980.
4. W. Zhang, J.H. Schneibel, The sintering of two particles by surface and grain-boundary diffusion - a two dimensional study, Acta Metall.Mater. **43** [12] 4377-86 (1995).
5. J. Svoboda, H. Riedel, New solution describing the formation of interparticle necks in solid state sintering, Acta Metall. Mater., **43** [1] 1-10 (1995)

MANUFACTURE AND CHARACTERIZATION OF LOW TEMPERATURE SINTERED Co_2Z CERAMICS

Xiao-Hui Wang, Longtu Li, Zhilun Gui, Shuiyuan Su , Ji Zhou
State Key Lab of New Ceramics and Fine Processing
Department of Materials Science and Engineering
Tsinghua University
Beijing, 100084, P. R. China

ABSTRACT

Z-type ferroxplanar ferrite Co_2Z ($Ba_3Co_2Fe_{24}O_{41}$) presents a gyromagnetic permeability whose resonance frequency stands around 2GHz. In this paper, ultrafine Co_2Z powders were synthesized by a novel citrate precursor method. Single phase Co_2Z was formed at a relatively low temperature between 1150-1200°C. The grain size of the powder derived at 1150°C was about 300nm. By adding a small amount of sintering aids, Co_2Z ultrafine powders can be sintered below 900°C to form ceramics with high density of 5.10g/cm^3. The low temperature fired ceramics were in Z-type phase, which exhibited good hyper frequency properties such as high initial permeability of 5, high quality factor about 50 and cut-off frequency as high as 1.5GHz. The results show this material can be used to produce multilayer chip inductor used in hyper frequency region.

INTRODUCTION

As well known, from the crystallographic point of view, Z-type ferrites are among the most complex compounds in the family of hexagonal ferrites with magneto-plumbite structure. The chemical formula of these ferrites is $M_3Me_2Fe_{24}O_{41}$, where M is a large divalent ion (Ba,Sr) and Me is a small divalent transition metal (Ni, Zn, Co etc.)[1,2]. Co_2Z, like $Co_2Y(Ba_2Co_2Fe_{12}O_{22})$, and $Co_2W(BaCo_2Fe_{16}O_{27})$ [3] is known as ferroxplanar ferrites, so called because its preferred direction of magnetization is at an angle to the c-axis. Co_2Z has a cone of magnetization at an angle to the c-axis below -53℃, above which temperature it has a preferred plane of magnetization perpendicular to the c-axis until 207℃, and then it undergoes a further change to magnetization parallel to the c-axis from 207 ℃ to the curie point at 400℃ [4]. This means that at room temperature, Co_2Z is magnetically soft, because although a large amount of energy is needed to move out of this plane, the magnetic vector can easily rotate within the preferred plane [5]. Co_2Z material has a much higher permeability and ferromagnetic resonance up to the 1.3~3.4GHz region [6], this brings it into the microwave region useful for inductor cores and ultra high frequency (uhf) communications.

It is well known, chip inductors made from Ni-Zn-Cu ferrites sintered below 900℃, have been frequently applied as one of the most important surface mounting devices (SMD) below 300MHz[7]. But there is few ideal materials to be used to produce chip inductors in the frequency range about 300MHz. How to decrease sintering temperature of Co_2Z to realize the co-firing of ferrite and Ag electrode material under 900℃ has been the main problem, because Co_2Z ceramics has a high sintering temperature of 1300 ℃ in a conventional ceramic route. Consequently, the aim of this study is to approach a better method to obtain novel Co_2Z material with good properties and low sintering temperature, in order to produce chip inductors applicable in the high frequency region.

In this paper, a novel and economical method--citrate precursor method was utilized to synthesize Co_2Z. Using this method, ultrafine Co_2Z ferrite powders were obtained at 1150℃, and could be sintered below 900 with suitable amount of sintering aids. The microstructures and electron-magnetic properties of the ceramics were investigated.

EXPREMENTAL PROCEDURE

Sample Preparation

The starting materials were iron citrate, cobalt acetate, barium acetate and citric acid. Stoichiometric amounts of the barium and cobalt salts were dissolved into hot iron citrate solution (1.0mol/L) at 70℃ with a small amount of polyethylene glycol to give the required Co_2Z composition. Then the pH value of the sol was adjusted to 7 ~ 8, by adding ammonia solution into it. The above sol was dried at 135℃, resulting in the citric precursor gel. After calcining the gel above 1150℃, single phase Co_2Z powders were obtained.

The powders were mixed with a suitable amount of sintering aid Bi_2O_3 or BPb glass, ground for 4 hours in water, dried, and pressed into toroidal (inner diameter of 10mm, outer diameter of 20mm and thickness of 3mm) and disk (diameter of 10mm and thickness of 1mm) samples. Sintered at temperature 870~900℃, ceramic specimens with high density were achieved.

Characterization and Property Measurements

The decomposition temperature of the citrate precursor was determined using a differential thermal analyzer. The structures of the samples were determined by a Rigaku diffractometer using Fe Kα radiation. Transmission electron microscope (TEM) was used to study morphopogy of powders. The microstructures of the ceramic samples were observed using a HITACHI S-450 Scanning electron microscope (SEM). The magnetization measurements were carried out on a LDJ 9600 vibrating-sample magnetometer. The permeability spectra of the ceramic samples were measured on an HP4291B RF impedance analyzer from 1MHz to 1.8GHz. The DC resistivity of ceramics was measured by an HP 4140B using silver contact.

RESULTS AND DISCUSSION

DTA and TG studies

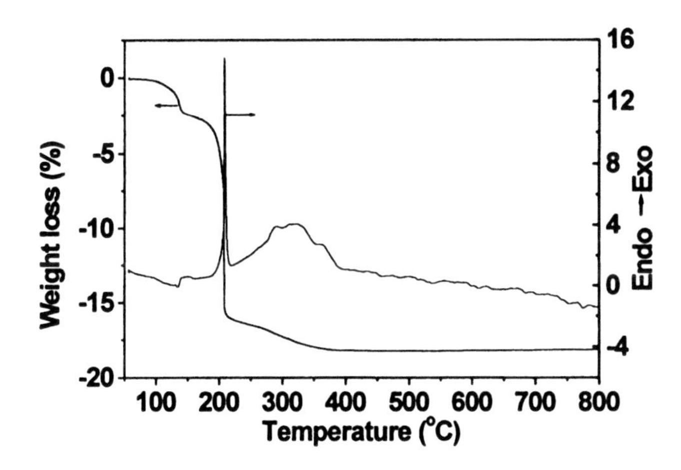

Figure 1 DTA and TG curves of the citric precursor

The decomposition processing of citrate precursor gel was recorded using

DTA and TG analyzer. Figure 1 shows DTA and TG curves of the gel. From DTA curve, two major peaks were observed and the whole reaction ended at about 450 ℃. Beyond 500℃, there is not any endothermic or endothermic peak and no weight loss, which indicates all the organic materials have burned out completely. Therefore, the gel was heat treated above 600℃ to remove organic compounds, then powders were obtained at various calcining temperature from 600℃ to 1250 ℃. Crystalline structures of the powders were determined by XRD.

X-ray Diffraction Studies

Unlike the preparation of $BaFe_{12}O_{19}$ (BaM) by citrate precursor method [8], in that case BaM powders can be directly obtained from precursor decomposition at the right temperature, the pure Z-type phase powder can't be obtained directly due to complexity of the crystal structure, but gradually transited from simple oxide or spinel to a mixture of BaM and Co_2Y to pure Co_2Z phase finally. As seen in figure 2, hematite (α-Fe_2O_3) and ferrite ($CoFe_2O_4$) had formed at 600℃as identified crystalline phases. With increasing temperature (at 800℃), these peaks are replaced by the appearance of BaM as the major phase. By 1000℃ the spinel had been lost due to the formation of Co_2Y, and BaM and Co_2Y appeared to co-exist in equal proportion. At 1100℃, the formation of Co_2Z phase seemed to coincide with the loss of BaM, but not Co_2Y, which co-existed with Co_2Z. Only when there is a further increase in temperature to 1150℃, Co_2Z seemed to be majority phase with a small amount of Co_2Y retention. Between 1150℃ to 1250℃, the majority phase Co_2Z kept stable. Due to the sharing of many peak positions the signals of small amounts of these hexaferrite compounds are extremely difficult to distinguish from the major phases in the very convoluted and similar XRD pattern of the hexagonal ferrites (refer to reference spectra in Figure 2). There was no evidence of Co_2Z decomposing to give Co_2W at temperatures even up to 1270℃. The above complex changes were confirmed further by magnetization measurements as the following discusses.

Magnetization Measurements of Powders

Magnetization of powder samples calcined at various temperatures were measured using vibrating-sample magnetometer at room temperature. Figure 3 shows the curves of magnetic hysteresis for these samples. The values of specific saturation magnetization, specific remanent magnetization and coercive force strength are given in table 1.

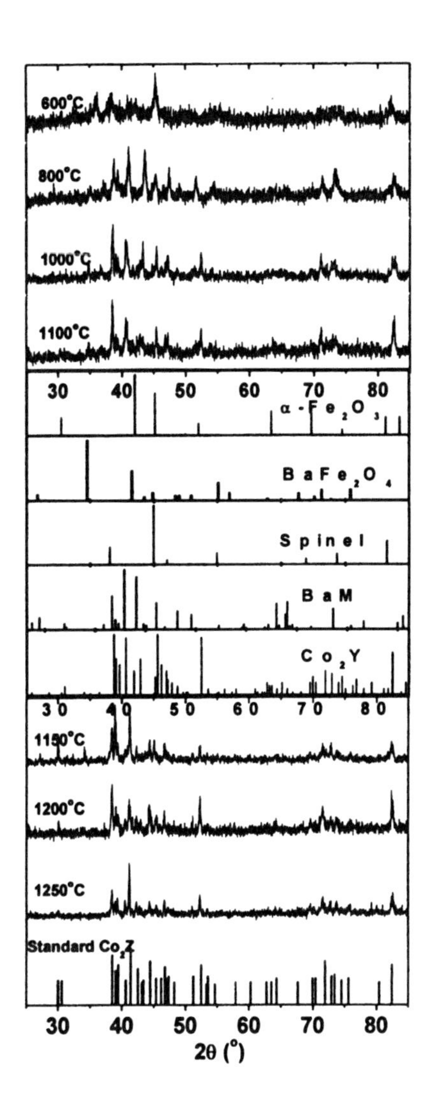

Figure 2 XRD patterns of powders calcined at different temperatures

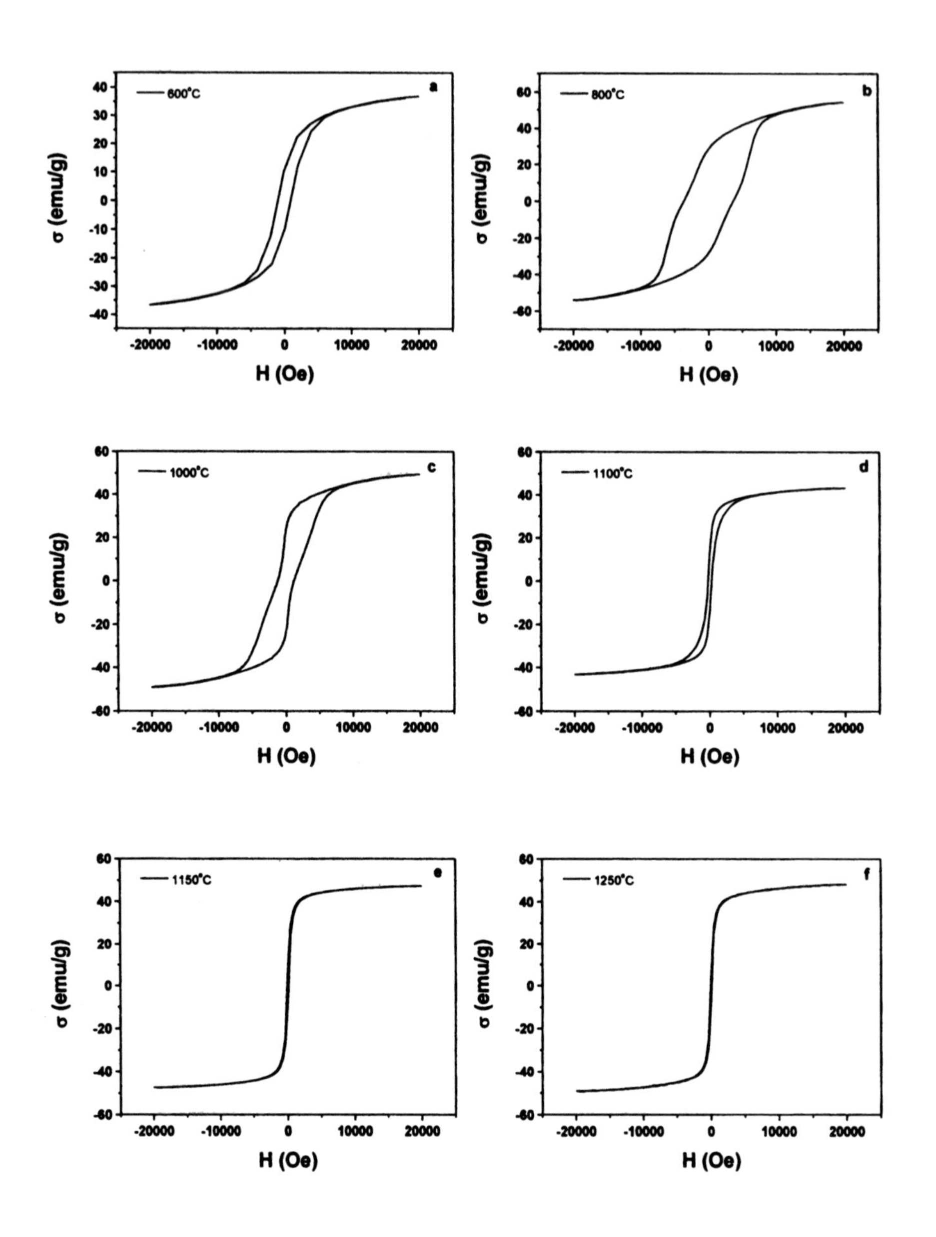

Figure.3 The magnetization curves of the samples caclcined at (a) 600 (b) 800℃ (c) 1000℃ (d) 1100℃ (e) 1150℃ (f) 1250℃

Table I Magnetization parameters for powders calcined at different temperatures

Sample number	Calcining condition	σ_s (emu/g)	σ_r (emu/g)	H_c(Oe)
1	600℃/6h	36.84	11.18	9444.24
2	800℃/6h	54.16	28.21	3663.90
3	1000℃/6h	49.44	23.43	1214.32
4	1100℃/6h	43.3	14.06	267.80
5	1150℃/6h	47.24	8.64	108.26
6	1250℃/6h	48.62	7.03	67.25

The 800℃ sample, whose XRD pattern is characteristic of BaM ferrite, does not saturate at 10KOe, as expected, because barium ferrite has a very high anisotropy field, of the order of 16KOe (as seen figure 3b). Obviously, the powder obtained at 1000℃ is a mixture of magnetically hard phase BaM and soft phase Co_2Y (figure 3c), which is consistent with XRD data. As seen in table 1, the values of coercive force strength and specific remanent magnetization decrease further with increasing calcining temperature due to the disappearance of BaM and occurrence of major Co_2Z phase. As the 1150℃ sample, the magnetization hysteresis curve appears typical features of magnetically soft materials (as shown in Figure 3e). Thus, it can be identified also from the results of magnetization measurement that pure Co_2Z can be achieved only after calcining above 1150℃.

Morphology of Co_2Z Powders

Figure 4 TEM image of Co_2Z ultrafine powder calcined at 1150℃

TEM was used to characterize Co_2Z powders obtained at various temperatures. Figure 4 shows hexagonal plates for Co_2Z ultrafine powders calcined at 1150℃. The average grain size is about 300nm in diameter and 50nm in thickness. When calcining temperature increases to 1250℃, platelet grains grow up to about 3 μ m in diameter.

Properties of Low -Temperature Sintered Co_2Z Ceramics

The ultrafine Co_2Z powders prepared by this method show excellent sintering property due to highly reactive particles, which could be sintered at 1150-1200℃ for 6hrs to form fully dense ceramics. This means the sintering temperature could be decreased to 1200℃, lower than 1300℃ used in a classical ceramic method. However, in order to realize the co-firing of Co_2Z hexaferrite and Ag electrode for MLCI application, the sintering temperature of Co_2Z must be decrease further below 900℃. Therefore, suitable amount of sintering aids Bi_2O_3 and BPb glass were introduced into the above Co_2Z powders to realize sintering below 900℃.

The magnetic properties such as initial permeability, quality factor and cut-off frequency were investigated for the ceramic samples doped with Bi_2O_3 as well as BPb glass.

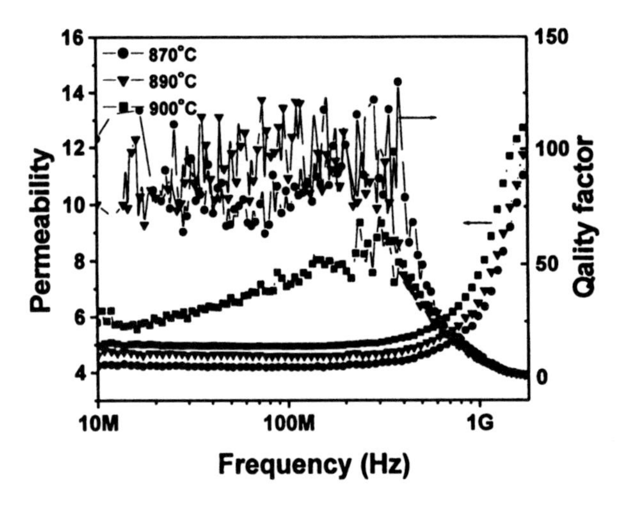

Figure 5 Frequency dependence of properties for Co_2Z ceramics with 2wt % Bi_2O_3 sintered at different temperatures

Figure 5 shows frequency dependent properties of ceramic samples with 2wt% Bi_2O_3. It is obvious, that the ceramics sintered at 900℃ exhibits good magnetic properties with the initial permeability as high as 5 and high quality factor above 50 at the frequency of 300 MHz. Moreover, the electrical resistivity of the sintered bodies reached 10^8~10^9ohm•cm. For comparison, the results of ceramics doped with appropriate amount of BPb glass are given in figure 6. Obviously, the initial permeability of the ceramics doped with BPb glass is much lower than that of ceramics doped with Bi_2O_3. Therefore, the optimum sintering aid for Co_2Z in this case is Bi_2O_3.

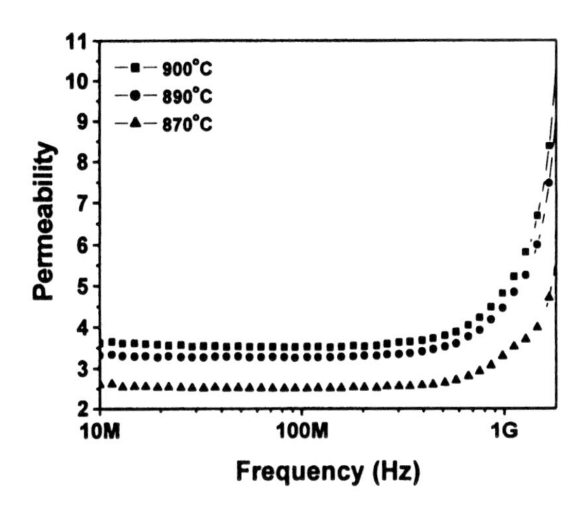

Figure 6 Frequency dependence of properties for Co_2Z ceramic with BPb glass sintered at different temperatures

The microstructure of Co_2Z ceramics with 2 wt% Bi_2O_3 sintered at 890℃ is shown in figure 7. The sintered body is of high density with few pores in it. The bulk density is 5.10g/cm^3 determined by the Archimedes method, more than 95% of theoretical density (d_{th}=5.35 g/cm3). The platelet grains were uniform with average grain size of 3μm.

CONCLUSIONS

In summary, ultrafine Co_2Z powders with hexagonal planar structure were synthesized by the citrate precursor method at temperature of 1150℃. By doping some sintering aids, sintering temperature of Co_2Z ceramic has been decreased below 900℃, which can meet requirement for multi-layer chip inductors. With sintering aid of Bi_2O_3, highly dense Co_2Z ceramics were achieved below 900℃, showing excellent high frequency properties such as high initial permeability of 5.0, quality factor of over 50 and cut-off frequency of 1.5 GHz. This soft magnetic material is a promising media applicable to produce high frequency MLCIs.

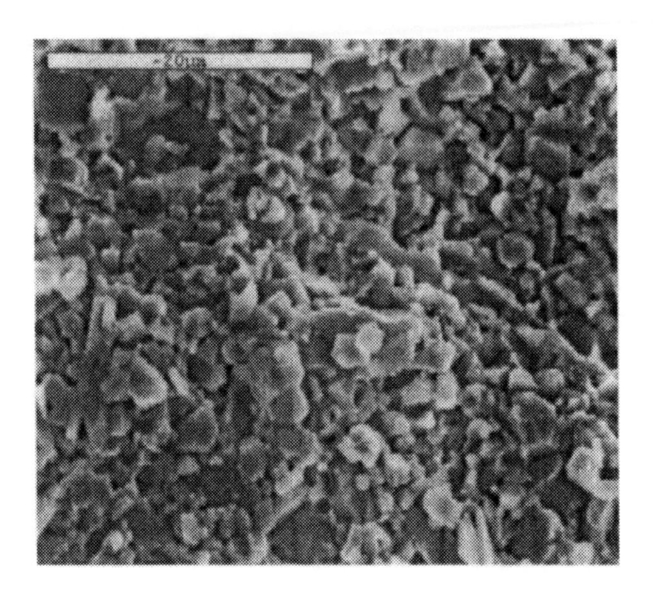

Figure 7 SEM micrograph of the Co_2Z ceramic sintered at 890℃

ACKNOWLEDGEMENT

This work was supported by the National Natural Science Foundation of P. R. China under grant No.59995523.

REFERENCES

[1]J. H. Hankiewicz, Z. Pajak and A.A. Murakhowski. "Nuclear magnetic resonance in $Ba_3Co_2Fe_{24}O_{41}$ ferrite", *J. Magn. Magn. Mater.*, **101** 134-136 (1991).

[2]S. Nicolopoulos, M. Vallet-Regi and J. M. Gonzalez-Calbet, "Microstructural study of hexaferrite related compounds: $Z(Ba_3Cu_2Fe_{24}O_{41})$ and $BaFe_2O_4$ phase", *Mater. Res. Bull.* **25** 567-574 (1990).

[3]R.C.Pullar, S. G. Appleton, M. H. Stacey, M. D. Taylor, A. K. Bhattacharya, "The manufacture and characterization of aligned fibres of the ferroxplana ferrites

Co_2Z, 0.67% CaO-doped Co_2Z, Co_2Y and Co_2W", *J.Magn. Magn. Mater.*,**186** 313-325 (1998)

[4]J. Smit, H.P.J.Wijn, *Ferrites,* Philips Technical library, Eindhoven, pp204-207, 1959.

[5]C. Heck, *Magnetic Materials and their Application,* Buterworths, London, pp. 511-517, 1974.

[6]J. Smit, H.P.J.Wijn, *Ferrites*, Philips Technical library, Eindhoven, PP.278-282. 1959.

[7]M. Fujimoto, "Inner stress induced by Cu metal precipitation at grain boundaries in low temperature fired Ni-Zn-Cu Ferrites", *J. Am. Ceram. Soc.*, 77(11) 2870-2878(1994)

[8]V. K. Sankaranaryanan, D. C. Khan, "Mechanism of the formation of nanoscale M-type barium hexaferrite in the citrate precursor method", *J. Magn. Magn. Mater.*, **153**(3) 337-346 (1996)

FABRICATION AND COFIRING BEHAVIORS OF LOW-SINTERING MONOLITHIC PIEZOELECTRIC TRANSFORMERS

Longtu Li, Ruzhong Zuo, Zhilun Gui
State Key Lab. of New Ceramics and Fine Processing, Department of Materials Science and Engineering, Tsinghua University, Beijing, 100084, P.R. China

ABSTRACT

Low-sintering monolithic piezoelectric ceramic transformers (MPT) were successfully fabricated. Low-firing PZT-based quarternary system piezoelectric ceramics were cofired with Ag-Pd inner electrodes to form the monolithic structure. Line shrinkage profile measurements of the ceramics and internal electrode paste showed little mismatch and favorable cofiring characteristics. Silver migration during cofiring was found by energy-dispersive X-ray spectroscopy (EDS). The interfacial microstructure of prototype devices by scanning electron microscopy (SEM) demonstrated that the MPT had desired physical integrity and high reliability. The effects of cofiring defects on the performance of the MPT were discussed.

INTRODUCTION

In recent years, the rapid spread of portable electric gadgets, such as notebook-type computers and digital video cameras, has promoted the demand for miniaturized and highly efficient transformers, which can be applied in the backlight power for liquid crystal displays (LCDs) and convert low dc battery voltage to high frequency and high voltage.

Multilayer piezoelectric ceramic transformers (MPTs) which have many advantages over conventional piezoelectric ceramic transformers, such as low driving voltage and high voltage step up ratio, have been extensively investigated.[1-4] However, the preparation of these MPTs adopts process technology similar to monolithic multilayer ceramic capacitors. The cofiring between piezoelectric ceramics and inner metal electrodes must be involved. Thus, the interaction between the electrode and the ceramic has been a major concern for the reliability of multilayer piezoelectric transformers. In the past, platinum or palladium was considered as adequate electrode materials, which allowed almost unchanged sintering conditions and a free ceramic composition. However, from a commercial viewpoint is it necessary to develop MPTs having superior electrical properties, a high reliability and a possibility of mass production at low cost.

In this paper, we reported low-firing high-performance multilayer piezoelectric ceramic transformers based on quarternary PMN-PNN-PZT ceramics and 70Ag-30Pd electrodes, which have outstanding electrical response, high voltage step up ratio and low sintering temperature. In view of the reliability of MPTs, the associated interfacial problems during cofiring were discussed. The ceramic-doped electrode technology was used to modify the cofiring behaviors and the favorable interfacial integrity was obtained.

EXPERIMENTAL PROCEDURE

Preparation of Multilayer Piezoelectric Transformers

$Pb(Mg_{1/3}Nb_{2/3})O_3$-$Pb(Ni_{1/3}Nb_{2/3})O_3$-$Pb(ZrTi)O_3$ (abbreviated as PMNNZT) was prepared by conventional oxide calcination processing using the chemical grade and analytical grade raw materials Pb_3O_4, TiO_2, Nb_2O_5, $Ni(AC)_2 \cdot 4H_2O$, $Mg_5(OH)_2(CO_3) \cdot 6H_2O$, ZnO, Li_2CO_3, etc. The as-calcined PMNNZT ceramic powders was mixed with binder, solvent, dispersant and plasticiser to obtain homogeneous slurries for tape casting. Green tapes were screen-printed with 70Ag-30Pd electrode paste and then laminated into green prototype MPT samples, whose structure is shown in Fig. 1.

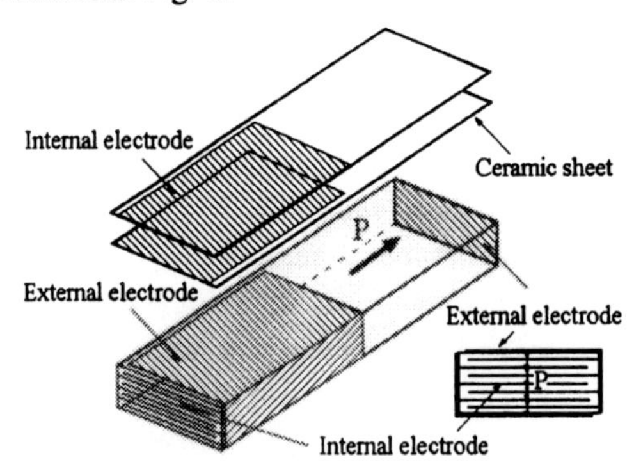

Fig. 1 Schematic image showing the inner structure of MPTs

In order to minimize problems caused by shrinkage and expansion of the specimens during the burn-out process, binder burn-out was carried out at 400°C for 12 h at heating rate of 5°C/h and cooling rate of 30°C/h. Afterwards, the MPT samples were sintered at 1050°C for 3h in a closed Al_2O_3 crucible in air. The external electrodes were formed by printing silver paste. Finally, the MPTs were poled by applying 30 kV/cm dc field at 90°C for 30min.

Measurements

The electrical property parameters of the MPTs were measured by HP4194A impedance analyzer and d_{33} meter. The sintering densification behavior of the ceramic and the electrode was measured by thermal mechanical analysis (TMA92, SETARAM, France) at a heating rate of 8°C/min. The microstructure at the surface and the cross section of the MPTs was observed by scanning electron microscopy (SEM, JSM 6301F) equipped with an energy dispersive X-ray spectroscopy detector (EDS, Link ISIS300). EDS was used to analyze the interdiffusion of different compositions near the interface.

RESULTS AND DISCUSSION

Table 1 lists the fundamental properties of the MPT sintered at 1050°C. The high-performance MPTs would be successfully produced in mass production at low cost and well satisfy the requirement of the backlight power source of LCDs of portable computers. This may be attributed to not only high property PMNNZT piezoelectric ceramic materials, but also excellent preparation processing.

Table Ⅰ. The properties of the MPT sintered at 1050°C

Sizes l x w x t (mm^3)	C_i (nf)	C_o (pf)	d_{33}(i) (C/N)	d_{33}(o)(C/N)	A_{vo}	$f_{r(\lambda)}$(kHz)
33 x 6.1 x 2.7	2000	40	7000×10^{-12}	450×10^{-12}	2000	104-110

As shown in Fig. 2, the surface of the fabricated MPT has a grain size of 2-3 μm and a dense structure with no open pores. It is clearly revealed from Fig. 3 that the cofired MPTs with the 70Ag-30Pd electrode show good sinterability without shortage of electrodes and the interfacial separation between the electrode and the ceramic caused by the mismatch of sintering densification behavior between the metallic electrode and the ceramic during cofiring. The electrode line is straight and the thickness of the ceramic layer is uniform. And also, the MPT does not exhibit the delaminations and cracks in the ceramic bodies, the gapping phenomena induced by thermal expansion coefficient difference between Pd internal electrodes and ceramic parts or protection phenomena by thermal diffusion of Ag external electrodes. Therefore, the 70Ag-30Pd electrode paste used in this study shows good matchability with PMNNZT piezoelectric ceramics.

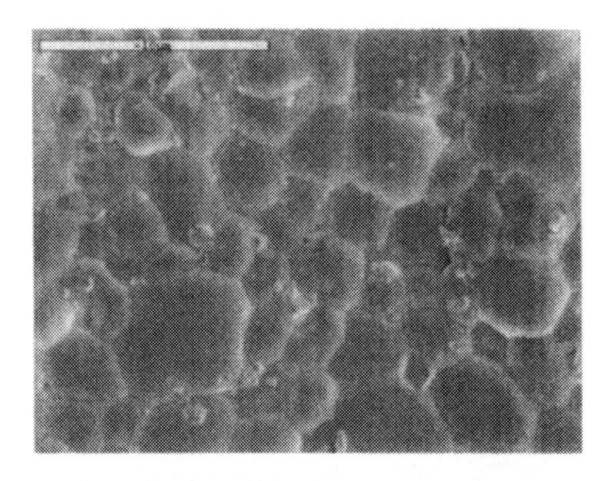

Fig. 2 SEM photograph of the MPT surface

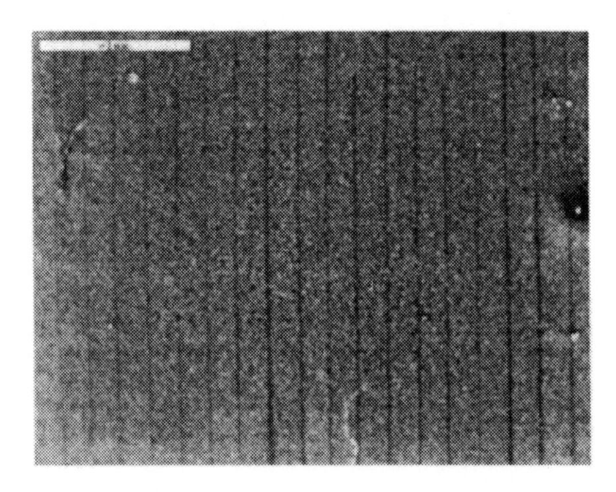

Fig. 3 Cross-sectional microgragh of the MPT showing interfacial integrity

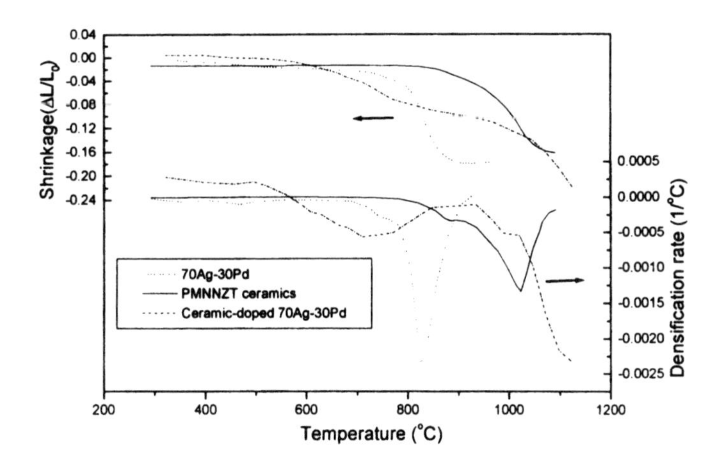

Fig. 4 Sintering densification behavior of PMNNZT ceramics and dried 70Ag-30Pd electrode paste at heating rate of 5 °C/min

Conventionally, the mismatched sintering shrinkage of the internal electrode and the ceramic is a major reason for producing cofiring interface defects. Fig. 4 shows the sintering characteristic of dried 70Ag-30Pd electrode paste and PMNNZT ceramics. It is evident that a certain sintering shrinkage difference occurs between the undoped Ag-Pd electrode paste and PMNNZT ceramics during cofiring, including sintering densification temperature and sintering shrinkage rate. The mismatched sintering behavior will inevitably cause the formation of cofiring defects, which may further degrade the reliability of the MPTs. Therefore, ceramic-modified 70Ag-30Pd electrode paste was used for the fabrication of the MPTs. The shrinkage of the internal electrode, by anti-sintering effect of the ceramic additive, was considerably reduced, as shown in Fig. 4. Thus, the good physical integrity of the as-fabricated MPT samples was greatly attributed to the ceramic-doped inner electrode technology, which is particularly important for preparing high-reliability MPTs. This is because the inner electrode paste used for the MPTs has no glass frit, generally different from the electrode paste used in the multilayer ceramic capacitors, considering the adverse effect of alkali metal ions in glass frit diffusing into the ceramic on the electrical properties of the piezoelectric ceramics. However, the weak interfacial bonding lacking of glass frit will lead to the invalidation of the MPTs in service. So, the addition of the ceramic powder also promotes the combination of the electrode and the ceramic at the interface.

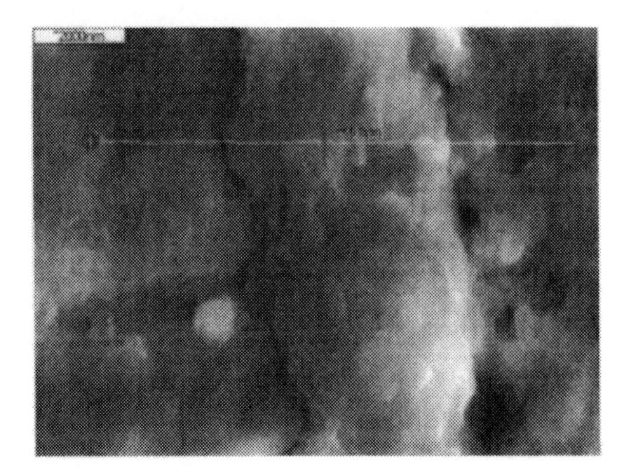

Fig. 5 Interfacail microstructure between PMNNZT and ceramic-doped 70Ag-30Pd electrode

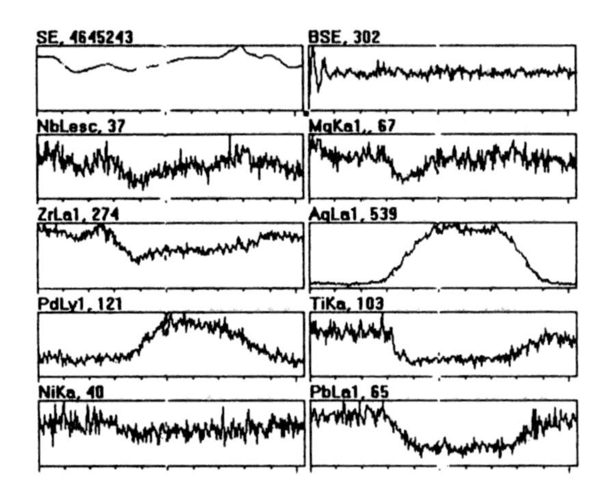

Fig. 6 Linear profile of different elements between PMNNZT ceramics and 70Ag-30Pd electrode (scanning distance is 10.3 μm)

Fig. 5 shows the microstructure between ceramic-doped 70Ag-30Pd electrode and PMNNZT ceramics. At the interface, the ceramic can be well cofired with the

Ag-Pd electrode. Except for the modified sintering densification, it is considered that the good combination of the ceramic particles in the electrode with the PMNNZT ceramic is another reason for the interfacial physical integrity. As shown in Fig. 5, an EDS line scanning analysis was made along the direction perpendicular to the interface. The linear distribution of different elements was indicated in Fig. 6. A little interdiffusion was observed including both silver ions and other metallic ions. The silver diffusion is a conventional phenomenon in the cofired systems based on the silver-rich electrode and the ceramic. However, the effect of silver migration on the electrical properties and reliability of multilayer piezoelectric ceramic devices could not be ignored. Moreover, It is considered that Pd and Pb interdiffusion should be correlated with Pd-Pb-based interfacial chemical reactions. Interestingly, metallic ions with considerable content were found in the internal electrode, for example, Zr, Mg, and Nb, etc. We think that except the contribution of PMNNZT ceramic composition, these metallic ions mainly come from doped ceramic additive. Seen from the measured properties of the MPTs, it seems that a little silver migration has no great effect on the properties. However, it is necessary to effectively restrain the interfacial diffusion of silver in view of further electromigration under the load of electric field. Currently, this work is in progress on the basis of previous work.[5]

ACKNOWLEDGEMENT

This work was supported by the National Natural Science Foundation of China under Grant No. 59995523.

CONCLUSIONS

High-performance low-sintering multilayer piezoelectric ceramic transformers were fabricated and successfully applied in the backlight power source of LCDs of portable computers. Low-firing quarternary system PZT-based piezoelectric ceramics shows excellent compatibility with ceramic-modified 70Ag-30Pd inner electrode paste. The good interfacial physical integrity of the MPTs was attributed to favorable cofiring characteristics. The experimental results demonstrate that no-glass frit ceramic-doped 70Ag-30Pd electrode has great significance for reliable multilayer piezoelectric devices.

REFERENCES

[1]T.R Shrout, W.A. Schulze, J.V. Biggers and L.T. Bowers, “Resonance

Behavior of Internally Electroded PZT Devices," *Materials Research. Bulletin*, **15** 551-559 (1980).

[2]H. Tsuchiya and T. Fukami, "Multilayered Piezoelectric Transformer," *Ferroelectrics*, **63** 299-308 (1985).

[3]H.Tsuchya, "Design Principles for Multilayer Piezoelectric Transformers," *Ferroelectrics*, **68** 225-234 (1986).

[4]L.T. Li, W.T. Deng, J.H. Chai, Z.L. Gui, and X.W. Zhang, "Lead Zirconate Titanate Ceramics and Monolithic Piezoelectric Transformer of Low Firing Temperature," *Ferroelectrics*, **101** (1990) 193-200

[5]R.Z. Zuo, L.T. Li, and Z.L. Gui, "Vapor Diffusion of Silver in Cofired Silver/Palladium-Ferroelectric Ceramic Multilayer," *Materials Science and Engineering B-Solid State Materials for Advanced Technology*, in press (2001).

FUNCTIONALLY GRADIENT RELAXOR DIELECTRIC COMPOSITES WITH X7R CHARACTERISTICS

Zhilun Gui, Ruzhong Zuo, Chengxiang Ji, Longtu Li
State Key Lab. of New Ceramics and Fine Processing, Department of Materials Science and Engineering, Tsinghua University, Beijing, 100084, P.R. China

ABSTRACT

The dielectric composites with X7R-type temperature dependence of dielectric constant (k) and high dielectric constant (k>9000 at room temperature) were prepared successfully by cofiring multilayer ceramic green bodies, each layer having different Curie temperature (Tc). During cofiring, the controllable interfacial interdiffusion made the composite possess functionally gradient characteristics with a continual change of Tc and flat dielectric-temperature response. The sensitivity to the stacking process increased the flexibility of structural design for the dielectric composite. Dielectric measurements showed that the layer-like dielectric was a multiphase ceramic composite. Favorable physical integrity from SEM observations demonstrated that the dielectrics were expected to be commercially used for high-property capacitors.

INTRODUCTION

The demand of high-performance ceramic capacitor for dielectric materials with low temperature coefficient of capacitance (TCC) and high dielectric constant greatly promoted the research and development in this field. Although the dielectric-temperature feature of modified $BaTiO_3$-based and/or Pb-containing relaxor-based ferroelectric ceramics can realise the Electronics Industries Association (EIA) X7R specification, the dielectric constant is not sufficiently large,[1,2] i.e., below 5000. Moreover, the heterogeneous relaxor materials synthesized by mixed-sintering[3] have a poorly commercial prospect because of low process reproductivity. Therefore, to date no great breakthrough in this field was realized in view of satisfying high properties and excellent process stability simultaneously.

In fact, a composite route in the preparation of functional or structural materials is generally considered as an effective method for obtaining high-property ceramic materials. Functionally gradient materials and layer-like structural composite materials are typical examples of obtaining excellent functional or structural characteristics by cofiring the composites of different types of materials with a suitable structural design. These materials are extensively applied to piezoelectric devices, composite coating, and high-intensity structural components because of their macroscopic inhomogeneity and gradually changed compositional distribution.

Recently, the "composite route" has been adopted to prepare a novel dielectric composite with high K and X7R characteristics, having an electrically parallel structure. It merits in the combination of high dielectric constant, flexible design, low TCC, and low sintering temperature. Thus, the capacitors made of the dielectric composite easily realize larger capacitance and highly thermal-stable property. Further, the composite structural design may be expected to apply in X7R-type high voltage capacitors because the high dielectric constant allows the increased active thickness of the dielectric composite in the same capacitance, which further improves breakdown tolerance. These possible practical applications become the main driving force for investigating the dielectric composite.

Therefore, the main objective of this work is to clarify the preparation of the dielectric composite and its dielectric-temperature response characteristic. The associated mechanism of obtaining outstanding properties was explained by microstructural observation and compositional analyses.

EXPERIMENTAL

Preparation of Dielectric Composites

Four kinds of lead-based relaxor ferroelectric ceramics used in this study were prepared by using reagent-grade raw materials PbO, MgO, ZnO, NiO, TiO_2, Nb_2O_5. Their main compositions were $Pb(Mg_{1/3}Nb_{2/3})O_3$-$Pb(Ni_{1/3}Nb_{2/3})O_3$-$PbTiO_3$ (A), $xPb(Mg_{1/3}Nb_{2/3})O_3$-$(0.96-x)Pb(Zn_{1/3}Nb_{2/3})O_3$-$0.04PbTiO_3$ (B, C, D, with different x value). The columbite precursor method[4] and Pb protection atmosphere were used to synthesize various ceramic powders. The dielectric properties measured from ceramic disks sintered at 1050°C are presented in Table 1.

Table I . Dielectric properties of different compositions A-D

Properties	A	B	C	D
$T(\varepsilon_{max})$/°C	-45	25	65	120
ε_{max}	7900	19500	23500	21000

Ceramic tapes of various compositions (A-D) were prepared by a conventional tape-casting technique, with a thickness of 150 μm. Green multilayer composites were fabricated by laminating these tapes according to a certain volumetric ratio and a suitable order. Organic binders in the green composite were removed slowly by the oxygenolysis at lower temperature up to 450°C, as confirmed by DTA data of organic mixtures, followed by a sintering schedule at 1050°C for 2h.

Measurements

The as-sintered multilayer composites were cleaned up and then were pasted with silver paste into the electrode for electrical measurements. The dielectric properties of multilayer composites were measured pseudo-continuously at 1 V_{rms} and 1 kHz at the temperature of –60°C-130°C using an impedance analyzer (HP4192A), a Delta 2300 automatic temperature control box, and a computer-monitoring system.

The interfacial microstructure and compositional line profile at the cross section of these composites were confirmed by field-emission scanning electron microscopy (SEM, JEOL 6301F) equipped with an energy dispersive X-ray spectroscopy (EDS, Link 2300, Oxford).

RESULTS AND DISCUSSION

Phase Confirmation and Microstructure

Fig. 1 shows the phase formation of compositions A-D. The individual ceramic sample sintered at 1050°C almost has complete perovskite structure, except those having higher-content Zn and Ni. Also, it is seen that perovskite stabilizer $PbTiO_3$ effectively restrains the formation of pyrochlore phase

The cross-sectional morphology of as-sintered dielectric composite is shown in Fig. 2. The brightness and darkness along the cross section evidently indicates layer-like distribution of different compositions inside the dielectric, showing that the dielectric has a composite structure even though it was sintered at high temperature.

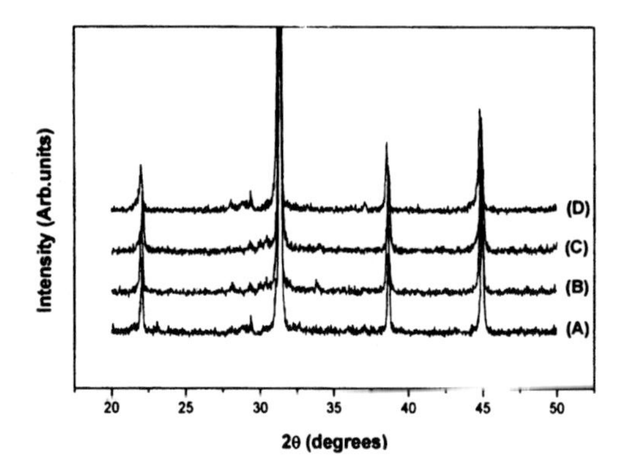

Fig. 1 XRD spectrum of as-sintered different composition samples making up dielectric composites

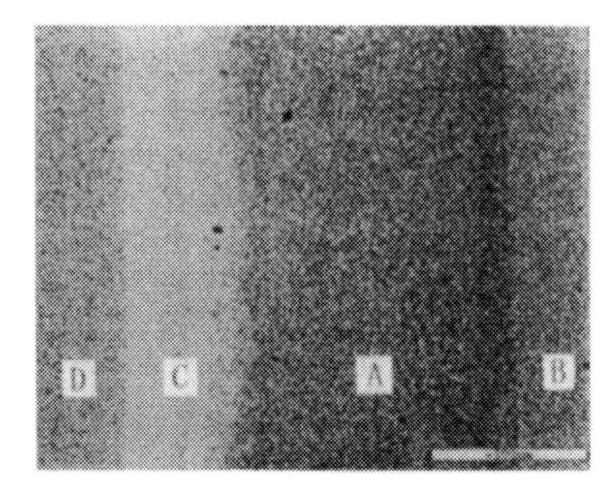

Fig. 2 Cross-sectional backscattered SEM image of a multilayer composite.

The grain morphology of different compositions near the interface is shown in Fig. 3. It is evident that there is a certain difference among the grain sizes of different compositions. A monotonous increase in grain sizes of B, C, and D is due to the improved sintering caused by ZnO addition,[5] which is based on the formation of the eutectics between PbO and ZnO at lower than 900°C. Importantly, there are no clear pores appearing at the heterogenous interfaces. The physical integrity at the interface and dense microstructure are shown.

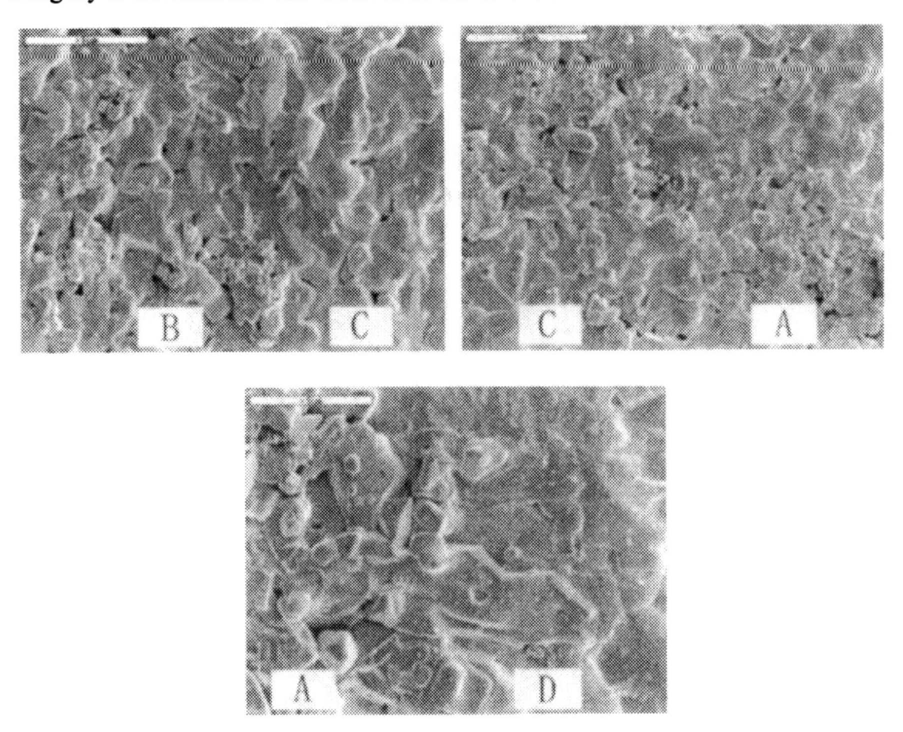

Fig. 3 Interfacial grain morphology in the as-sintered dielectric composite at 1050°C

Dielectric Response and Compositional Profile

Fig. 4 shows the temperature dependence of the dielectric constant of different dielectric composites. It is indicated that the laminating order of various compositions greatly affects the dielectric response of dielectric composites. Most of these dielectric composites can realize low capacitance-temperature variation up to X7R characterization (-55~+125℃, ±15%) with dielectric constant of above 9000 at room temperature. Fig. 5 simply accounts for the change of Tc of different phases in the dielectric composites before or after sintering. It is inevitable that interdiffusion between different phases will occur during cofiring. Just because of the interdiffusion, the compositional distribution along the cross section presents continuous gradient change, which finally leads to the continuous change of Tc.

This assumption is based on the peak-moving characterization of doped complex perovskite relaxors. Particularly, in PMN-PZN complex systems, the difference in the ratio of PMN and PZN will cause the continuous change of dielectric properties. Hence, the interdiffusion between layers makes the dielectric composites characteristic of functionally gradient materials. That's to say, four kinds of dielectric compositions with discontinuous T_c before sintering was changed into monolithic samples with continuous T_c along the direction perpendicular to the layers.

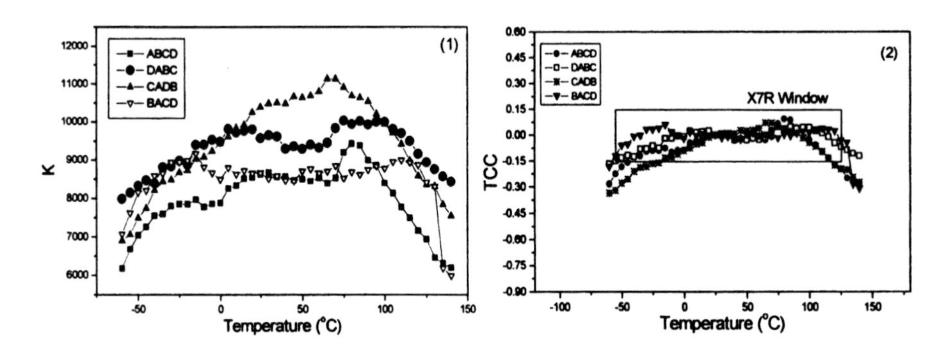

Fig. 4 Dielectric properties of different dielectric composites sintered at 1050°C: (1) dielectric-temperature property; (2) TCC.

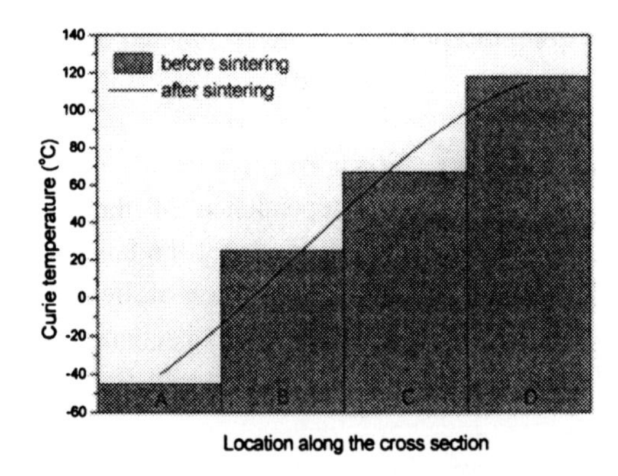

Fig. 5 Schematic image showing the change of the dielectric property for layer-like composites along the cross section before and after sintering.

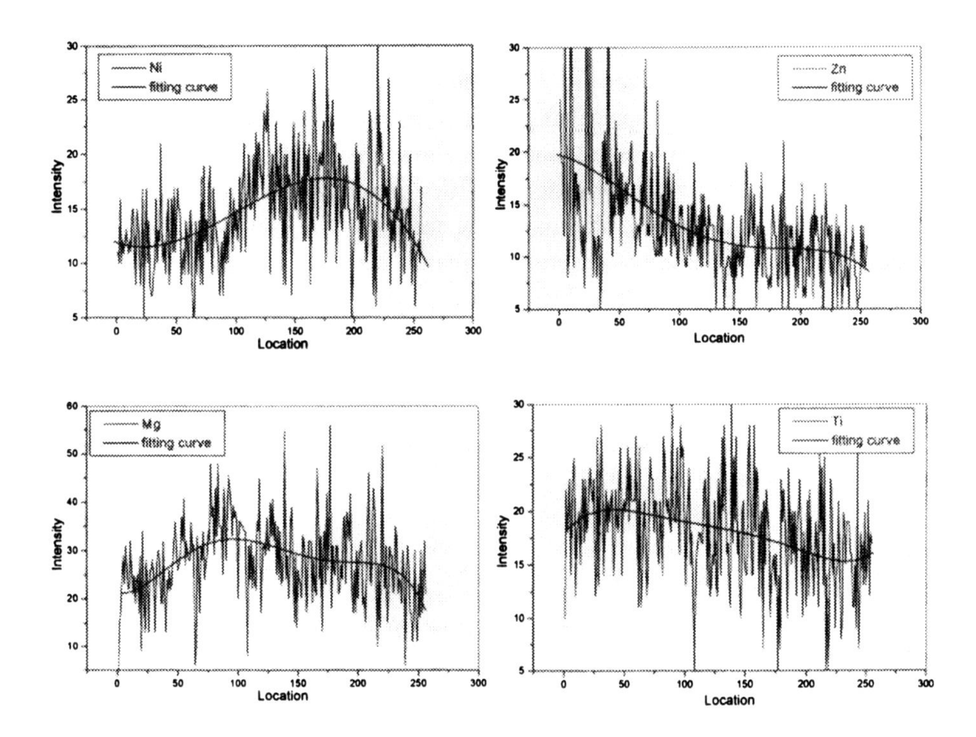

Fig. 6 EDS linear distribution of different elements in DCAB dielectric composite

The linear profile of the composition in DCAB composite (Fig. 3) was shown in Fig. 6. Different elements evidently present gradient distribution, particularly for Ni and Zn elements. In the initial composition, there is no Zn element in composition A and there is no Ni element in compositions B, C, and D. After cofiring at high temperature, interdiffusion of different elements between layers occurs. As discussed above, the gradient change of the compositions will cause the gradient change of the properties, which is the gradient distribution of the temperature corresponding to maximum dielectric constant for the dielectric composite. Hence, this dielectric composite is also called as functionally gradient relaxor ferroelectric composite, even though it is different from conventional gradient materials, which is to reach continuous functional change. This composite utilizes the continuous change of T_c to realize low TCC value.

CONCLUSIONS

High-performance layer-like dielectric composites were successfully fabricated by mature tape-casting and cofiring technique. Microstructural observation and compositional analyses show that multiphase structure and continuous distribution of dielectric properties are responsible for obtaining both high dielectric constant and thermal-stable features. The dielectric composite is hoped to be applied in high-performance plate capacitors, and even high-voltage capacitors because high dielectric constant increases the active thickness of capacitors under the condition of the same capacitance.

ACKNOWLEDGEMENT

This work was supported by the National Natural Science Foundation of China under Grant No. 59995523.

REFERENCES

[1]D. Hennings and G. Rosenstein, "Temperature-Stable Dielectrics Based on Chemically Inhomogeneous $BaTiO_3$," *Journal of the American Ceramic Society*, **67**[4] 249-52 (1984).

[2]F. Uchikoba and K. Sawamura, "JIS YB Lead Complex Perovskite Ferroelectric Material," *Japanese Journal of Applied Physics*: Supplement, **31** part 1, No. 9B, 3124-27 (1992).

[3]L.J. Ruan, L.T. Li, and Z.L. Gui, "Temperature Stable Dielectric Properties for PMN-BT-PT System Ceramics," *Journal of Materials Science Letters*, **16**, 1020-1022 (1997).

[4]T. R. Shrout and A. Halliyal, "Preparation of Lead-Based Ferroelectric Relaxor for Capacitors" *American Ceramic Society Bulletin*, **66**[5] 704-711 (1984).

[5] Z.L. Gui, L.T. Li, and X.W. Zhang, "Study on Dielectric Properties of PMN-PZN-PT Ferroelectric Relaxor Ceramics," in Proceedings of the 3rd International Conference on Properties and Applications of Dielectric Materials, Tokyo, Japan (1991) pp. 816-819.

DIELECTRIC, PIEZOELECTRIC AND FERROELECTRIC PROPERTIES OF PMN-PNN-PZT QUARTERNARY SYSTEM

Xiaobo Guo, Jinrong Cheng and Zhongyan Meng
School of materials science and engineering, Shanghai University, Shanghai, 201800, China

Haiyan Chen
Institute of composite materials, Shanghai Jiaotong University, Shanghai,200030, China

ABSTRACT

In this paper, the dielectric, piezoelectric and ferroelectric properties of $Pb_{0.95}Sr_{0.05}(Mg_{1/3}Nb_{2/3})_a(Ni_{1/3}Nb_{2/3})_b(Zr_mTi_n)_cO_3$(aPMN-bPNN-cPZT) +0.05wt%CeO_2+0.02wt%Bi_2O_3 were investigated as a function of PNN content. With the increasing PNN content, the temperature of the dielectric maximum (T_{max}) decreases rapidly and the crystal structure transforms from tetragonal or rhombohedral phases to cubic phase. A transition from a fat hysteresis loop to a linear hysteresis curve was observed with PNN content increasing. The ceramics reveal some characteristics of relaxor. The dielectric constant ε, piezoelectric constant d_{33} and electromechanical coupling factor Kp appear a maximum in the composition 0.1PMN-0.4PNN-0.5PZT which is near the morphotropic boundary (MPB) determined by XRD.

INTRODUCTION

It is well know that lead zirconate titanate (PZT)-based perovskites have excellent piezoelectric properties, many researchers have focused on the pseudoternary solid solution compounds PZT, for example: (I) It's reported that $PbNi_{1/3}Nb_{2/3}O_3$-$PbTiO_3$-$PbZrO_3$ (PNN-PZT) has a large electric-field-induced displacement and a electromechanical coupling coefficient[1-4]. (II) $Pb(Mg_{1/3}Nb_{2/3})O_3$-$PbTiO_3$-$PbZrO_3$ (PMN-PZT) has been investigated by many researchers[5-8]for their large dielectric constants, high electric-field-induced strains, and quick response time. Because of these excellent properties, PNN-PZT and PMN-PZT ceramic are widely used in many electromechanical coupling devices such as displacement transducers, precision micropositioners and actuators.

Since both PMN-PZT and PNN-PZT have these good properties and usage, it is expected that $Pb(Mg_{1/3}Nb_{2/3})O_3$-$PbNi_{1/3}Nb_{2/3}O_3$-$PbTiO_3$-$PbZrO_3$ (PMN-PNN-PZT) quarternary system, which is the solid solution of the both, also shows similar behavior. Zhong, Z. Gui et al[9] studied the doping effect of Bi_2O_3 and La_2O_3 on the PMN-PNN-PZT quarternary system ceramic. Zhilun et al[10] also found that the Cd^{2+} ion has a key role in the low-temperature sintering of PMN-PNN-PZT ceramics by lowering the sintering temperature and also improving the piezoelectric properties.

However there is lack of report on the influence of PNN content on the electrical properties of PMN-PNN-PZT quarternary system. As reviewed above, the purpose of this paper was to study the dielectric, ferroelectric and piezoelectric properties of the PMN-PNN-PZT system as a function of PNN content.

EXPERIMENTS

The selected composition were $Pb_{0.95}Sr_{0.05}(Mg_{1/3}Nb_{2/3})_a(Ni_{1/3}Nb_{2/3})_cO_3$+ 0.5wt% Bi_2O_3+0.2wt%MnO_2 (where a=0.1, $0.2 \leq b \leq 0.6$, $0.3 \leq c \leq 0.7$, a+b+c=1) The analytical grade raw materials PbO, , ZrO_2, MgO, NiO, Nb_2O_5, $SrCO_3$, Bi_2O_3, MnO_2 and chemical grade TiO_2 were used. In the first step, the precursor magnesium niobate ($MgNb_2O_6$) and nickel niobate ($NiNb_2O_6$) were prepared using the columbite process[11], then the powder of starting materials of $MgNb_2O_6$ $NiNb_2O_6$, PbO, ZrO_2 TiO_2were mixed by ball milling and dried. In the next step, the mixture was calcinated at 820~860℃ for 3h, and then ball-mixed and dried to fine powders. The powders were pressed into pellets of 12mm in diameter and 1mm in thickness at 150Mpa pressure. Disk samples were sintered at 1200~1260 ℃ for 2h. The density was measured using the Archimede's method, the phase structure was performed by X-ray diffraction (XRD) in the range of $43° \leq 2\theta \leq 47°$.

The specimens were metallized with Ag paste and were poled under a DC field of 2~3kV/mm at 100 ℃ in a silicone oil bath. The Curie temperature were determined from the temperature dependence of the dielectric constant measured at 1kHz, 10kHz and 100kHz. Dielectric constants were measured with HP4294A. Using resonance-antiresonance method, the electromechanical coupling factor was obtained by HP4294A. P-E hysteresis loops were measured with RT6000HVS.

RESULTS AND DISCUSSION

Phase Transition

Fig.1 shows the change of XRD patterns of (200) peaks of the 0.1PMN-bPNN-(0.9-b)PZT (b=0.3, 0.4, 0.5, 0.6) in the range of $43° \leq 2\theta \leq 47$

° . The T(002), T(200) and R(200)peak indicates that 0.1PMN-0.3PNN-0.6PZT is in the ferroelectric tetragonal phase (F_T), while 0.1PMN-0.5PNN-0.4PZT is in ferroelectric rhombohedral phase(F_R), and 0.1PMN-0.4PNN-0.5PZT is in the morphotropic boundary (MPB). 0.1PMN-0.6PNN-0.3PZT shows a cubic phase because the Curie temperature(see Fig.2) is lower than room temperature. This is to say, with the increaing of PNN content, the crystal structure transforms from tetragonal and rhombohedral phase to cubic phase.

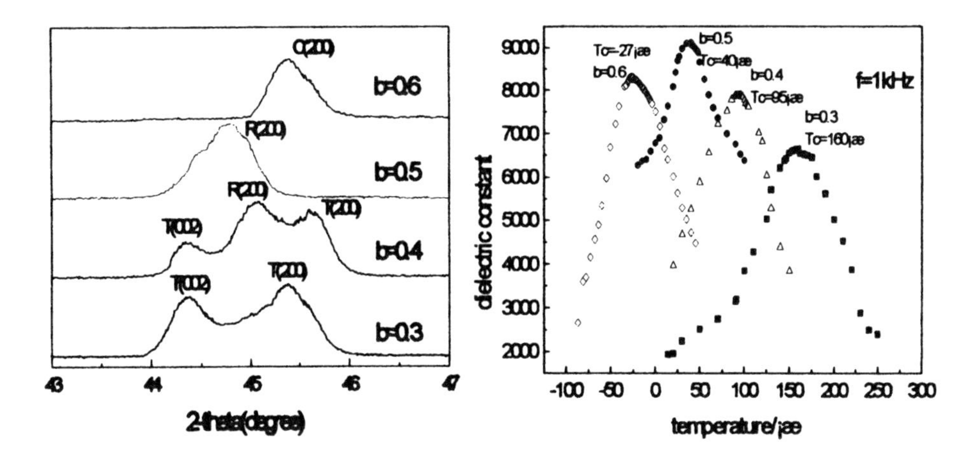

Fig1. Change of XRD patterns of (200) peaks of the 0.1PMN-bPNN-(0.9-b)PZT (b=0.3, 0.4, 0.5, 0.6)

Fig.2 Temperature dependence of dielectric constant measured at 1kHz.

Dielectric Properties

The temperature dependence of dielectric constants (measured at 1kHz) for the composition 0.1PMN-bPNN-(0.9-b)PZT (b=0.3, 0.4, 0.5, 0.6) is shown in Fig.2. It is found that the temperature of the dielectric maximum (T_{max}) decreased with the increasing of PNN content. It is reasonable that PNN has a comparatively low Curie temperature (Tc= −120 ℃). From Fig.2, it also can be seen that the dielectric constant at T_{max} increased with the increasing of PNN content and then decreased after the PNN content is 0.5. The dielectric constant for the composition 0.1PMN-0.6PNN-0.3PZT measured at 1kHz, 10kHz,100kHz as a function of temperature is shown in Fig.3, a frequency dispersion can be observed at temperature below T_{max}. Similar phenomenon was observed in other 0.1PMN-bPNN-(0.9-b)PZT (b=0.3,0.4,0.5). It reveals that this quarternary system ceramic had some characteristics of relaxor.

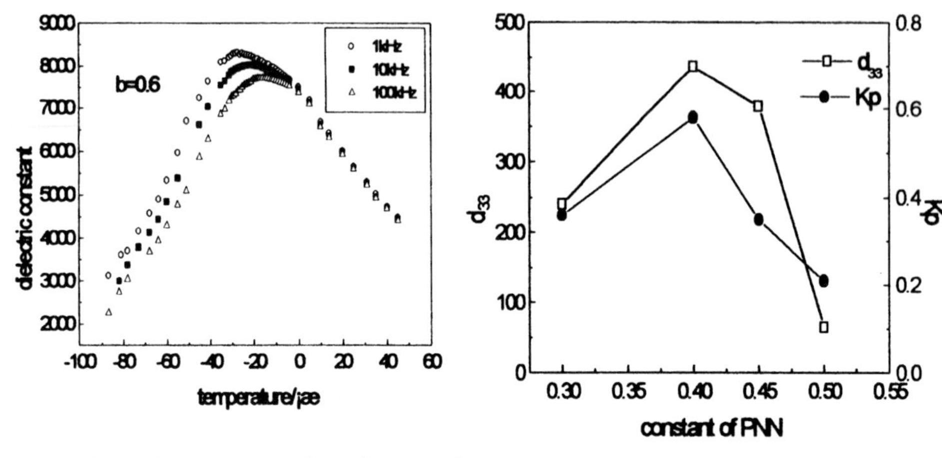

Fig.2 Temperature dependence of dielectric constant of the composition b=0.6 measurec at 1kHz,10kHz and 100kHz.

Fig.4 Piezoelectric constant d_{33} and electromechanical coupling factor Kp versus the content of PNN

Piezoelectric Properties

Fig.4 showed the piezoelectric constant d_{33} and electromechanical coupling factor Kp versus the content of PNN. The curves illustrated that the largest d_{33} and Kp was obtained in the composition 0.1PMN-0.4PNN-0.5PZT, it can be explained that this sample is near the MPB, which is in a mixed phase region coexisting ferroelectric tetragonal phase (F_T) and rhombohedral phase (F_R). For the composition near MPB, the two ferroelectric phases coexist and their structure energies are nearly equal. Their crystal structure can be transformed from rhombohedral phase to tetragonal phase when electric field or mechanical stress is applied, and vice versa. This is beneficial for the mobility and polarization of ferroelectric domain. Therefore in such structural states, the pizoelectric activity was enhanced by oscillation of boundaries between coexisting rhombohedral and tetragonal regions, and the electromechanical coupling factor become higher.

Ferroelectric Properties

Fig.5 showed the P-E hysteresis loops of the specimens 0.1PMN-bPNN-(0.9-b)PZT (b=0.3, 0.4,0.5,0.6). With increasing PNN content, the transition from a fat hysteresis loop representing a typical ferroelectric to a slim hysteresis curve was observed, almost linear P-E behavior was observed for the composition of b=0.6. This indicated that a transition from ferroelectric phase to paraelectric phase has taken place, which was consistent with the results of Fig1. With the increasing of PNN content, the structure transform from tetragonal or rhombohedral phase to pseudo-cubic phase, and because the coercive field Ec of pseudo-cubic phase is comparatively low, so Ec decreased with increasing PNN content.

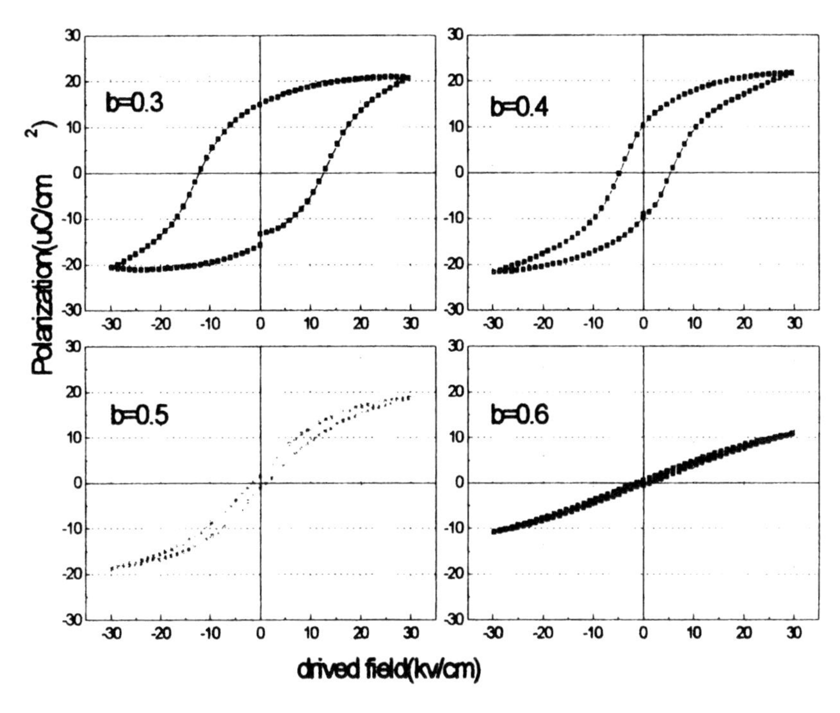

Fig.5 P-E hysteresis loops of the specimens 0.1PMN-bPNN-(0.90-b)PZT(b=0.3, 0.4, 0.5, 0.6) measured at room temperature.

CONCLUSIONS

The dielectric, piezoelectric and ferroelectric properties of PMN-PNN-PZT were systematically studied as a function of PNN content. It was found that the crystal structure transforms transforms from tetragonal or/and rhombohedral phase to cubic phase. the temperature of the dielectric maximum (T_{max}) decreased rapidly with increasing PNN content. The transition from a fat hysteresis loop to a linear hysteresis curve was observed. This quarternary system ceramic had some characteristics of a relaxor. The piezoelectric constant d_{33} and electromechanical coupling factor Kp are 437×10^{-12}C/N and 0.58 respectively in the composition 0.1PMN-0.4PNN-0.5PZT which was near the MPB.

REFERENCES

[1]D. Luff, R. Lane, K. R. Brown and H. J. Marshallsay, "Ferroelectric Ceramics with High Pyroelectric Properties", *Trans. J. Br. Ceram. Soc.*, 73 251-264 (1974)

[2]Masao Kondo, Mineharu Tsukada and Kazuaki Kurihara, "Temperature Depedence of Piezoelectric Constant of 0.5 $PbNi_{1/3}Nb_{2/3}O_3$-$PbTiO_3$-0.5$Pb(Zr,Ti)O_3$ Ceramic in the Vicinity of Morphotropic Boundary," *Jpn. J. Appl. Phys. Vol.* 38(1999) pp.5539-5543

[3]XINHUA ZHU, ZHONGYAN MENG, "The influence of the morphotropic phase boundary on the dielectric and piezoelectric properties of the PNN-PZ-PT ternary system," *J. Mater. Sci.*, 31(1996) 2171-2175.

[4]Sheng-yuan CHU etal., "Doping effect on the piezoelectric properties of low-temperature sintered PNN-PZT-based ceramics," *J. Mater. Sci. Lett.*, 19(2000) p609-612.

[5]Ki Hyun Yoon and Hong Ryul Lee, "Effect of Ba^{2+} Substitution on Dielectric and Electric-Field-Induced Strain Properties of PMN-PZ-PT Ceramics," *J. Am. Ceram. Soc.*, 83 [11] p2693-98 (2000).

[6]M. Kiyohara, K.Kato, K. Hayama, and A. Funamoto, "Actuator Characteristics of $Pb(Zr,Ti)O_3$ Composition near Morphotropic Phase Boundary," *Jpn. J. Appl. Phys.*, 30 [9B] 2264 (1991).

[7]S. Nomura and K. Uchino, "Electrostrictive Effect in $Pb(Mg_{1/3}Nb_{2/3})O_3$-Type Materials," *Ferroelectrics*, 41, 117 (1982).

[8]K. Uchino, "Piezoelectric and Electrostrictive Actuators," *IEEE Trans. Ultrason., Ferroelectr., Freq, Contro,* 33[6] 806 (1986)

[9]Zhong, Z. Gui, and L. Li, "Bi_2O_3 and La_2O_3 doped $Pb(Mg_{0.67}Ni_{0.33})_2O_3$-$Pb(Ni^{0.67}Nb_{0.33})_2O_3$- $PbZrO_3$-$PbTiO_3$ piezoelectric ceramic materials," *Proc. 1992 spring American Ceramic Society Annual Meeting, Minneapolis, MN, Apr. 1992*

[10]GUI ZHILUN, etal., "Low Temperature Sintering of Lead Magnesium Nickel Niobate Zirconate Titanate (PMN-PNN-PZT) Piezoelectric Ceramic, With High Performances," *Ferroelectric,* 1990, Vol.101, p93-99.

[11]Swartz, S.L. and Shrout, T.R., "Fabrication of perovskite leas magnesium niobate," *Mater. Res. Bull.*, 1983,7, 1245-1250.

OPTIMIZATION OF FERRITE POWDER PROCESSING BY CHARACTERIZATION OF SLURRY PROPERTIES

Jürgen Wrba and Ralph Lucke
P.O. Box 80 17 09
D-81617 Munich
Germany

ABSTRACT

In the fabrication of Mn-Zn ferrites -ferrimagnetic ceramic- the raw materials are mixed and calcinated. After wet milling and spray drying, ferrite granules are pressed and sintered. To obtain granules with good pressing behavior, the intensive characterization of the wet milled slurry is very useful to investigate the influence of the previous steps on the following spray drying and pressing. It is shown, that a careful adjustment of the rheological behavior and zeta potential of slurries with high solid loading is necessary to get highly sophisticated granules for a well-controlled pressing step. The influences on slurry properties will be discussed.

INTRODUCTION

Many processing steps are needed to get ferrite cores which are used as signal or power transformer in electronic devices [1,2].

The first part is the production of granulated ferrite powder (Fig. 1). The weighed raw materials Fe_2O_3, Mn_3O_4 and ZnO are mixed and then pelletized with a small amount of water. The red pellets with a pellet size of 3-5 mm are calcinated in a rotary kiln at about 1000°C. Here the oxides react partly to the magnetic ferrite with spinel structure. After that the black pellets, water and some inorganic additives (amount from 0.01-0.1 wt-%) are added into an attritor for fine milling. The inorganic additives are necessary to improve the sintering behavior and / or the magnetic properties. After attritor milling the slurry is conditioned by adding binder in order to get a granulate with a good pressing behavior, i.e. good bulk density and flow speed, low springback and high green strength.

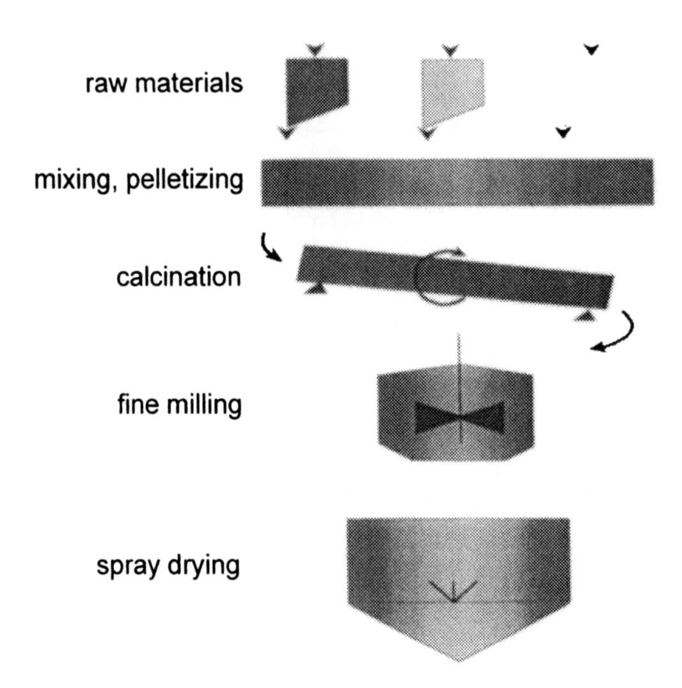

Fig. 1: Production processing of granulated ferrite powder

The second part of core production contains pressing, sintering and grinding. These steps and the magnetic properties of the ferrite core are influenced by the conditioning of the powder in considerable extent. Each processing step influences the following step.

In this work we will present only one aspect of the broad varieties of relationships between processing parameters and granulate properties. The bulk density is very important to get a high green density, especially for small cores. Each batch of a material should have the same bulk density with a good reliability. We will show the effects of rheology of ferrite slurries on the bulk density of granulated powder and how the rheology can be influenced.

Following parameters and their effects on slurry properties were investigated: composition, calcination temperature, milling time, the additives, binders and dispersants. The slurry were characterized by solid loading, pH, viscosity and zeta potential. It is the first time that the zeta potential of ferrite slurries with a solid loading up to 70 wt-% is measured without any dilution.

EXPERIMENTALS

In this work the production process of the standard EPCOS material N87 for power applications was investigated. Some tests were made with slurries from the laboratory (batch size: few kg). The other part of the used slurries came from the factory (batch size: tons), so the measured slurry properties could be compared immediately with the resulting granulate properties of the produced batch. Slurry properties of laboratory and production are strongly comparable. Because of the necessary production capacity the solid loading is about 62 wt-% and was not changed for the test.

For the measurement of zeta potential, the electrokinetic sonic amplitude (ESA) effect is used [3]. The electrokinetic charge (zeta potential) and the size of particles can be determined in concentrated suspensions. Applying an alternating electrical field to the suspension causes an oscillation of the particles because of the electrical charge they carry. The movement of the particles generates acoustic waves which are measured over a range of excitation frequencies. From this so-called dynamic mobility spectrum the charge and size can be calculated.

RESULTS AND DISCUSSION

In order to predict the bulk density from slurry properties, many slurries from the actual production and the resulting bulk densities were compared. As expected the viscosity of the slurry must be low in order to get high bulk density. The dependency on the pH is very important because an increase of the solid loading from 65 to 75 wt-% (e.g. during drying) causes an decrease of the pH from 9 to 8.4. Figure 2 shows a strong dependency of the viscosity where the rheological behavior of the slurry is changed. But the viscosity is only a first condition.

The other condition is a high zeta potential. Figure 3 shows that the bulk density increases with a higher absolute value of the zeta potential of the slurry. The shown slurries contain dispersant and binder. For the bulk density, the course of viscosity and zeta potential is decisive during drying. The drying process can be simulated by changing the pH. For high bulk density, a high zeta potential, i.e. high slurry stability is necessary. The small particles in the slurry should not agglomerate before most of the water is gone. So more spherical particles and less agglomerated particles (with rather complicated shapes) stay till the end of drying, and no bridges between particles can be formed which increase the resulting bulk density. It is easier to reach high density with spherical particles.

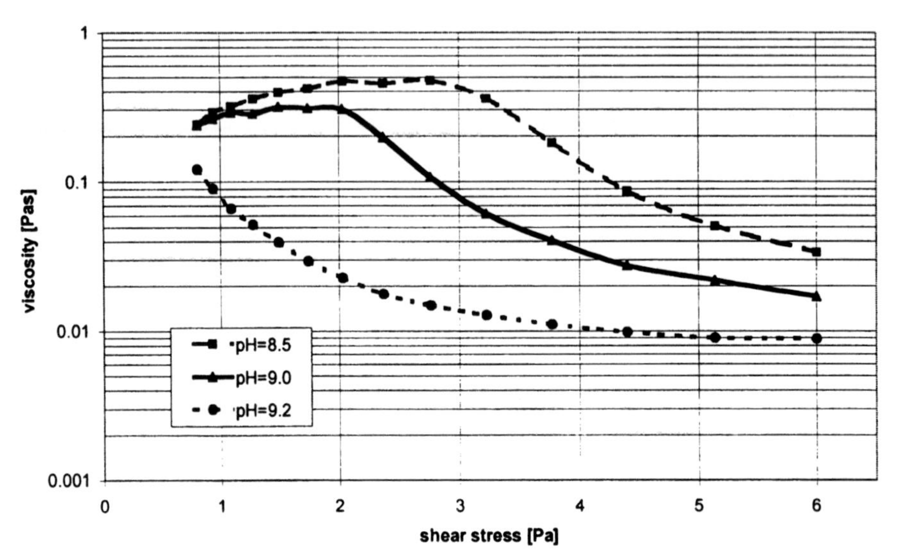

Fig. 2: Dependency between pH and viscosity of a ferrite slurry for different shear stresses.

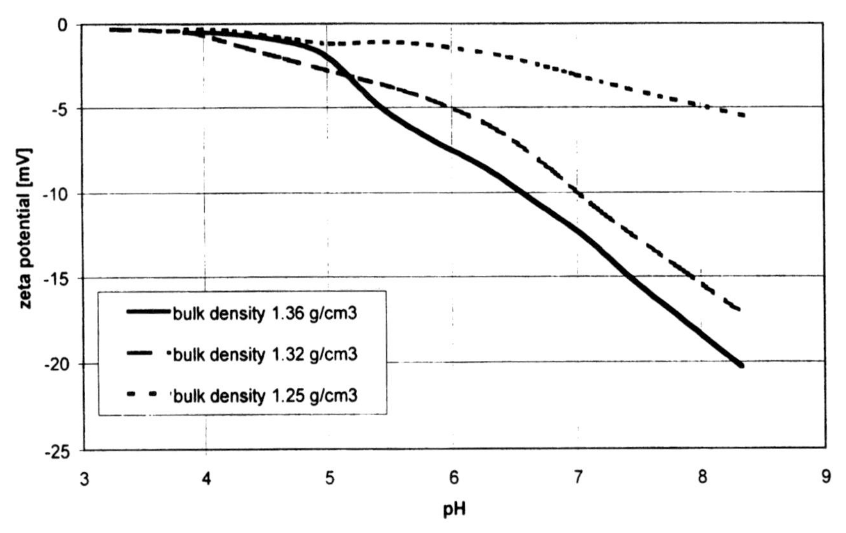

Fig. 3: Zeta potential of different slurries as a function of the pH value and the resulting bulk density of the granulate.

The variation of the composition of Mn-Zn-ferrites has no significant influence on the zeta potential. In Figure 4, the effects of the calcination temperature, the milling time and the inorganic additives are shown. The curve 1-4 has the normal additive set of N87. The slurries 1-3 were calcinated at different temperatures. For a discrete pH, a decrease of zeta potential can be seen with increasing calcination temperature. Particles with higher specific surface area could be better stabilized in the suspension. The milling time has no significant effect on the zeta potential (slurries 2 and 4).

Figure 5 shows that the addition of an dispersant changes the sign of the zeta potential. Ferrite particles have a positive potential, which will be negative after adding dispersant to get higher solid loadings. If binder is added to a slurry, the zeta potential decreases (Fig. 6). The type of binder can have a strong effect on the isoelectric point. It is important to know the isoelectric point in order to evaluate the right pH range for the start of spraying. By adding acid or base the right range can be adjusted.

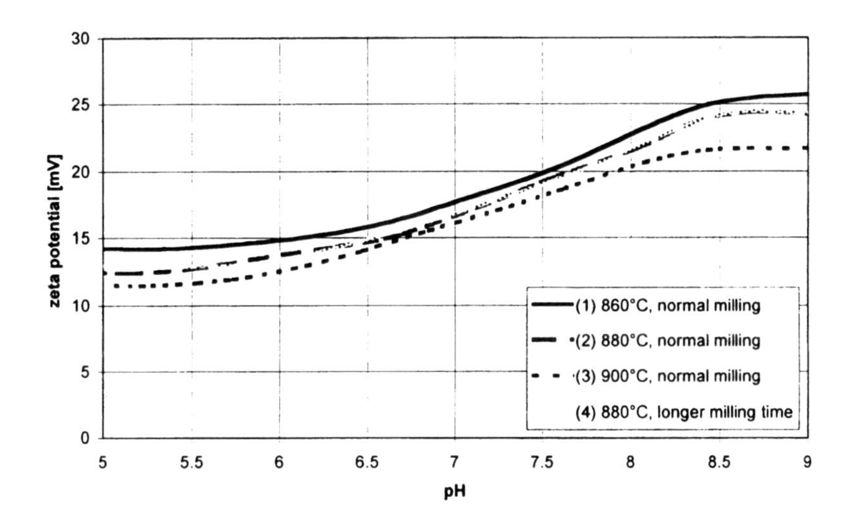

Fig. 4: Influences of calcination temperature, milling time and inorganic additives on the zeta potential for the pH range from 5 to 9.

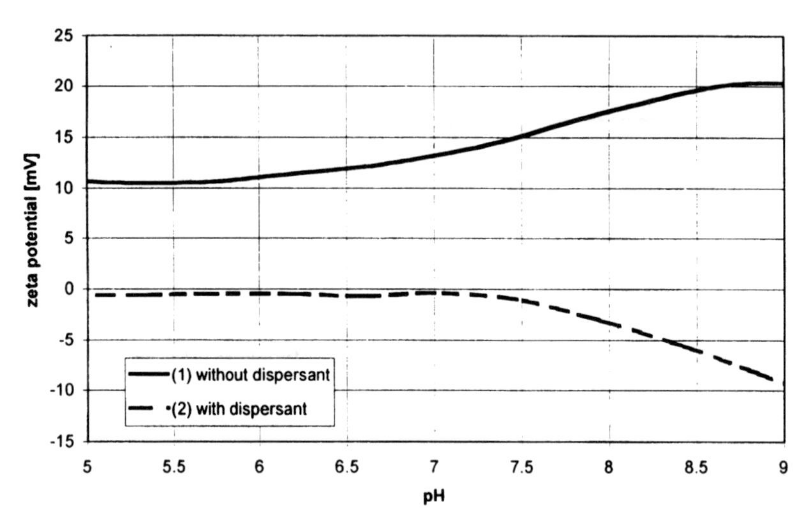

Fig. 5: Zeta potential as a function of the pH value without dispersant (curve 1) and with dispersant (curve 2).

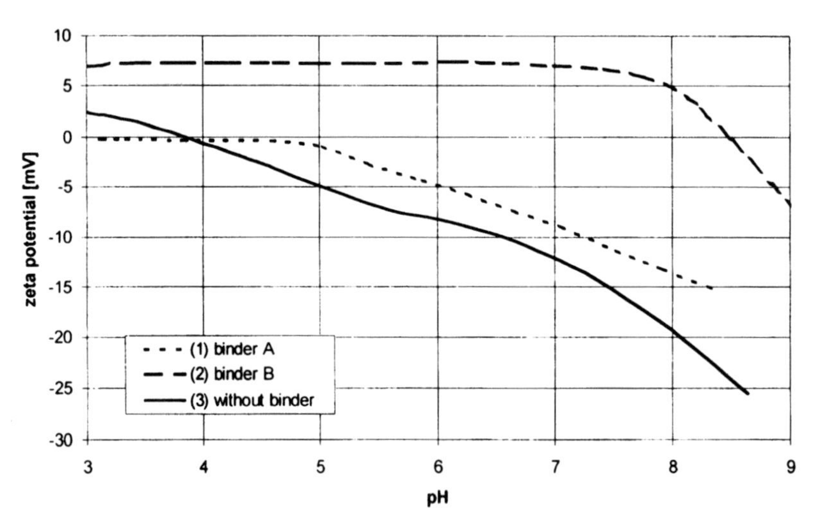

Fig. 6: Zeta potential as a function of the pH value of slurries without binder and with 2 chemically different binders.

Another application of this knowledge and measurement method of zeta potential is the search for the right pH for adding binder. Figure 7 shows the optimization of the binder adding process by using zeta potential measurements. For one curve (starting pH=8.4), only binder was added into the slurry. The viscosity increases with factor 48. The other curves show different pH values, adjusted by adding NH_3 prior binder addition. It can be seen that the change of the potential by the binder is smaller with NH_3 than without NH_3. The viscosity increases only with factor 8. In Figure 7 can be also seen that only a small change of pH (small amount of NH_3) is necessary before adding the binder. Apparently, this binder needs a defined particle surface, prepared by the base, to work in the right way.

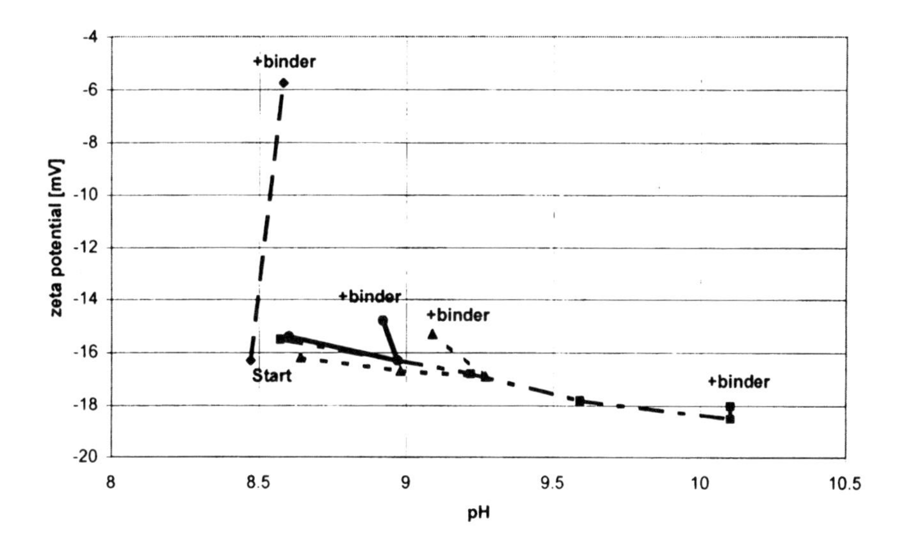

Fig. 7: Change of zeta potential by variation the binder adding process

CONCLUSION

Both viscosity and zeta potential influence the bulk density after spray drying. They are strongly dependent on the pH value. Using ESA method to measure the zeta potential, the slurry properties can be optimized without carrying out spraying tests. The working range of new organic additives, the optimal concentrations and the adding process can be easily evaluated. Time and money can be saved.

So a measurement method is established which can improve the production process to get high quality granulated powder with a good reliability, i.e. stable properties even if the quality of raw materials is changed.

REFERENCES

[1]M. Sugimoto, "The Past, Present, and Future of Ferrites", *Journal of the American Ceramic Society*, **82** [2] 269-80 (1999).

[2]R. Lucke, J. Wrba, "Weichmagnetische Ferritkeramik für elektrotechnische Anwendungen"; pp. 43-53 in *Das Keramiker-Jahrbuch 2001*. Edited by H. Reh. Goeller-Verlag, Baden-Baden, 2001.

[3]T. Oja, G.L. Petersen and D.C. Cannon, U.S. Pat. No. 4497208, 1985.

MANUFACTURING OF ADVANCED DIELECTRIC COATINGS BY THERMAL SPRAYING

A. Killinger
Institute for Manufacturing Technologies of Ceramic Components and Composites
University of Stuttgart, Baden-Württemberg,
Allmandring 7b
70569 Stuttgart
Germany

ABSTRACT

Thermally sprayed dielectrics have been successfully inserted in numerous industrial applications throughout the last years. Besides specified values for breakthrough voltage, permittivity and volumetric resistance, the sprayed coatings have to withstand thermal load and mechanical stresses without delamination. Therefor the appropriate mechanical and thermophysical properties of the coatings are also of importance.

The presentation gives an overview of selected thermophysical and electrophysical material properties of various thermally sprayed oxide coatings and discusses some aspects of the thermal spray process and their impact on the physical and mechanical properties of the acquired coatings.

INTRODUCTION

Oxide ceramics play an important role as dielectric materials for multiple purposes in industrial applications. They serve as insulators, dielectric discharge barriers, as dielectric coatings in corona treatment devices and printer rollers etc.

In many applications, coating solutions have successfully replaced expensive bulk ceramic composites. The processing of refractory oxides with high melting points using thermal spray technologies is a performing and cost effective way to manufacture coating systems in the range of 30 µm up to several 1000 µm. Figure 1 compares common coating technologies and correlates their achievable ranges of thickness.

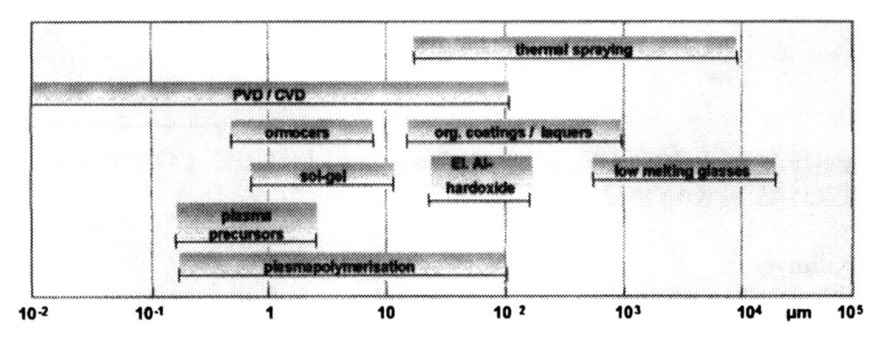

Figure 1. Comparison of different coating processes and their typical achievable coating thicknesses for depositing insulating coatings

With growing film thickness thermophysical incompatibilities of the ceramic coating and the substrate material (for instance metal, glass or polymer) are becoming crucial and can lead to failure by crack formation or complete delamination of the coating composite under thermal and/or mechanical load. The problem can be solved by thoroughly adjusted spray parameters and the utilisation of simultaneous cooling techniques for the substrate to obtain an appropriate residual stress distribution in the layer composite. Two thermal spray procedures are basically suitable to process oxide ceramics: (1) atmospheric plasma spraying (APS) and (2) hypersonic oxyfuel flame spraying (HVOF). Both methods feature their own benefits and disadvantages regarding process costs and resulting coating properties.

Further coating attributes like thickness, porosity as well as chemical and phase composition can be influenced and optimized by the variation of thermal spray parameters and finally determine the resulting electrical behaviour of the coating material.

Examples for industrial applications like thermally sprayed dielectrics for use as dielectric barrier discharge material in tubular ozonizers and multilayer composites with different electric properties for applications in the printing industry were presented in an earlier work [1], [4].

THERMAL SPRAY TECHNOLOGY

Functional coatings can be applied by various techniques. For the deposition of thick coatings consisting of high melting metals and ceramics thermal spraying techniques are a cost effective and promising technology. One of the main advantages, compared to many other coating processes is the possibility to use a wide range of materials ranging from low melting polymers, metals and metal alloys to high melting oxide ceramics [2], [3].

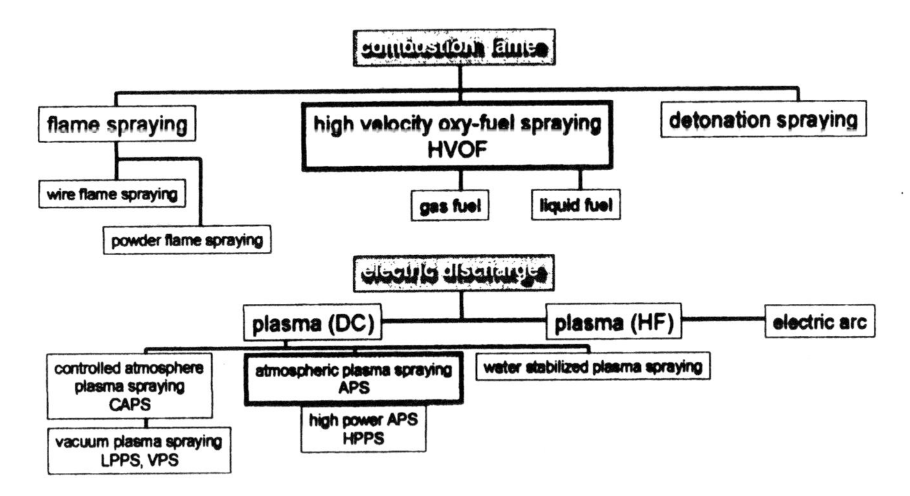

Figure 2. General survey of the various thermal spray techniques classified by the kind of heat source (combustion or electric discharge), spray techniques used in this work are outlined

An outstanding feature of the thermal spray technique is the substrate's low thermal load during the coating process compared to many other methods. Using simultaneous air or liquid CO_2 cooling techniques, the substrate temperature can be controlled, for most applications a low substrate temperature between 80 – 150°C is desired, however some substrates like glass need to be preheated before the coating process is initiated. During thermal spraying the spray powder, suspended in a carrier gas, is injected into the heat source of the torch. In electric arc spraying a metal wire is used instead of powder. After being totally or at least partially molten and accelerated, the powder particles impact the substrate, cool down very rapidly (quenching) and build up the coating. The substrate surface is not fused during the process unless a post-processing is performed. There exist various thermal spraying techniques e.g. arc-, plasma-, or flame-spraying, a general survey is illustrated in Figure 2. Within all these variations two thermal spray processes are suitable to manufacture high melting oxides, the Atmospheric Plasma Spraying (APS) and the High Velocity Oxygen Fuel spraying (HVOF).

In APS a plasma serves as the energy source, an electric arc discharge between a water cooled copper anode and a tungsten cathode dissociates and ionises the processed gas and builds up a plasma, that expands into the atmosphere forming a plasma gas jet (see figure 3 a). Generally a mixture of Ar and H_2 is used as the plasma gas. The process supplies extremely high gas temperatures combined with a medium range particle velocities, being able to

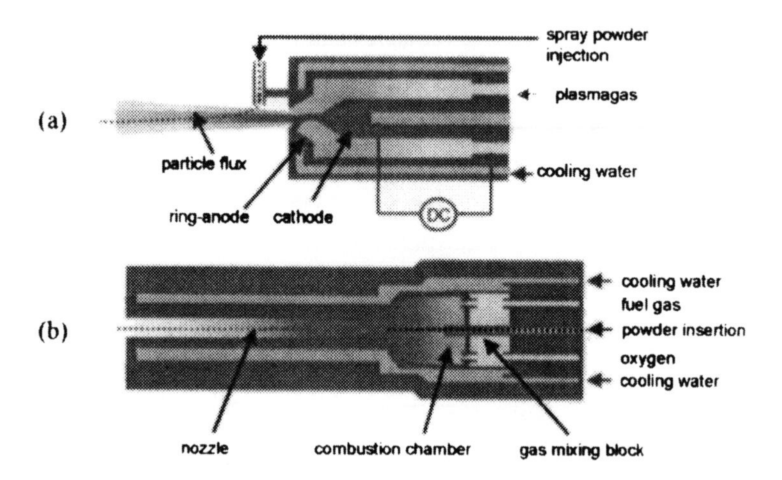

Figure 3. (a) Schematic sketch of an Atmospheric Plasma Spray torch with axial powder injection. (b) Schematic sketch of HVOF torch used for thermal spraying operated by diffusion flame, with combustion chamber and laval shaped nozzle. In experiments a Metco F4 torch was used as a plasma gun and a TopGun gas system (GTV) for HVOF spraying. For details refer to text.

melt virtually every material. Therefor APS is the favoured method to process oxides.

The HVOF process uses liquid fuels or fuel gases to perform a high energetic combustion with oxygen in a combustion chamber. A subsequent relaxation and acceleration of the combustion gas in an expansion nozzle leads to hypersonic gas and particle velocities at the nozzle exit (see figure 3 b). The powder is inserted either directly into the combustion chamber or into the nozzle entrance. HVOF performs highest particle velocities but reduced process temperatures and is commonly used for materials having melting points below 2000°C. Therefore the APS is preferentially used for the deposition of refractory oxides, the HVOF system for metal and cermetic materials.

However, it is possible to spray oxides with HVOF, therefor the process needs to be operated with acetylene as the fuel gas, because it reveals highest gas temperatures and is able to melt oxide ceramic powders.

Precedent to the coating process a surface cleaning, degreasing and grit blasting of the substrate surface is performed. The roughening of the surface with corundum of defined size improves the micromechanical adhesion of the coating and induces compressive stresses into the substrate material. However there exists substrate materials like glass where a grit blasting is not necessary because the sticking mechanism is of chemical nature [4]. Following to the coating deposition a mechanical (grinding, polishing) or thermal post-treatment of the coating surface is optional.

As mentioned earlier, the physical properties of the thermally sprayed coatings, i. e. the residual stress distribution, deformation, microstructure, porosity etc., can be tuned to a certain extend by adjusting the spray parameters. In APS there are a couple of process parameters involved, most important ones are: plasma energy and gas mixture (Ar/H_2 ratio), spraying distance, powder morphology, powder flux, torch kinematic, simultaneous cooling or selective heating of the substrate, etc.

For many applications APS sprayed coatings will meet all requirements especially when a higher coating thickness is desired. Compared to other coating methods thermal spraying and spray lacquer techniques belong to the most cost effective deposition methods to manufacture thick films in the range of 50- 1000 µm. However, to fulfill additional requirements like low porosity, low surface roughness, or a limitation of the coating thickness, HVOF spraying or a combination of the APS and HVOF method (multilayer coating) can be the appropriate method to solve the problem (see table 1).

The differences between APS and HVOF coatings in their respective morphological structure is visible in Figure 4. The APS coating features a pronounced slit pore structure whereas the HVOF coating has only small spherical pores. The overall porosity typically lies in the range of 7 – 10% for APS and 1 - 4% for HVOF coatings respectively. Responsible for these structural differences are the different particle velocities that occur in the APS and HVOF process respectively. The difference in porosity also has an important impact on the mechanical properties of the coating. Dense HVOF

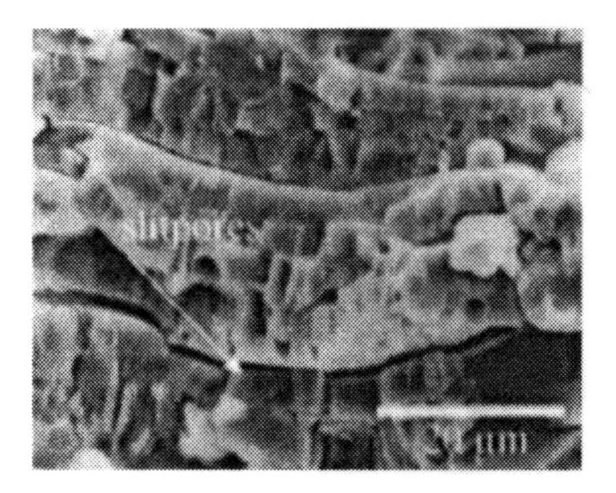

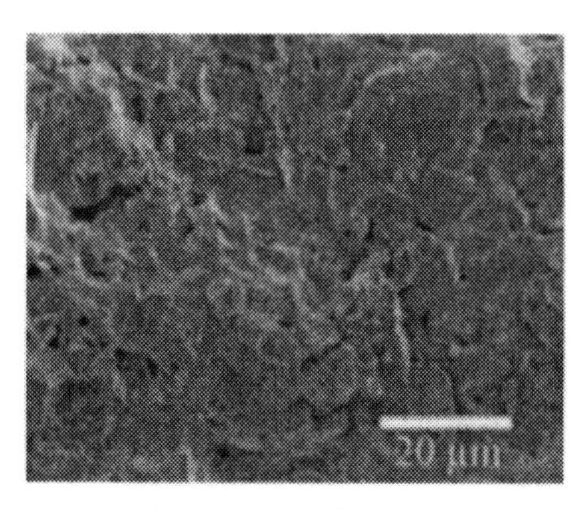

Figure 4. Comparison of SEM images of fracture surfaces of an APS and a HVOF sprayed titania coatings showing the differences in coating morphology. The APS coating exhibits a distinctive slit pore structure partially spacing individual particles.

Table 1. Comparison of thermal spray methods concerning dielectric manufacturing

HVOF properties	Resulting benefits and disadvantages
Higher particle velocity	dense coatings, higher breakthrough voltage but reduced elasticity of the coating
Lower gas temperature	less efficient in oxide powder processing
Higher heat impact	extensive cooling and reduced coating thickness
Oxidising flame possible	reduced oxygen loss for titania and zirconia
Running costs	more expensive

APS properties	Resulting benefits and disadvantages
Lower particle velocity	higher porosity, lower breakthrough voltage but higher elasticity of the coating
Higher gas temperature and higher gas enthalpy	Very efficient in oxide powder processing
Reduced heat impact on workpiece	Moderate cooling and coating thickness up to several mm
Running costs	Less expensive

coatings are limited to a few 100 μm due to their reduced elasticity (higher Young´s modulus) whereas it is possible to manufacture APS coatings having a thickness up to 1000 μm.

It should be pointed out that both techniques lead to different phase compositions in the coating that may differ from the initial phase composition of the powder material. Depending on the temperature / time history of the particle during flight and solidification, different phases are formed. The rapid cooling process promotes the formation of metastable high temperature phases. Within the coating complex spatial distributions of different phases can be formed.

POWDER TECHNOLOGY FOR THERMAL SPRAYING

Depending on the kind of material that is processed thermal spray powders have distinct morphologies that strongly affect their flowability in the powder feeder device (the powder is injected into the torch with the help of a feeding gas), their melting behaviour and thereby the spray-coating efficiency. The microstructure of the deposited film, i. e. its porosity and phase composition, is strongly influenced by the powder morphology and its thermal history. Figure 5 summarises the common morphology types for ceramic powders. Fused and crushed qualities are the standard powders and they are quite cost effective. These powders are electro-thermally fused and milled to the appropriate grain size. More expensive are sintered qualities based on submicron primary grains because they include an additional sintering step after the agglomeration process. Granulate size distributions typically range

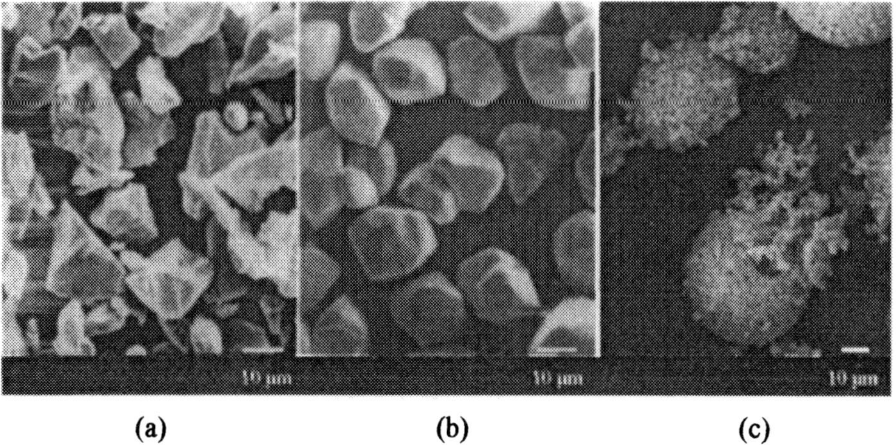

(a) (b) (c)

Figure 5. (a) Fused and crushed alumina powder; (b) Chemically synthesized alumina powder, (c) agglomerated titania powder

from 5 – 20 µm, 25 – 40 µm and 45 – 90 µm respectively. Narrower size ranges can be achieved by additional sieving or separating techniques.

Mixed oxides are mechanically mixed powders (consisting of two separate phases) or alloyed species (mixed crystals). High end quality oxide ceramic powders are obtained by redox-reactions and sol-gel techniques based on organometallic compounds or hydrothermal synthesis. They exhibit a very narrow size distribution and can be scaled down to a particle size of a few 100 nm. Figure 5 shows SEM images of alumina particles, manufactured in

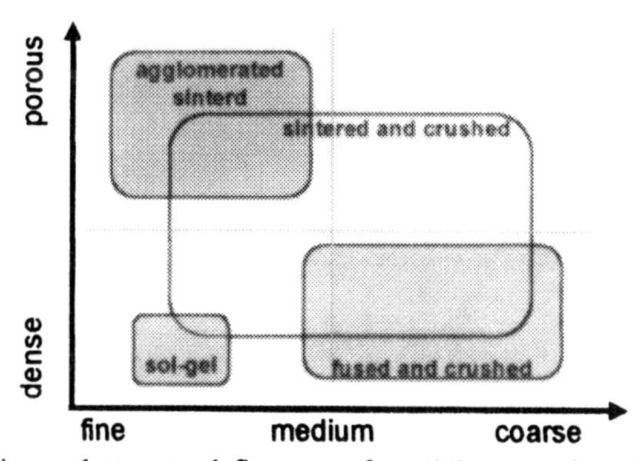

Figure 6. Porosity and structural fineness of particles as a function of powder production method

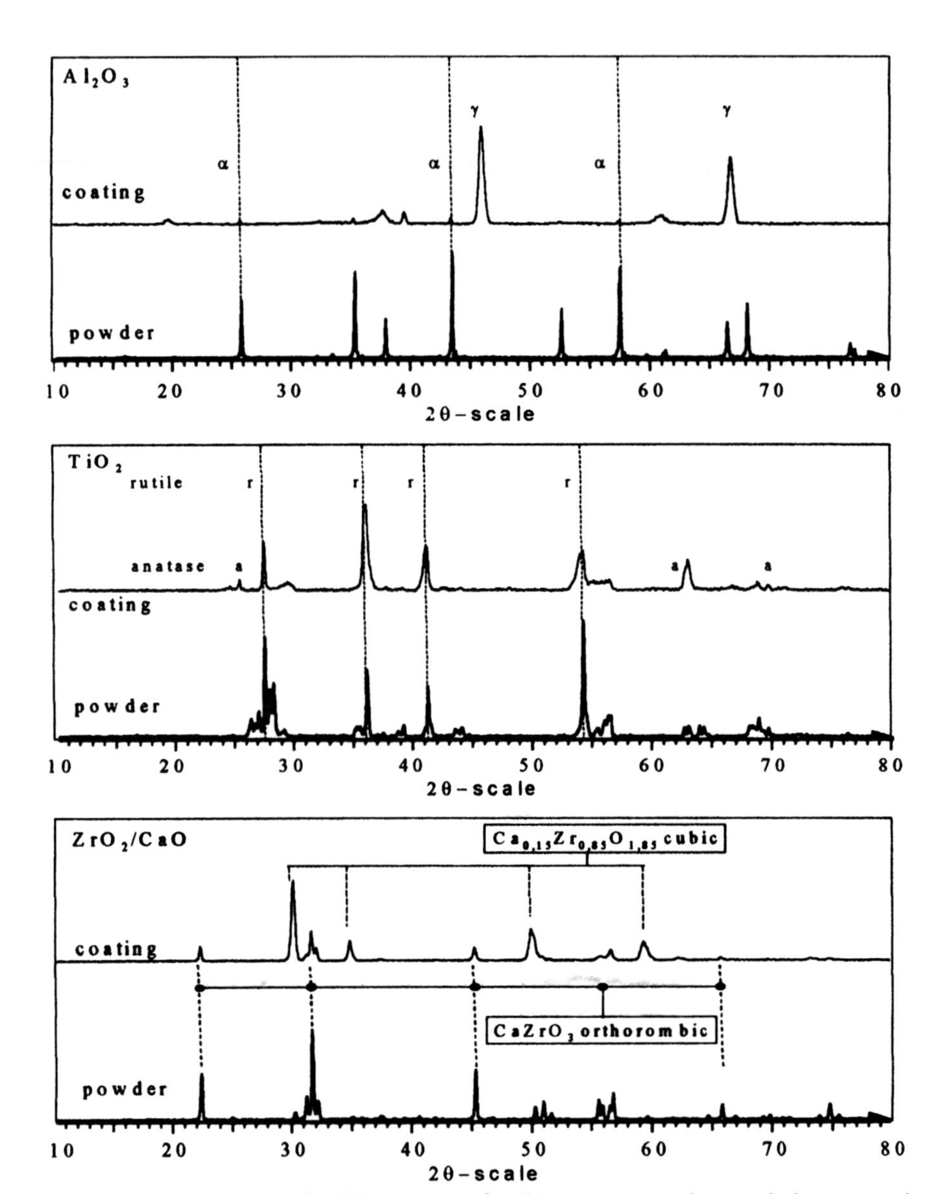

Figure 7. Comparison of XRD pattern of oxide spray powders and the respective plasma sprayed coatings. (a) Spraying hexagonal α–alumina (corundum) leads to a phase transitions into the metastable cubic γ–phase; (b) spraying titania (rutile) leads to the formation of anatase and substoichiometric species; (c) spraying of partially Ca-stabilized zirconia (mainly orthorhombic) leads to Ca depletion and the formation of cubic species. Strongest peaks were identified according to the JCPDS-data base

different production routes: fused and milled (figure 5a), chemically synthesized (figure 5b) and agglomerated / sintered (figure 5c).

Figure 6 illustrates the correlation of the various production routes with their respective achievable porosities and grain sizes. In general it can be stated that very dense and fine powder qualities are most expensive to manufacture but are most sensitive in handling regarding their behaviour in the powder feed stream.Very crucial for the quality of the thermal sprayed coatings is their grain size distribution. For example powder qualities may have a bimodal distribution with a fine fraction band. During HVOF spraying, a fine fraction of the oxide powder will then partially evaporate and condense at the nozzle exit leading to a built up of undesired residuals. This will influence the continuous operation of the HVOF torch, lower the process stability and consequently the coating quality. A coarse fraction will not contribute to the coating build up because the particles cannot be fully molten. The particles will escape as overspray and lead to a decrease of the deposition efficiency of the coating process and therefor to an undesired increase of costs.

PHASE COMPOSITIONS OF THERMALLY SPRAYED OXIDES

As mentioned earlier phase compositions of the powder and the resulting coating will differ in most cases. Due to the high temperature conversion and the rapid cooling a number of different metastable phases are formed in the coating. XRD spectra of spray powders and their respective plasma sprayed coatings are shown in Figure 7 for three selected oxides (alumina, titania, zirconia) commonly used for electrical applications. Transformation to metastable phases (alumina, figure 7a), phase transitions (titania, figure 7b) and modification of the chemical composition (titania, figure 7b and zirconia figure 7c) may occur.

Well known is the behaviour of α-Al_2O_3 (corundum). During the rapid cooling a metastable high temperature phase is formed, the cubic γ-Al_2O_3. Compared to the hexagonal α-phase it features a reduced mechanical and chemical stability. Therefor a maximum amount of α-phase is desired in the coating. However, this is only possible by heating the substrate (and therefor lower the cooling rate) to several hundred degrees or by deposition of partially unmolten particles through the adjustment of the plasma parameters (Figure 8). Numerous oxides like TiO_2 and ZrO_2 exhibit a pronounced tendency to form substoichiometric species when thermally sprayed. The material releases oxygen in the liquid phase and due to the rapid cooling it will remain in a suboxide state. Figure 9 demonstrate the differences in phase distribution according to the varying spray processes for titania coatings. The APS coating (figure 9a) features an inhomogeneous structure of substoichio

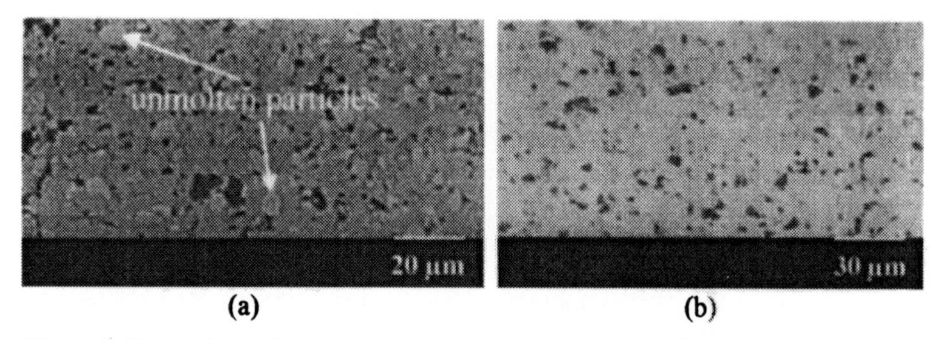

Figure 8. Comparison of microscopic cross section images of plasma sprayed alumina using different plasma powers. Partially unmolten particles reveal a higher fraction of α-Al_2O_3 in the coating but have higher porosity (a) than coatings sprayed with high plasma power (b).

metric areas. In contrast the HVOF coating shows a much more uniform structure (figure 9b). Of course the substoichiometry strongly affects the electrical properties of the coatings and will be discussed in the next chapter.

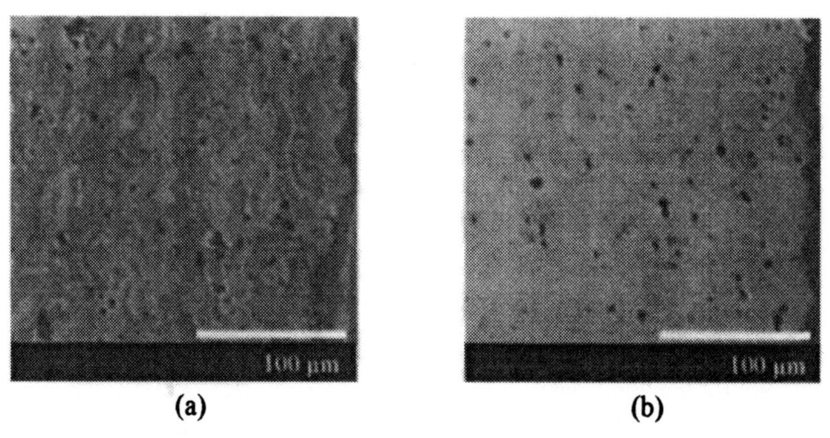

Figure 9. Comparison of microscopic cross section images of (a) APS F4 and (b) HVOF TopGun sprayed titania coatings. Variation in the color indicates the presence of different suboxide species in the APS coating

PHYSICAL PROPERTIES OF THERMALLY SPRAYED OXIDES

Thermophysical properties

A remarkable aspect of thermal spraying is the possibility to adjust the internal stress distribution of the layer composite thus compensating the thermophysical mismatches that arise when substrate and coating materials are different. Table 2 shows values for typical substrates and a selection of thermally sprayed oxides respectively. Most crucial is the ratio of heat expansion values of the substrate and the coating material. For substrate materials it can vary in a large range from 3 – 24x10^{-6} K^{-1} (f. e. glass and light metal alloys in table 2). For thermally sprayed oxide ceramics, it ranges from 2 - 11x10^{-6} K^{-1} (refer to table 2). Depending on the substrate / coating material combination of the coating, very opposed situations can emerge where either the CTE α of the substrate or the coating exhibits the lower value. This will contribute to an overall residual stress of the composite in form of tensile coating stresses for $\alpha_{coating} > \alpha_{substrate}$ and of compressive stresses for $\alpha_{coating} < \alpha_{substrate}$.

Substrate pre-processing, substrate cool-down after the coating process as well as mechanical surface post-treatment superimpose and form the residual stress of the whole composite. For a more detailed discussion see [5].

Table 2. Comparison of thermophysical properties of common substrate materials and plasma sprayed oxides

	Material	CTE α (10^{-6} 1/K)	Thermal conductivity λ (W/mK)	Young's modulus E (GPa)
Substrate materials	Al-alloy ($AlMg_3$)	23,8	134	70
	Mg-alloy (AM50)	26	65	45
	Cr-steel (X5CrNi11810)	16	15	200
	Borosilicate glass	3,3	1,12	63
Coatings materials	Al_2O_3	3-5	3-5 (20°-400°)	40-50
	TiO_2	7-9 (20°-400°)	5 (20°-400°)	-
	ZrO_2/Y_2O_3 92/8	9-11	2,5	97
	Mullite: $3(Al_2O_3)\ 2(SiO_2)$	4,3-5,0	-	-
	Cordierite: $2(MgO)\ 2(Al_2O_3)\ 5(SiO_2)$	2,2-2,4	-	-

Electrical properties

To use a material as an insulator a high dielectric strength and a low electrical conductivity are the most important properties. Furthermore dielectric applications require a high polarisation ability of the material, i.e. a high permittivity value. Besides the crystallographic structural requirements for dielectric materials it can be stated that oxides containing elements with a high atomic weight like Zr, Ba or Hf also exhibit high permittivities. Aluminium oxide possesses excellent dielectric strength and is the preferred material for typical insulator applications.

Table 3 summarizes all relevant electrophysical data of thick plasma sprayed samples (1000 μm). It should be noted that on thin coatings a lower porosity and therefor higher dielectric breakthrough voltages can be achieved, however, many application require thick coatings to meet the desired dielectric strength.

Porosity plays the major role when dealing with the electric properties of the material. Its influence on the dielectric strength is clearly detectable and was measured for thermally sprayed alumina coatings (see Figure 12).

Applications where the oxide coating serves as a dielectric often require a high coating thickness of several 100 μm up to 1 mm. An example is given in Figure 13. A metal ceramic composite consisting of a 30 μm Al/Si electrode and a zirconia dielectric of 600 μm thickness on borosilicate glass substrate was developed for ozonizer tubes. For the same application we have manufactured alumina coatings of about 1mm thickness that withstand a 13 kV voltage [4].

When dealing with thick coatings (d>300μm) the internal stresses become critical. The risk of crack formation rises and will lead to the destruction of the entire dielectric if present. An electric breakdown normally will form a pinhole going through the entire coating layer (Figure 10). Eventually the

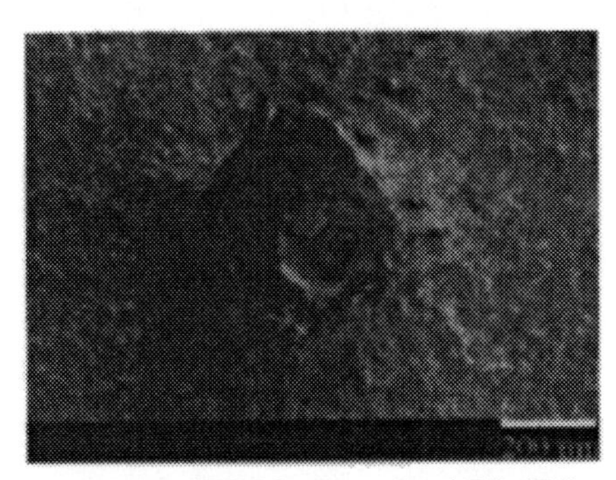

Figure 10. SEM image of a APS (F4) alumina coating with a pin hole produced by dielectric breakdown. (Top view)

Figure 11. Microscopic cross section image of the plasma (F4) spray coated metal ceramic composite on borosilicate glass substrate. For details see text.

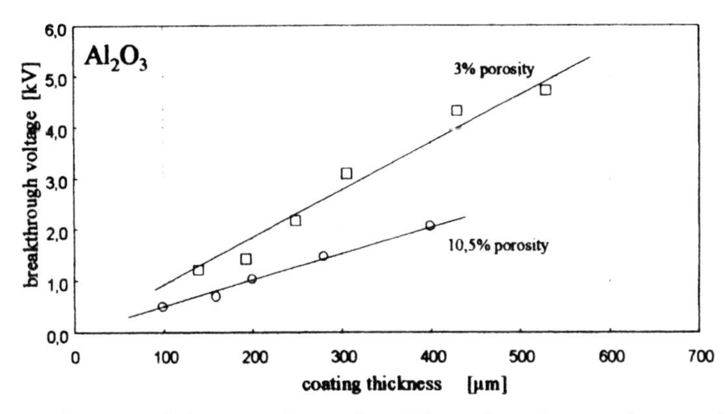

Figure 12. Influence of the porosity on breakthrough voltages of sprayed alumina coatings measured on samples with varying thickness and porosity [7]

molten material will heal out but in general this means the destruction of the coating and therefor failure of the entire component.

Titania is very wellknown for its ability to form suboxides with unusual tribological properties (Magnélli phases, lubricious oxides [6]). When thermally sprayed, titania will always form suboxides, there is no way to deposit stoichiometric TiO_2 unless the coating is thermally treated afterwards. Zirkonia is also able to form suboxides, nevertheless it is possible to deposit (nearly) stoichiometric ZrO_2 by plasma spraying on preheated substrates

Table 3. Measured electrophysical values of selected plasma (F4) sprayed coatings. Permittivity and volume resistance was determined on planar samples via impedance spectroscopy [7]

Chemical Composition	Volume resistance at 50 Hz	Typical porosity	Dielectric breakthrough voltage	Permittivity at 50 Hz
	ρ [Ωcm]	P [%]	E_d [kV/mm]	ε_r [-]
Al_2O_3	10^7 - 10^{11}	5,4 - 7,4	13,0	4
Al_2O_3/TiO_2 97/3	$2*10^8 – 3*10^9$	5,1 - 7,4	11,1	10–14
Al_2O_3/SiO_2 70/30	$10^8 – 8*10^8$	5,0 - 5,2	9,9	8
ZrO_2/SiO_2 67/33	$2*10^9$	7,6 - 8,6	11,2	12
ZrO_2	$2*10^8 – 3*10^9$	6,0 - 9,1	7,3	18-51
ZrO_2/Y_2O_3 93/7	$3*10^7 – 10^9$	4,3 - 6,6	11,5	23-26
ZrO_2/CaO 95/5	$2*10^8 – 3*10^{11}$	5,4 - 6,5	9,7	22-41
ZrO_2/CaO 70/30	$10^9 – 3*10^{11}$	3,4 - 4,2	9,9	21-23
HfO_2/Y_2O_3 95,5/4,5	$7*10^8 – 8*10^{10}$	3,0 - 4,8	10,3	16-22

(150°C). As zirkonia is an oxygen ion conductor the diffusion of oxygen into the crystal is much more efficient than in titania.

Mixing and spraying oxides with varying electrophysical properties allows the adjustment of permittivity and volume resistance of the coating in a certain range. An example is given in figure 13. Alumina / titania oxides have been plasma sprayed in various compositions starting from pure alumina to pure titania. With the increase of titania content the volume resistivity can be lowered from 10^{11} Ωcm to 10^{7} Ωcm. Approximately at a alumina / titania mass ratio of 60/40 there is a gap and the volume resistivity drops to a value of 10^{5} Ωcm. The permittivity behaves in an analogous manner. Pure titania finally behaves like a semiconductor with volume resistivity values in the region of 10^{2} Ωcm. This behaviour offers interesting possibilities of tailoring electric properties of thermally sprayed oxides for various new industrial applications.

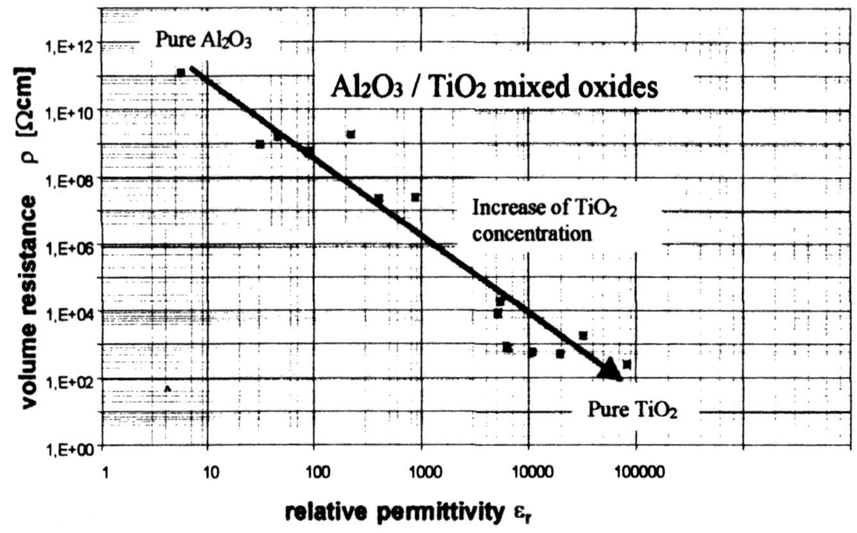

Figure 13. Measrured volume resistance and permittivity values of plasma sprayed Al_2O_3/TiO_2 mixed oxide samples., Results based on impedance spectroscopical measurements on planar samples, film thickness approx. 300 μm at 10 V and 50 Hz

SUMMARY

Thermal spraying of ceramic oxides offers a flexible and cost effective technique to produce functional coatings for electrical applications. APS and HVOF techniques are suitable to manufacture coatings with varying mechanical and electrophysical properties. Film thickness in the range of 50 μm up to 1mm can be realised. Porosity plays an important role with regard to the electrophysical and mechanical properties of the coating. Residual stress effects become critical when the film thickness exceeds approximately 300 μm (exact values depend on the material composition of substrate and layer respectively and the coating porosity). Thick coatings therefor require a minimum porosity to increase the materials elasticity . Manufacturing of coatings with specified volume resistivity and permittivity is possible by spraying mixed oxides species.

REFERENCES

[1] Friedrich, C., Gadow, R.; Killinger, A.Plasma and hypersonic flame sprayed ceramic coatings for dielectrical applications CIEC 7, 2000, Genova

[2] Matejka D., Benko B.,"Plasma spraying of metallic and ceramic materials", John Wiley & Sons, Chichester UK ,1989, ISBN 0-471-91876-8

[3] L. Pawlowski; "The Science and Engineering of Thermal Spray Coatings", John Wiley & Sons, 1995, ISBN 0-471-95253-2

[4] Gadow R.; Killinger A.; Friedrich C.; "Thermally Sprayed Multilayer Coatings as Electrodes and Dielectrics in High Efficiency Ozonizer Tubes", UTSC`99, Düsseldorf, Germany, 1999, ISBN 3-87155-656-X, pp. 676 - 682

[5] Buchmann, M.; Gadow, R., "High speed circular micro milling method for the determination of residual stresses in coatings and composites", 24th Annual Cocoa Beach Conference&Exposition, 23. – 28.01.2000, Cocoa Beach, USA

[6] Buchmann, M.; Gadow, R.; Killinger, A. „Thermally sprayed coatings for solid lubricant applications", Proceedings of the 10th Intern. Metallurgy and Materials Congress a. Trade Fair, 2000, Istanbul, Vol. II, ISBN 975-395-382-8

[7] Friedrich, C.,"Atmosphärisch plasmagespritzte dielektrische Oxidschichten für Ozongeneratoren, dissertation, IFKB, University of Stuttgart, Germany, 2001, in print

ELECTRICAL PROPERTIES OF BARIUM TITANATE THICK FILMS

Cesar R. Foschini, B.D. Stojanovic
José A. Varela
Instituto de Química-UNESP
Araraquara, São Paulo, Brazil
14801-970

V.B. Pavlovic, V.M. Pavlovic,
V. Pejovic
University of Belgrade and IRITEL,
Belgrade,Yugoslavia

ABSTRACT

The barium titanate thick films were prepared from powders mixture based on $BaCO_3$ and TiO_2. After homogenization and milling in high-energy vibro mill, the as prepared $BaCO_3$ and TiO_2 mixed powders were calcined at 700°C for 2 hours, ground in a agate mortar to effect comminution of the layer aggregates and screened through a 200 mesh nylon screen before preparing the paint. The fine dry powder is extremely active because of the high surface area of the particles. A thick film paste was prepared by mixing BT fine powders with small amount of low temperature sintering aid and organic binder. The thick films were screen-printed on alumina substrates and electroded with Ag-Pd. The BT films were sintered at 850°C for 1 hour. The film thickness, measured by SEM, was between 25 and 75 μm depending on the number of layers. The microstructure of thick films and the compatibility between BT layers and substrate were investigated by SEM. Depending on the number of layers, the permittivity reached values of 2600 with dielectric losses lower than 2%.

INTRODUCTION

The $BaTiO_3$ compound is a technologically important electroceramic material widely used as dielectric, thermistor, and ferroelectric. It is even today the most significant ceramic material to manufacture multilayer ceramic capacitors (MLC)[1]. Indeed, the difficulty in preparation of thin films in various thickness range is the primary reason for their notable lack of availability. Not withstanding the difficulties, thick films still hold promise as a interesting area of research since it is generally recognized that certain phenomena such as electrooptic effects, piezoelectricity or capacity can be profitably applied to devices within them [2].

Barium titanate ($BaTiO_3$) is used extensively as the dielectric in ceramic capacitors, particularly due to its high dielectric constant and low loss characteristics. As advanced miniaturization requires smaller circuit area, the multilayer ceramic capacitors with higher volumetric efficiency were developed. Conventionally, $BaTiO_3$-based materials are manufactured at high temperature, by solid-state reaction or from chemically derived intermediates. These methods typically produce large, agglomerated particles that consequently are milled to obtain mean particle sizes between 0.5 and 1.5 μm. The large, non-uniform particle size of conventionally prepared $BaTiO_3$, generally limits the ability to fabricate reliable MLC with various dielectric layer thickness [3].

Many researchers have been reported on the deposition mechanisms with different powders [4-7]. As the film thickness increases, the difficulty in preparing thick film is well recognized. The sub-micron powders are widely available, and an effective method to prepare thick film of barium titanate is by using reliable ceramic powders. One possibility to obtain required grains size is by the mechanical activation of raw materials during powder process preparation using high energy milling process. The mechanical activation is one of the most effective method for obtaining highly disperse system due to mechanical action stress fields form in solids. This effect results of changes of free energy, leading to release of heat, formation of a new surface, formation of different crystal lattice defects and initiation of solid state chemical reaction [8, 9]. Energy parameters of system and the amount of accumulated energy during mechanical activation have been changed during process [10]. Generally, plastic deformation of surface layers of particles due to high local stress on particle contacts is caused by the energy accumulated in the deformed material. The accumulated deformation energy determines irreversible changes of the crystal structure and smaller particles size [11]. As the consequence of deformation, the change of crystal and microstructure of material leads to change of their properties. Usually sub-micron grain size of ceramic powders could exceedingly advance their applications in thick and thin films technology, requiring for application in electronics.

It is known that thin films may be deposited on substrates by various techniques, for example, sputtering, evaporation, sol-gel processes, etc [12, 13]. Typical film thickness are between 0.1 and 1,0 μm. Meanwhile, layers thicker than a few micrometers could not be achieved by these methods.

Thick film technology is a method whereby resistive, conductive and dielectric pastes are typically applied to ceramic substrates [14]. The technology of producing thick films of various types involves a number of steps are common to all of them. Unlike thin films, the number of processes capable of producing high quality thick films are rather limited to specifically mentioned to sol-gel, MOD and screen printing. Previous investigators have employed dip coating, spin coating and chemical vapor deposition technique with limited success on films up to 25 μm in thickness. However, by experience, it was shown that films thicker than 2 or 3 μm, which could be prepared by previously mentioned deposition techniques, had a tendency to undergo cracking, debonding from substrates, increasing the roughness [15, 16]. Otherwise, the deposition of thick film pastes by screen printing is a relatively simple and convenient method to produce thicker layers with thickness up to 100 μm.

Screen printing is a relatively mature technology. This process is useful to accommodate the demands of miniaturization, circuit complexity, multilayer assembles, or high frequencies [17, 18]. Generally, the characteristics of thick film ferroelectric materials are similar to the characteristics of bulk materials.

The circuits defined by screen printing are fired typically at 850°C to fuse the films to the substrates and to produce the desired functions. This temperature is higher than for thin films, thus increasing the possibility of interactions with either the electrodes or substrates, and consequently to their possible degradation. Adhesion between support and film, as similar temperature expansion coefficients of the thick films and substrate are also important. Low-melting glasses were used to isolate the top layer conductor from the rest of the circuit below. However, because of undesirable interaction between the glass phase and the overlying conductor, crystallisable glasses with low dielectric constants could appear.

In the present work, $BaTiO_3$ thick films were prepared, from mechanically activated powders based on $BaCO_3$ and TiO_2 to obtain sub-micron grain sized powders. A thick film paste was prepared by mixing BT fine powders and organic vehicle screen-printed on alumina substrates and with electrodes based on Ag-Pd and sintered at 850°C for 1 hour. The microstructure and dielectric properties were determined and the effect on the layers of thick films was determined.

EXPERIMENTAL PROCEDURE

Mixture of $BaCO_3$ (Merck PA 99%) and TiO_2 (Ventron PA 99.8%) powders with ratio Ba/Ti=1 and 1.5 wt % LiF (Merck PA 99.3%) were first homogenized in planetary ball mill for 120 min and after that mechanically activated in high energy vibro-mill with rings (TM MN 954/3) in air for 90 min. The BT powders were dried, milled, sieved and pressed at 400 MPa and after calcination at 800°C for 1 hour, pellets were crushed, milled and sieved. Paste was prepared from the mixture of the suspension of organic material (resin, organic solvent and some additive to improve rheological behaviour of paste) and BT calcined powders, with a 30:70 ratio. Alumina substrates were used as a commercial product (Alcoa). The electrode was the silver/palladium mixture (Ag/Pd 70/30). Over the fired electrode, barium titanate layer was screen-printed and sintered at 850°C in air flow during 1 hour. The obtained films usually had thickness that range from about 25 to 100 μm, depending on the number of layers.

Specimens for SEM examination were prepared in planar and cross-sectional view of the films and the thickness of the films was determined. The microstructure was analyzed using a scanning electron microscope (TOPCON SM 300). The temperature dependence of the dielectric constant and dielectric loss were measured at 1 kHz, 10 kHz and 100 kHz using an LCR meter (HP 4291A) in the temperature range of 30° ~ 160°C. The dielectric constant and Curie temperature were determined for thick film samples with up to 3 layers.

RESULTS AND DISCUSSION

A major change in the past decade is the increase in the number of chips and multilayer dielectrics. Chips and myltilayers still use thick film technology, but the complexity is demonstrated in precisely deposition onto substrates. The difficulty in fabricating films in needful thickness range is the primary reason for their notable lack of availability. When thick films are being deposited, the differences in thermal coefficients of dielectric materials, electrodes and support can lead to stress (compressive or tensile) much more expressed than in a thin one. Furthermore, the elastic module of the substrate as well as the film must also be taken into consideration during preparation of thick films since their different module have to determine stress levels [22]. Taken all together, these stresses, specially tensile stresses, can produce disastrous results such as delaminating of the film from the substrate and when the films become thicker, may appeared the macro-cracking of films, or can lead to appearance of micro-cracks or pores in film. In general, it should be noted that with thick films most of the substrate and processing consideration are magnified in comparison to thin films because the extended processing times at temperature and the desire to deposit a maximum film thickness per layer [23].

The substrates selected for our investigation was based on pure Al_2O_3. Al_2O_3 substrates have a good chemical compatibility with the film, so the inter-diffusion is minimized. Previous experiments have been shown that the substrate do not put the barium titanate thick films neither in compression nor in tension. The most common problems in thick films relating to appearance of cracks was not observed in investigated case.

A typical microstructure of a BT MLC cofired with Ag-Pd electrodes is shown in Fig. 1. The thickness of one, two and three layers thick film, measured by SEM, was approximately 25, 50 and 75 μm, respectively. The adhesion of Ag-Pd/BT layer was rather strong, and it was not possible easily peeled off the alumina substrate. In the same time, the adhesion of the bottom electrode layer overprinted with alumina, and upper electrode overprinted with BT was good; it could not be peeled off with Scotch tape. The results indicate that the adhesion of the electrode film to alumina substrate and to barium titanate are quite well and that the firing process does not decrease the stability of interconnections.

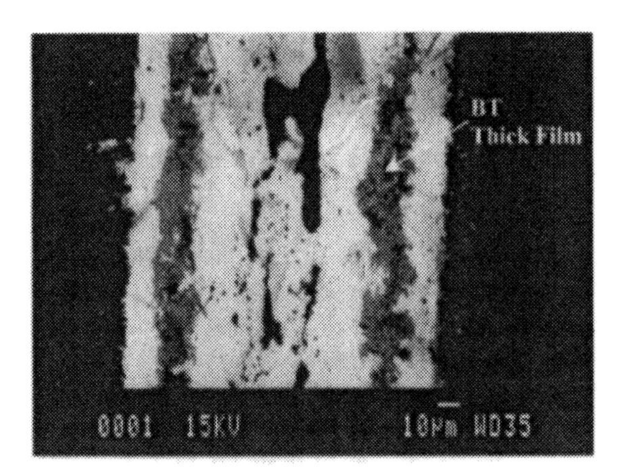

Figure 1: Micrograph of a two layers BT MLC deposited on Al_2O_3 substrates and cofired with Ag-Pd electrodes at 850°C, 2 hours (3000X).

The grain size of BT thick film is rather regular with rounded grains and withouth obvious presence of secondary phases (Fig. 2). The grain size was less than 1 μm, approximately 800 nm. The small grain size is caused not only due to low sintering temperature and rather short time for firing process but also resulted from tribophysical activation process that increases the surface area of starting powders.

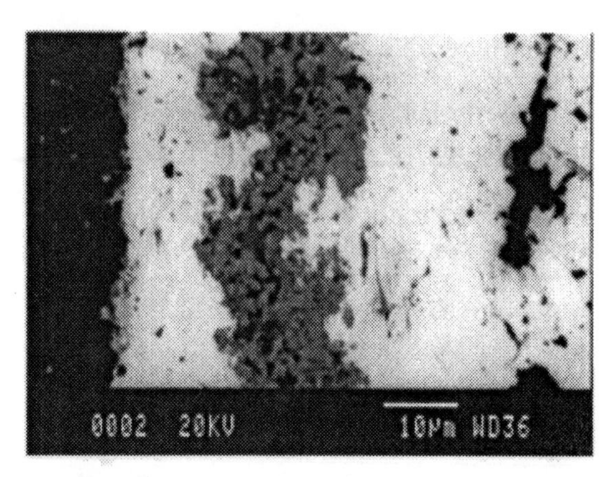

Figure 2: SEM cross-section analysis of BT layer presenting the small grain sized barium titanate.

The dielectric properties of barium titanate thick films are presented in Fig.3. The dielectric constant and dissipation factor of BT thick films with up to 3 layers as a function of temperature and frequency is shown. It could be noticed that the dielectric losses slightly change in frequency region from 1 kHz to 100 kHz. The dielectric losses values ranged from 0.1×10^{-2} to 2.5×10^{-2} depending on the number of layers. At the lowest frequency it was very difficult to obtain the exact results, because of the noise caused by the system used for measurements. Otherwise, the decreasing in the dielectric losses with the increasing of film thickness could be noticed. This observation is comparable with reported data that the thinner dielectric layer, the higher is dielectric loss [24]. On the other hand, the rather lower values obtained for dielectric losses compared with the literature data [13] related to BT thick film prepared by electrophoresis method resulted not only due to higher thickness of films but also due to the microstructure and grain size of the dielectric layer.

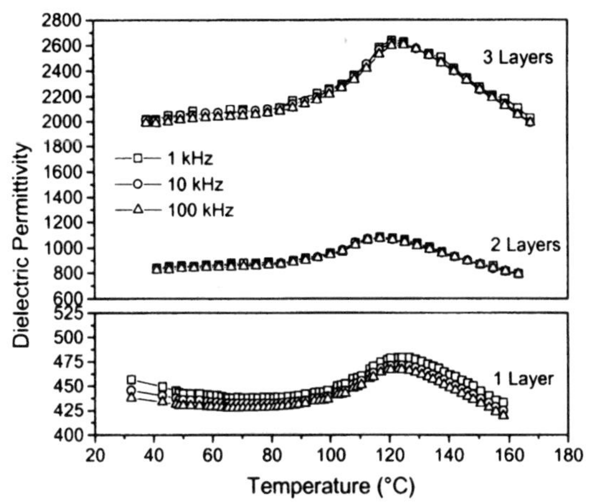

Figure 3. Dielectric permittivity and dielectric loss *vs* temperature and frequency, for BT thick films with 1, 2 and 3 layers.

The dielectric constant at room temperature (25°C, in the frequency region from 1 kHz up to 100 kHz) changes from 400 to 2000, depending of the number of layers. Dielectric constant at the Curie temperature is well expressed in Fig. 3. The Curie temperature of ~120°C was observed for all BT thick films independently of number of layers. The dielectric constant at the Curie temperature was 480, 1100 and 2600 for MLCs with 1, 2 and 3 layers respectively, and no significant difference during heating and cooling cycles was noticed. It is believed that the position of dielectric peak without shifting in comparison with the $BaTiO_3$ bulk is due to known phase transition from tetragonal to cubic.

Dielectric properties indicate that the barium titanate thick films show the similar characteristic with bulk material. This means that the screen-printed barium titanate thick films have good opportunity to evaluate their potential application as multilayer dielectrics.

CONCLUSION

The high surface activity associated with the prepared powders through mechanically activated technique which translate to overcoming the kinetic constraints by rapid atomic diffusion and high activities in grain boundaries, makes possible the preparation of BT thick films at low sintering temperature such as 850°C, with desired electrical properties for the applications in microelectronics. The most common problem in thick films such as cracking was not observed. The adhesion electrode/support and electrode/barium titanate layer was strong and no delamination was noticed. The microstructure of barium titanate thick film deposited by screen-printing is uniform and dense with rounded grains and withouth presence of secondary phases.

The barium titanate thick films show the dielectric constant ranging from 480 to 2600 at a Curie temperature of ~120°C. The dielectric losses range from 0.1-$2.0x10^{-2}$ depending on the number of layers.

ACKNOWLEDGEMENTS

This research was supported by CNPq, PADCT and FAPESP-Brazil (98/13678-3).

REFERENCES

[1] G.H. Haertling, in "Integrated Thin Films and Applications", Ed. by R.K. Pandley, D.E. Witter, U. Varshney, Ceramic Transactions, Vol. 86, 235-261, 1999.

[2] W.Borland, "Thick Film Hybrids" Vol.1.,ASM International, 1989.

[3] J.L. Larry, et all, "Thick Film Technology: In Introduction to the Materials", IEEE Trans, CHMT-3, 1980.

[4] D. Brown, F. Salt, "Mechanism of Eletrophoretic Deposition", J. Appl. Chem. 15 (1963) 40-48.

[5] K. Yamashita, M. Nagai,T. Umegaki, "Fabrication of Green Films of Single and Multi-Component Ceramic Composites by Eletrophoretic Deposition Technique", J. Mater. Sci, 32 (1997) 6661-64.

[6] K. Hasegawa ,H. Nishimori, M. Tatsumisago, T. Minami, "Effect of Polyacrilic Acid on tje Preparation of Thick Silica Films by Sol-gel Deposition",J. Mater. Sci, 33 (1998) 1095-98.

[7] P.S. Nicholson, P.Sarkar, S.Datta, "Producing Ceramic Laminate Composites by EDP", Am. Ceram. Soc. Bull., 75, 11 (1996) 48-51.

[8] G. Gomez-Yanez, C. Benitez, H. Balmori-Ramirez, "Mechanical activation of the synthesis reaction of $BaTiO_3$ from a mixture of $BaCO_3$ and TiO_2 powders", Ceramic International, 26 (2000), 271-277.

[9] E.M. Kostic, S.J. Kiss, S.B. Boskovic, S.P. Zec, Am. Ceram. Soc. Bull, 76 (1997) 60-64.

[10] J. Xue, J. Wang, D. Wan, "Nanosized Barium titanate Powder by Mechanical Activation", Amer. Ceram. Soc., 83 (2000) 1235-1241.

[11] L.J. Lin, "Implication of fine Grinding in Processing-Mechanochemical Approach", J. Therm. Anal.,57 (1998) 453-461.
[12] J. Xue, D. Wan, S. Lee, J. Wang, "Mechanochemical Synthesis PZT of Mixed Oxides", J. Amer. Ceram. Soc., 82, 7 (1999) 1687-92.
[13] J. Zhang, B. Lee, "Electrophoretic deposition and Characterization of Micrometer-Scale $BaTiO_3$", J. Am. Ceram. Soc., 83, 10 (2000) 2417-22.
[14] K. Adachi, H. Kuno, "Effect of Glass Composition the Electrical Properties of Thick Films", J. Am. Ceram. Soc., 83, 10 (2000) 2441-48.
[15] M. Prudenziati, R. dell'Acqua , "Thick Film Resistors", Ed. by M. Prudenziati, Elsevier Science, Amsterdam, 1994.
[16] R.W. Vest, "A Model for Sheet Thick Film", IEEE Trans.Compon., Hybryds, Manufac. Techol., CHTM-14 (1991) 396-406.
[17] R. Bouchard, "Thick Film Technology: an Historical Perspective", In: "Dielectric Ceramic materials", Ed. by K.M. Nair, A.S. Bhalla, Publ. by The American Ceramic Society, Westerville, Ohio, USA, 1999.
[18] J. Holc, M. Hrovat, M. Kosec, "Interaction between Alumina and PLZT Thick Films", Materials Research Bull., 34, 14/15 (1999) 2271-78.
[19] J. Holc, M. Hrovat, J. Mater. Sci. Lett., 8, (1989) 636.
[20] V.B. Pavlovic, B.D. Stojanovic, Lj. Zivkovic, G.O. Brankovic, M.M. Ristic "Grain Growth during Sintering of $BaTiO_3$ with LiF", Ferroelectrics, 186, (1996), 165-168.
[21] V.B. Pavlovic, B.D. Stojanovic, G.O. Brankovc, M.M. Ristic: "The Effect of LiF on the Dielectric and Microstructural Properties of Low Temperature Sintered $BaTiO_3$ Ceramics", Sci. Sint., 28, Spec. Issue, (1996), 143-148.
[22] B. Morten, G.De Cicco, A-Gandolfi, C.Tonelli, in Proc.of 8th Europian Hybrid Microelectronics Conference, ISHM-Europe 91, Rotterdam,(1991) 392-399.
[23] M. Prudenziati, In: "Thick Film Sensors", Ed. by M. Prudenziati, Elsevier, Amsterdam, 1995.
[24] B.C. Foster, W.I. Symes, E. A. Davis, "New Dielectric Compositions for Capacitors", Ceram.Ind., 12 (1998) 29-34.

MICROWAVE DIELECTRIC PROPERTIES OF Al_2O_3-MgO-ReO_x (Re:RARE EARTH) SYSTEMS AND THEIR APPLICATION TO NEW LTCC.

Hiroshi Kagata
Device Development Center, Matsushita Electric Industrial Co.,Ltd

Hidenori Katsumura
Corporate Components Development Center, Matsushita Electronic Components Co.,Ltd.
1006 Kadoma, Kadoma-shi, Osaka,571-8501 Japan

ABSTRUCT

Dielectric properties of Al_2O_3-MgO-ReO_x (Re: rare earth and Y) systems have been studied in the microwave region. We found that magnetoplumbite phases in the MgO poor region of $MgReAl_{11}O_{19}$(Re: Y and La~Tb) compositions had positive TCF(temperature coefficients of resonance frequencies) values though they had low dielectric constants less than 20. By mixing a lead-free glass to the above system, a new LTCC (that we call an AMSG) with a low dielectric constant (<10), a near zero TCF, and a high bending strength was obtained. The AMSG can be applied to laminated RF modules for various mobile communications.

INTRODUCTION

The dielectric material used for high frequency filters for mobile communication, etc. needs to possess an appropriate dielectric constant, small dielectric loss and small temperature coefficient of resonance frequency (dielectric constant). Dielectric ceramic materials with small temperature coefficients and a variety of dielectric constants have been developed for commercial use. The material having the smallest dielectric constant among them is $MgTiO_3$-$CaTiO_3$-based material, given a value of 20. A compound with a dielectric constant less than 20 usually has a negative temperature coefficient of resonance frequency but a positive temperature coefficient for its dielectric constant[1]. For this reason, there is demand for the development of a dielectric ceramic material having a dielectric constant less than 20 combined with a small temperature coefficient. A small temperature coefficient is also needed for the increasingly prevalent low temperature cofired ceramics (LTCC). The temperature coefficients of their dielectric constants in many LTCCs used today are large positive values of around +100 ppm/℃.

On the other hand, $MgReAl_{11}O_{19}$, a ternary system composed of Al_2O_3 - MgO-ReO_x, possesses a magnetoplumbite structure and, in single crystal form, has been investigated for use as a tunable laser[2]. Certain ceramic materials having similar composition have been found to possess large positive temperature coefficients of resonance frequency combined with dielectric constants less than 20. This article reports on their dielectric characteristics in the microwave range and their potential application to LTCCs.

EXPERIMENTAL

The specimens were fabricated using a conventional ceramic process. Al_2O_3, MgO and ReO_x (where Re is replaced by La, Ce, Pr, Nd, Sm, Eu, Gd, Tb, Dy, or Yb while x is a number determined stoichiometrically by the valence number of Re) were weighed for predetermined concentration and mixed in a ball mill. After drying the mixture, it was calcined at 1300 °C and crushed again in the ball mill. The crushed powder was then granulated and made into 17 mm diameter disks and sintered at 1650 to 1680 °C. The dielectric constant of the sintered body was measured using a TE_{011} mode resonator[3] while the temperature coefficient of the Q value and resonance frequency were measured using the resonance of the $TE_{01\sigma}$ mode. The temperature coefficient of resonance frequency was determined from the values at 20 °C and 85 °C. The produced phases were examined by using an X-ray diffraction (CuKα ray).

RESULTS AND DISCUSSION

(1)Al_2O_3-MgO-ReO_x-based Produced Phases

Shown in Fig. 1 is the relationship between the composition and produced phases of a ternary system when Sm was used in place of Re. By using the composition of the magnetoplumbite phase reported in the formula $MgReAl_{11}O_{19}$, a spinel phase ($MgAl_2O_4$) was produced in addition to a magnetoplumbite phase. Use of a poor-MgO composition, however, produced almost a pure magnetoplumbite phase. Reference 2 below also reports the vaporization of Mg and Re components under a high temperature melted conditions during single crystal production using the predetermined composition of $MgReAl_{11}O_{19}$. It is presumed that either the magnetoplumbite phase can be easily obtained under poor-MgO conditions or the stoichiometrical composition of magneto-plumbite phase is shifted away from that for $MgReAl_{11}O_{19}$.

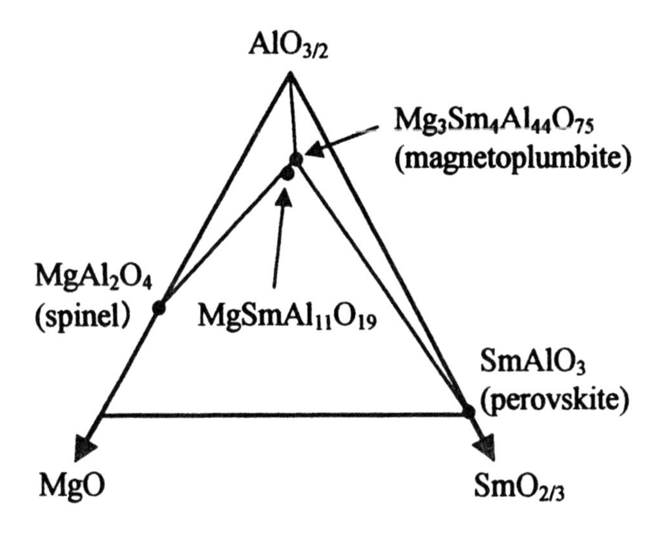

Fig.1 Crystalline phases sintered at 1650℃-1680℃ in the Al_2O_3-MgO-Sm_2O_3 system.

Table I. Crystalline phases in the composition of $Mg_3Re_4Al_{44}O_{75}$ sintered at 1650-1680℃.

Re	Primary phases	Other phases
La	MP	PE, uk
Ce	MP	uk
Pr	MP	uk
Nd	MP	uk
Sm	MP	uk
Eu	MP	uk
Gd	MP	uk
Tb	MP	uk
Dy	GN	α-alumina, SP, MP
Yb	GN	α-alumina, SP

MP:magunetoplumbite, PE:perovskite, GN:garnet, SP:spinel, uk:unknown phase

The produced phase of $Mg_3Re_4Al_{44}O_x$, which produced an almost entirely single phase, was further examined using a different Re component other than Sm. The results are summarized in Table I . The produced structure using La through Tb in order of atomic number was primarily a magnetoplumbite phase, but only a small quantity of this type of phase was produced with Dy. When Dy and Yb were used in place of Re, the primary structure was garnet phase ($Re_3Al_5O_{12}$), which was a mixture of Al_2O_3 and spinel phase ($MgAl_2O_4$). No ternary system compound was present in the area examined. The interesting observation was made that the magnetoplumbite phase was generated as a primary structure when a perovskite phase ($ReAlO_3$) was present in a system of Al_2O_3-ReO_x, while magnetoplumbite phase was produced in only very limited quantities when garnet phase was present. The X-ray diffraction pattern of the magnetoplumbite phase we obtained is shown in the Figure 2. It was found that a compound which showed an X-ray diffraction pattern resembling a magnetoplumbite phase could be obtained with Tb in place of Re. The peaks of the X-ray diffraction pattern of the magnetoplumbite phase we obtained were able to be identified for the most part as that of $MgLaAl_{11}O_{19}$, but some peaks near 30° of 2-theta could not be identified. Full equilibrium may have not been reached under the sintering conditions used. Analysis of the detailed stoichiometric composition of the magnetoplumbite phase produced and its crystalline structure will be left for a future study.

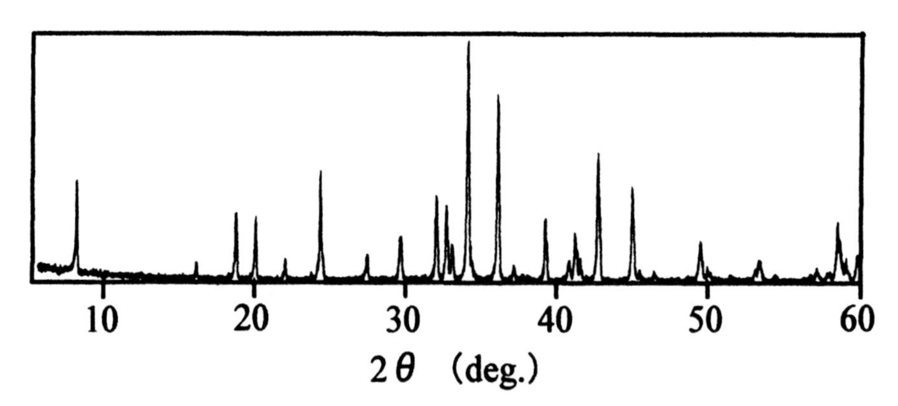

Fig.2 XRD pattern of $Mg_3Gd_4Al_{44}O_{75}$ composition.

(2)Microwave Dielectric Characteristics of $Mg_3Re_4Al_{44}O_x$-based Material

The microwave dielectric characteristics of $Mg_3Re_4Al_{44}O_x$-based materials are shown in Table II. A comparison with Table I shows that when magnetoplumbite phase is the primary structure, the dielectric constant increases and the temperature coefficient of resonance frequency increases with increasing atomic number of Re. This material is significantly different from well-known compounds in that its TCF value has a high positive number even though the dielectric constant is less than 20. Compounds having a magnetoplumbite structure have only rarely been studied as dielectric materials, but new compounds with unique temperature coefficients may now be found based on the compositions investigated in this study. Concerning potential commercial application, the unique dielectric characteristics of the material with La in place of Re, with a dielectric constant of 13 and almost zero TCF, are of particular interest. Future applications of this material are expected to be coaxial resonators, etc. used in dielectric filters.

Table II. Microwave Dielectric Characteristics of $Mg_3Re_4Al_{44}O_x$ ceramics.

Re	εr	Qf (GHz)	TCF(ppm/°C)
La	13.0	7700	+3
Ce	14.0	9000	+11
Pr	14.5	10000	+23
Nd	15.0	11000	+35
Sm	16.5	11000	+93
Eu	17.3	11000	+147
Gd	18.3	4800	+175
Tb	18.3	5900	+200
Dy	10.3	28000	-49
Yb	10.3	41000	-57

(3) Applications to LTCCs

Using the composition examined above, Sm was selected in place of Re and low temperature sintering was studied by adding glass. Lead-free glass with the composition of SiO_2-Al_2O_3-B_2O_3 was selected. The material needs to be sufficiently sintered at approximately 900 °C in order for the LTCC to be cofired with silver. The base material composition was determined to be $Mg_3Sm_4Al_{44}O_x$, and the oxide mixture of the base material was calcined at 1500 °C. This base material and glass were mixed in equal quantities. A sintering body was then produced by using the same method as that used in the experiment, and its characteristics were evaluated. The LTCC produced will be called AMSG hereafter. The characteristics of AMSG are shown in Table III in comparison with those of a conventional Al_2O_3-glass-based LTCC. It can be seen from Table III that AMSG matches the characteristics of conventional LTCCs with respect to relative dielectric constant, Q value, bending strength, etc., but has a temperature coefficient of relative dielectric constant close to zero. By taking advantage of its good temperature characteristics, future applications of this material are likely to include highly integrated RF modules, including narrow-band devices such as band-pass filters and notch filters.

Table III. Characteristics of New LTCC, "AMSG".

Materal Code	Composition	ε r	Qf (GHz)	TCC (ppm/℃)	Thermal Exp.Coeff. (ppm/℃)	Bending Strength (MPa)
AMSG	$Mg_3Sm_4Al_{44}O_{75}$ + B-Si-O glass	7.8	10000	-9	+11	250
Conventional	Al_2O_3-glass	7.8	10000	+100	+6	250

REFERENCES

[1] A. J. Bosman and E. E. Havinga, "Temperature Dependence of Dielectric constants of Cubic Ionic Compounds," Pys. Rev., 129, 1593 (1963).

[2] D. Gourier, L. Colle, A. M. Lejus, D. Vivien, and R. Moncorge, "Electron-spin Resonance and Fliorescence Investigation of LaMgAl11O19: Ti3+, a Potential Tunable Laser Material," J. Appl. Phys., 63, 1144 (1988).

[3] B. W. Hakki and P. D. Coleman, "Dielectric Resonator Method of Measuring Inductive Capacitance in the Millimeter Range," IRETrance. Microwave Theory Thech., MTT-8, 402(1960).

AN ULTRASONIC MOTOR FOR CATHETER APPLICATIONS

Serra Cagatay, Burhanettin Koc[†], and Kenji Uchino
International Center for Actuators and Transducers, Materials Research Laboratory, The Pennsylvania State University, University Park, PA, 16802
†Kirikkale University, Department of Electrical and Electronics Engineering, Turkey

ABSTRACT

Ultrasonic surgical devices typically operate at frequencies between 20 kHz and 60 kHz and have application in many surgical specialties including neurosurgery and general surgery. An ultrasonic motor that converts input electrical energy to ultrasonic vibrations may be fabricated from piezoelectric ceramics. This design consists of two hollow metal cylinders, one of which is inserted into the other. The outside surface of the outer cylinder is flattened on two sides at 90-degrees to each other and rectangular piezoelectric plates are bonded on these surfaces. This motor utilizes two orthogonal bending modes of the hollow cylinder. A wobble motion is generated on the outer cylinder when one of the piezoelectric plates is excited at a frequency between the two orthogonal bending modes. Performance of the motor can be estimated by the displacement mechanism and load characteristics of the motor.

INTRODUCTION

First investigation of piezoelectric motors using bending vibration of a square beam was proposed by Williams and Brown in 1942. Four piezoelectric rectangular plates were bonded to all faces of a square elastic beam forming a composite structure. A wobble motion was generated at one end of the bar upon exciting piezoelectric elements by two voltages such as sine and cosine at a bending mode resonance frequency of the square beam. The rotor was pressed against the stator's surface and a rotation was produced via frictional interaction between the touching surfaces of the stator and rotor [1].

Even though the principle and the structure of the piezoelectric motor proposed by Williams et al. is similar to today's piezoelectric motors [2,3,4], the original piezoelectric motor never became a commercial product due to the lack of high quality piezoelectric materials and high voltage-high frequency driving techniques.

Piezoelectric ultrasonic motors with their exceptional properties, such as high resolution of displacement control, absence of parasitic magnetic fields, frictional locking at the power-off stage, and high thrust to weight ratio, make them good candidates for use in precision micro mechanical systems.

Also medical applications such as, endoscopes and prosthetic devices could benefit from miniature motors. Endoscopy techniques are replacing many surgical procedures. Instead of a critical surgery, a thin tube is inserted through a small slit and fed to the appropriate place in the body. Miniature motors would add additional functionality to endoscopic procedure [5].

Miniaturization of piezoelectric motors can be accomplished if the structure and driving circuit of these motors are simplified so that they can be manufactured at low cost. One of the most suitable structures for miniaturization of piezoelectric motors for aforementioned applications is multi- (or mixed-) mode excitation type piezoelectric motor. Exciting at least two orthogonal resonance modes of the stator vibrator generates elliptical motion on the stator surface. According to the shape of the stator, these orthogonal resonance modes may generate longitudinal-torsional [6,7] radial-torsional [8], longitudinal-flexural [9], or flexural-flexural motions.

The motor presented in this paper is a multi-mode-single-vibrator excitation type, which uses two orthogonal bending modes of a hollow cylinder. Since the structure and poling configuration of the active piezoelectric elements used in the stator are simple, this motor structure is very suitable for miniaturization. Moreover, a single driving source can excite two bending modes at the same time, thus generate a wobble motion.

STRUCTURE AND OPERATION PRINCIPLE

A square beam has two orthogonal bending modes whose resonant frequencies are equal to each other. The first bending mode frequencies in any direction for circular cylinders are also equal to each other. The stator of the motor presented in this study combines the circular and square cross-sections.

The outside surface of a hollow metal cylinder was flattened on two sides at 90-degrees to each other and two uniformly electroded rectangular piezoelectric plates were bonded onto these two flattened surfaces (Figure 1(a)). The basic configuration of the motor is shown in Figure 1(b). Since the stator is symmetric with respect to the x'-axis, the area moment of inertia about the principal axis is on the x'-axis. The area moment of inertia about the other principal axis is on the y'-axis. This causes the stator to have two degenerated orthogonal bending modes, whose resonance frequencies are close to each other.

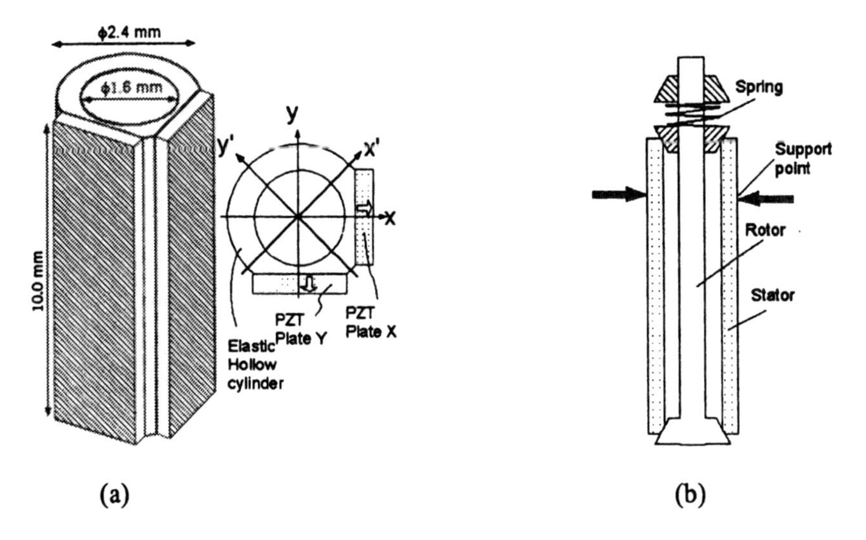

Figure 1. (a) Structure of the stator (b) Assembly of the motor. Two rectangular plates are mounted on the flattened surfaces of the hollow cylinder. The bending modes in x' and y' directions can be excited when only one plate is driven with an AC signal.

The split of the bending mode frequencies is due to the partially square/partially circular outside surface of the hollow cylinder. The orthogonal bending modes correspond to the minimum and maximum bending moment of inertia of the beam (stator). Since the piezoelectric plates are oriented by 45° to the direction of minimum and maximum bending moment of inertia, driving one piezoelectric plate (while short circuiting the other to ground) at a frequency between the two orthogonal bending mode frequencies excites both modes, thus, causing the cylinder to wobble. When the other piezoelectric plate is driven at the same frequency, the direction of wobble motion is reversed.

FINITE ELEMENT ANALYSIS

The behavior of the free stator was simulated using ATILA finite element software to verify the operation principle. The piezoelectric plates on the surface of the cylinder were placed in such a way that one piezoelectric plate can excite the two orthogonal bending modes of the stator. Figure 2(a) shows the two orthogonal bending mode shapes when only the plate on the x-axis (plate X) was excited, while the electrode of the plate Y was short-circuited. Wobble motion was generated on the cylinder when only one piezoelectric plate is excited at a frequency (72.7 kHz) between the two orthogonal bending modes frequencies (Figure 2(b)). When the other piezoelectric plate was excited at the same frequency, the direction of wobble motion was reversed.

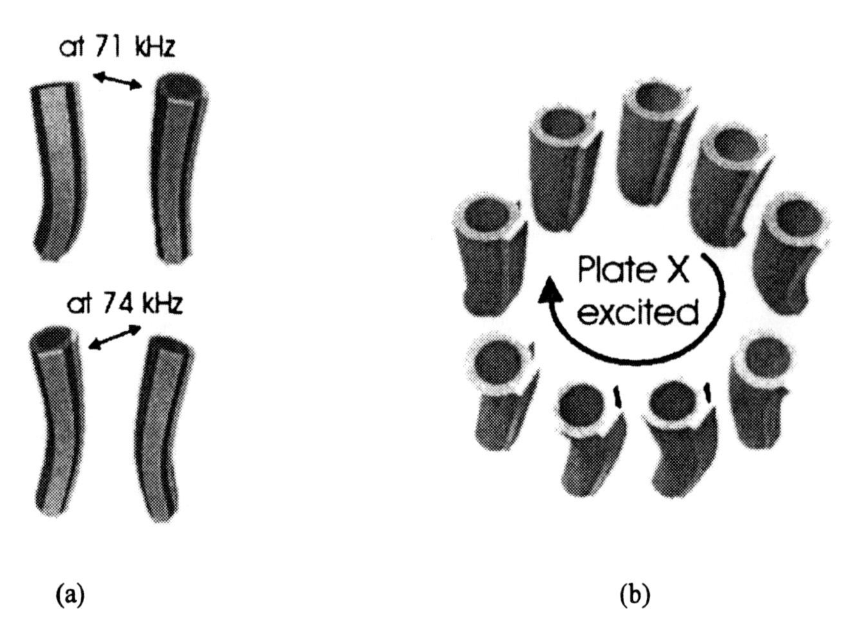

Figure 2. (a) Two orthogonal mode shapes at 71 and 74 kHz when plate in x-axis was excited, (b) Wobble motion in clockwise direction can be generated on the cylinder when plate in x-axis was excited in between the two orthogonal bending resonance frequencies (72.7 kHz). When the other piezoelectric plate is excited at the same frequency, the direction of wobble motion can be reversed.

EXPERIMENTAL PROCEDURE

A. Assembly of the Motor

The stator of the prototype motor consists of a hollow metal tube (brass) with an outer diameter of 2.4 mm, an inner diameter of 1.6 mm, a length of 10 mm and two rectangular piezoelectric plates with dimensions, 10 mm in length, 2 mm in width and 0.5 mm in thickness. The outside surface of the metal cylinder was ground on two sides at 90-degrees to each other to obtain two orthogonal flat surfaces. The distance from the flat surface to the center of the tube was 1.0 mm. The PZT plates (APC841, APC International Ltd.), which were electroded uniformly and poled in the thickness direction, were bonded onto the flat orthogonal surfaces of the cylinder using an epoxy and cured at 60^0C. The rotor was a cylindrical rod and it was pressed by a spring using a pair of stainless steel ferrules. The assembled motor is shown in Figure 3.

B. Characterization of the Stator

As a first step to clarify the behavior of the stator, the admittance spectra of the free stator were measured. Figure 4(a) and 4(b) show the magnitude and phase of the free stator when plate X or Y was excited. When plate X was excited while short-circuiting the electrode of plate Y to the ground, the stator had two degenerated bending mode resonance frequencies around 71.8 and 74.0 kHz. When plate Y was excited, the stator showed a similar behavior.

The peak-to-peak displacement spectrum around the two orthogonal bending mode frequencies in the x and y-axes directions was also measured. A function generator (HP33120A) and a power amplifier (NF4010) were used to excite the stator. By exciting either plate X or plate Y, the vibration velocity in x and y-axes directions was measured with two Laser Fiber Optic Interferometers (Politec OFV-3001/OFV-311). The displacement spectrum in the x and y directions when only plate X was excited is shown in Figure 5(a). When only plate Y was excited, similar behavior was obtained and it is shown in Figure 5(b).

The wobble motion in the xy-plane was also verified by measuring the magnitude of the vibration velocity in x and y directions at the same time. Firstly, plate X was excited at frequencies of 70.7, 71.6, 72.2, 73, 74, 75, 76 kHz, with an input voltage of 112.0 V p-to-p. The elliptical displacements in x and y directions were then calculated using the vibration velocity data. The results are plotted in Figure 6(a). The measurements were repeated by exciting only plate Y and the results are shown in Figure 6(b). The interesting points here are: i) the direction of wobble motion when only plate X was excited was clockwise, and counterclockwise when only plate Y was excited, ii) the wobble motion, when only plate X was excited, was almost identical to the wobble motion when only plate Y was excited at the same frequency. In conclusion, the designed motor can be driven with a single AC source and exciting either plate X or Y, the direction of the rotation can be reversed.

MOTOR CHARACTERISTICS

The performance of the motor was measured using a transient characterization method, which was initially proposed by Nakamura [10]. The principle of this method is mounting a load (usually a disk whose moment of inertia is known) onto the motor, running the motor, and, finally, analyzing the transient speed obtained as a function of time. More explicitly, the angular acceleration of the motor is calculated from the speed measurement by Newton's second law.

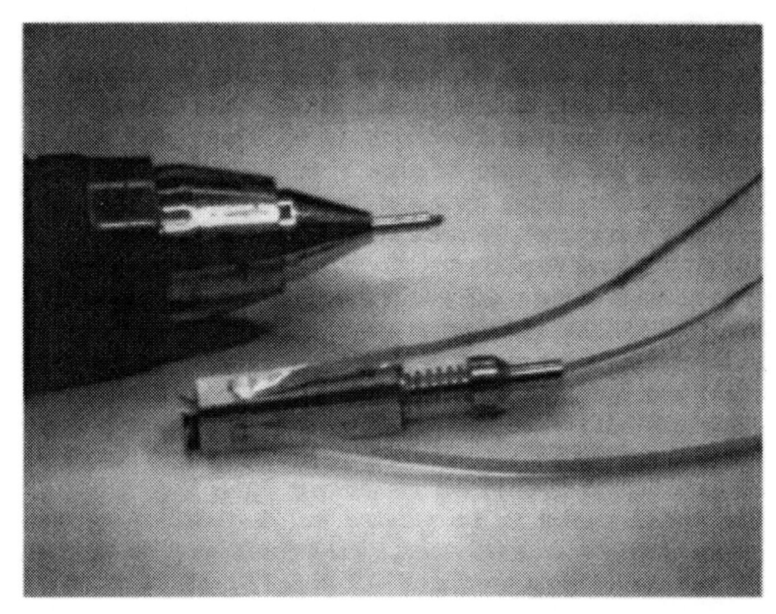

Figure 3. Photo of the assembled motor.

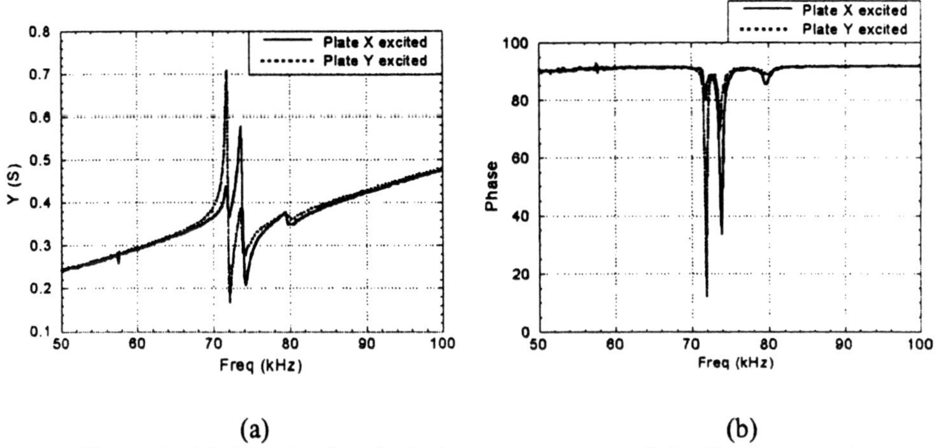

(a) (b)

Figure 4. (a) Magnitude of admittance spectrum of the free stator when plate X or Y was excited, (b) Phase of the admittance spectrum of the free stator when plate X or Y was excited

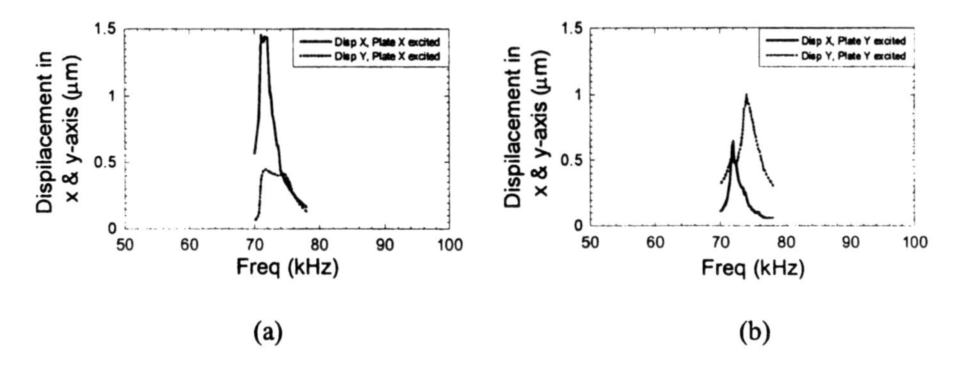

(a) (b)

Figure 5. Amplitude of the displacement in x and y directions around two orthogonal resonance frequencies, (a) when plate X was excited, (b) when plate Y was excited

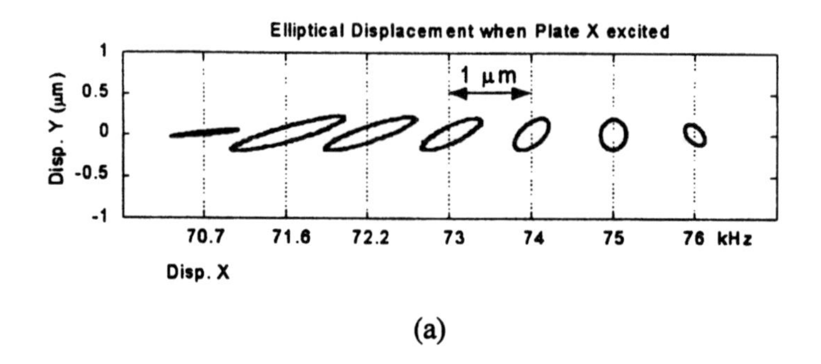

(a)

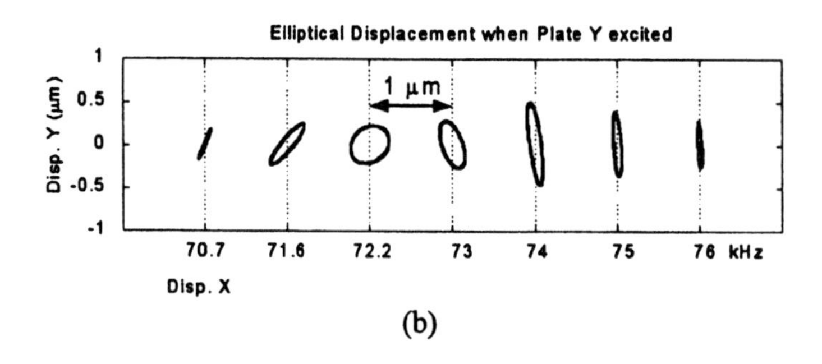

(b)

Figure 6. Elliptical displacements at frequencies 70.7, 71.6, 72.2, 73, 74, 75 and 76 kHz (a) when plate X was excited and, (b) when plate Y was excited. The direction of wobble motion is clockwise and counterclockwise, respectively

The transient torque is then calculated by multiplying the angular acceleration with the moment of inertia of the load. Using this method, the starting transient response of the motor gives the speed-torque relation. Similarly, the friction coefficient between the rotor and the stator can be estimated from the transient response for stopping the motor.

The load, a metal disk (60 g) with a diameter of 34 mm and moment of inertia (8.6 kg·mm^2) was mounted onto the stator as shown in Figure 7. The motor was then driven with an AC voltage of 120 V at 69.5 kHz. The position of the rotating disk was detected through an optical encoder (US Digital HEDS-9100-S00). The transient position data were then converted into voltage signal using a frequency-to-voltage converter, which is also shown in Fig. 7.

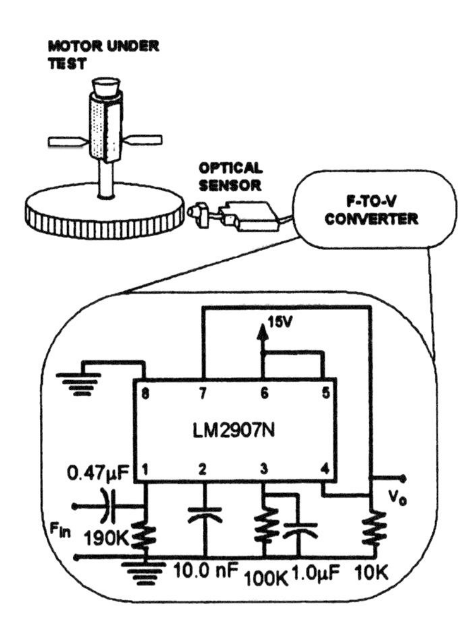

Figure 7. Experimental setup to obtain load characteristic of using transient response method.

Since the output voltage of the converter is proportional to the input frequency, the speed of the motor was obtained with a simple gain factor. The transient speed of the motor under loaded condition is shown in Figure 8 when plate X is excited. The angular acceleration of the motor was estimated using the derivative of the angular speed. Finally, the transient torque was calculated by multiplying the angular acceleration with the moment of inertia of the rotating disk. The steady state speed reached 86 rad/sec in 7 sec. The load characteristics of the motor were obtained by plotting the transient speed as a function of transient torque (Figure 9). Same results were obtained when plate Y was excited.

A starting torque of 1.8 mNm at 120 Volt is similar to other bulk piezoelectric cylindrical type micro motors in literature and almost one order of magnitude higher than that of a thin film motor with a similar size [11,12,13]. The product of output torque with output speed gives the output power. A maximum power of 60 mW was obtained at a speed of 60 rad/sec and a torque of 1 mNm. The efficiency curve that was obtained by dividing the output mechanical power to input electrical power is also shown in Figure 9. The maximum efficiency of 25 % at 120 V is similar to other bulk cylindrical type micro piezoelectric motor [13].

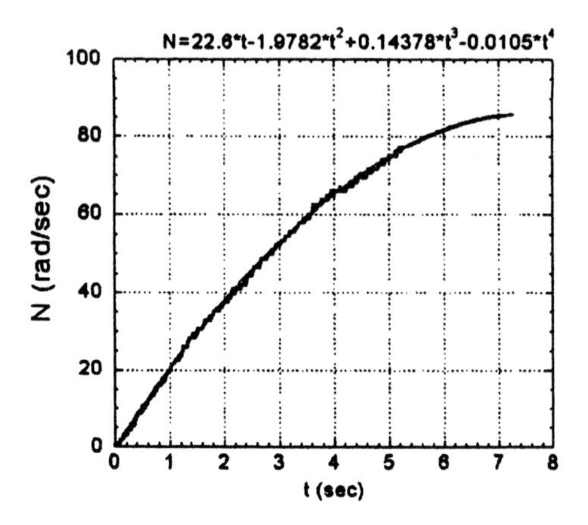

Figure 8. Transient response of the motor at 120 V.

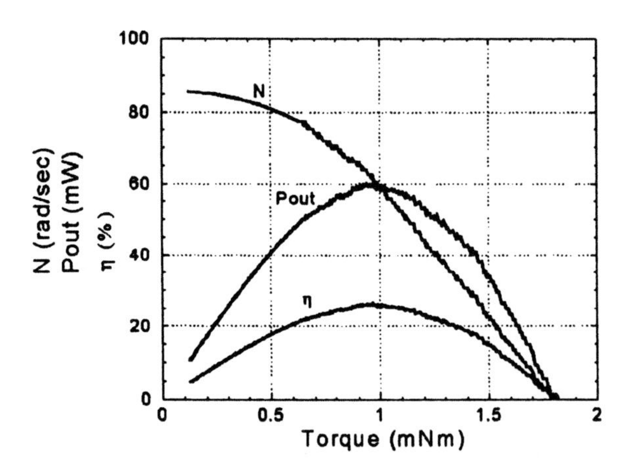

Figure 9. Load characteristics of the motor.

CONCLUSION

The ultrasonic motor presented is a new Multi-Mode-Single-Vibrator type piezoelectric motor. The rotation of this motor takes place by exciting two orthogonal bending modes of a hollow cylinder.

The structure of the stator was analyzed by using the ATILA finite element code, and its dynamic behavior was predicted. The motor in this paper provided the following motor characteristics: torque, 1.8 mNm, speed, 60 rad/sec, power, 60 mW and efficiency, 25%.

Significant advantages of our metal tube motors to the PZT tube motors [14] are; (1) lower manufacturing cost by 1/100, (2) simpler driving circuit (only one sine wave), and (3) higher scalability.

The proposed motor may find applications in medical industries such as catheter and other micromechatronic device applications.

ACKNOWLEDGEMENT

The authors would like to thank the Office of Naval Research for supporting this work through contract: N00014-96-1-1173.

REFERENCES

[1] W. Williams and W.J. Brown, "Piezoelectric Motor, " U.S. Patent 2439499, April 13, 1948.

[2] S. Ueha and Y. Tomikawa, *Ultrasonic Motors: Theory and Applications.* Oxford U.K.: Clarendon, 1993, pp.4-6.

[3] T. Sashida and T. Kenjo, *An Introduction to Ultrasonic Motors.* Oxford, U.K.: Clarendon, 1993, pp.6-8.

[4] J. Wallaschek, "Contact mechanics of piezoelectric ultrasonic motors," *Smart Mater. Struct.*, vol. 7, pp. 369-381, (1998).

[5] H. Flemmer, B. Eriksson and J. Wikander, "Control design and stability analysis of a surgical teleoperator," *Mechatronics*, vol. 9, iss.7, pp. 843-866, (1999).

[6] J. Tsujino, M. Takeuchi and H. Koshisako, "Ultrasonic rotary motor using a longitudinal-torsional vibration converter," in *IEEE Ultrasonics Symposium*, 1992, vol. 2, pp. 887-892.

[7] O. Ohnishi, O. Myohga, T. Uchikawa, M.Tamegai, T. Inoue and S. Takahashi, "Piezoelectric ultrasonic motor using Longitudinal-Torsional composite Resonance Vibrator," *IEEE Trans. Ultrason., Ferroelect., Freq. Contr.*, vol. 40, pp. 687-693, (1993).

[8]B. Koc, A. Dogan, Y. Xu, R.E. Newnham and K. Uchino, "An ultrasonic motor using a metal-ceramic composite actuator generating torsional displacement," *Jpn. J. Appl. Phys.* vol. 37, part 1, no.10, pp. 5659-5662, (1998).

[9]Y. Tomikawa, T. Takano, and H. Umeda, "Thin rotary and linear ultrasonic motors using double-mode piezoelectric vibrator of the first longitudinal and second bending modes," *Jpn. J. Appl. Phys.*, vol. 31, pp.3073-3076, (1992).

[10]K. Nakamura, M. Kurosawa, H. Kurebayashi, and S Ueha, " An Estimation of load characteristics of an ultrasonic motor by measuring transient response," *IEEE Trans. Ultrason., Ferroelect., Freq. Contr.*, vol. 38, pp. 481-485, (1991).

[11]M. Kurosawa, T. Morita and T. Higuchi, "A cylindrical ultrasonic micro motor based on PZT thin film," in *IEEE Ultrasonics Symposium*, 1994, vol.1, pp.549-552.

[12]T. Morita, M. Kurosawa and T. Higuchi, " A cylindrical micro ultrasonic motor using PZT thin film deposited by single process hydrothermal method," *IEEE Trans. Ultrason., Ferroelect., Freq. Contr.*, vol. 40, pp. 687-693, (1993).

[13]T. Morita, M. Kurosawa and T. Higuchi, " Cylindrical micro ultrasonic motor utilizing bulk Lead Zirconate Titanate (PZT)," *Jpn. J. Appl. Phys.*, vol. 38, pp.3347-3350, (1999).

[14]S.Dong, S.P.Lim, K.H.Lee, J.Zhang, L.C.Lim and K.Uchino, "Piezoelectric Ultrasonic Micromotors with 1.5 mm Diameter," *IEEE Trans. Ultrason., Ferroelect., Freq. Contr.*, to be published.

GRAIN SIZE DEPENDENCE OF HIGH POWER PIEZOELECTRIC CHARACTERISTICS IN A SOFT PZT

Chiharu Sakaki and Kenji Uchino
International Center for Actuators and Transducers
Materials Research Institute
The Pennsylvania State University
University Park, PA 16802

ABSTRACT

Grain size dependence of low and high power piezoelectric properties for Nb doped PZT-based ceramics were studied, including heat generation under high AC electric fields. The mechanical quality factor (Q_m) increased with decreasing grain size under low electric fields. The temperature rise decreased and the vibration velocity increased with a decrease in grain size. The temperature rise of the fine grain ceramic (0.9 μm) was about 40% lower than that of the coarse grain ceramic (3.0 μm) near the maximum vibration velocity level (0.30 m/s).

INTRODUCTION

In recent years, demands for piezoelectric devices used for high power applications such as actuators and transformers are increasing. These applications need to exhibit high vibration velocities in order to obtain large displacements for actuators and to obtain high voltage for transformers. Significant heat generation occurs in them under these conditions, which is one of the major problems in high power applications. Since it is difficult to suppress heat generation dramatically just by modifying the design of these devices, the use of materials with high vibration velocity and low heat generation is essential.

Heat generation is mainly caused by the loss originated from the domain wall motions in a piezoelectric ceramic. The pinning effect of the domain wall by doping acceptor ions, e.g. Fe^{3+}, Mn^{2+}, Co^{2+} etc., is one of the methods for reducing heat generation, via an oxygen vacancy generation. As the domain motion is suppressed by the oxygen vacancy generation, the mechanical quality factor (Q_m) and the heat generation are remarkably improved. Fe doped PZT and $Pb(Mn_{1/3}Sb_{2/3})O_3$-$PbTiO_3$-$PbZrO_3$ are used popularly for high power applications.[1) 2)] However, Hirose et al. reported that not all high Q_m materials did suppress heat generation.[3)] Doping acceptor elements is not always effective to the reduction of heat generation.

On the other hand, decreasing the grain size may be another way for suppressing the domain motion, because a grain boundary increases with reducing the grain size will contribute to additional domain pinning sites. Though there are several papers which discussed this issue under low electric fields[4)-9)], few papers have discussed heat generation under high electric fields. Therefore, We studied the grain size contribution to heat generation under high electric fields.

In this study, We chose $PbZr_{0.52}Ti_{0.48}O_3$+2.0 at%Nb, known as the low Q_m material and has no oxygen vacancy ideally: this is suitable in order to observe only the pinning effect by grain boundaries. We studied piezoelectric properties under low electric fields, temperature rise and vibration velocity under high electric fields as a function of grain size.

EXPERIMENTAL PROCEDURE

The starting powders used in this investigation were reagent grade and included PbO, ZrO_2, TiO_2, Nb_2O_5 (Alfa Aesar). The individual oxides were weighed to be the composition as $Pb(Zr_{0.52}Ti_{0.48})O_3$ + 2.0 at%Nb, mixed, vibratory-milled with alcohol and 5mmϕ YTZ-grinding media (Tosoh Co.), dried, passed through a 80-mesh sieve, and calcined at 850°C for 4.5 h in Al_2O_3 crucible. The calcined powders were vibratory-milled again for 4, 24, 72 h, dried, and passed through a 200-mesh sieve. The ground material was cold-isostatic-pressed.

The compacts were put on platinum sheet, covered with an Al_2O_3 crucible to prevent the evaporation of PbO, and sintered at different temperatures in the range of 1000°C-1150°C, and different periods between 10 min – 1000 min. Density of the sintered specimens was measured using Archimedes' method after sintering. The sintered specimens were polished and cut into a rectangular plate vibrator, $43 \times 7 \times 1$ mm^3, electroded with gold sputtering and finally poled under a field level of 30 kV/cm for 10 min.

The grain size was measured using SEM (Hitachi 3500N). The electromechanical properties under low electric fields were measured using an impedance analyzer (HP-4194A). The temperature rise and the vibration velocity under high electric fields were measured using a thermocouple and an interferometer (Polytech, OFV-511), respectively.

RESULTS

Figures 1 and 2 show the sintering temperature and period versus the density (ρ) relation as a function of the milling time, respectively. Although the density at 1000°C is slightly low, the density difference between the samples prepared at 1000°C and at the other temperatures are less than 5 %. Thus, the density change will not contribute significantly to the property. Figure 3 shows the grain size dependence of the density regardless of the milling time. Though the density slightly decreases below 1.0 μm, the density is considered to be almost the same for all the samples.

The grain size dependences of piezoelectric properties are shown in Figures 4, 5, and 6 for mechanical quality factor (Q_m), piezoelectric constant (d_{31}), dielectric compliance (s_{11}^T), respectively. The mechanical quality factor (Q_m) increases with decreasing the grain size. The piezoelectric constant (d_{31}) decreases with reducing the grain size in the range of less than 1.5 μm, and appears to be saturated for more than 1.5 μm. On the other hand, the elastic compliance (s_{11}^T) is almost constant in all grain size range.

Figure 7 shows the grain-size versus heat generation relation for various vibration velocities. The temperature rise from room temperature decreases with a decrease in the grain size in all vibration velocity range. In addition, the slope increases as the vibration velocity increases. The temperature rise of the fine grain ceramic (0.9 μm) is 40 % lower than that of the coarse grain ceramic (3.0 μm) near the maximum vibration velocity (0.30 m/s).

DISCUSSIONS

The vibration velocity is expressed by the following equation[2]:

$$v = \frac{4}{\pi}\sqrt{\frac{\varepsilon_{33}^{T}}{\rho}}k_{31}Q_{m}E = \frac{4}{\pi}\frac{1}{\sqrt{\rho s_{11}^{E}}}d_{31}Q_{m}E\,, \tag{1}$$

where k_{31}, ε_{33}^{T} and E are electromechanical coupling factor, dielectric constant and electric field, respectively. On the other hand, heat generation is expressed by mechanical loss factor (tan ϕ), which is inversely proportional to mechanical quality factor (Q_m). Under low electric fields, the piezoelectric constant (d_{31}) decreases slightly with reducing the grain size in the range of less than 1.5 μm. The density (ρ) and the elastic compliance (s_{11}^{T}) are constant in the all grain size range. However the mechanical quality factor (Q_m) increases dramatically with a decrease in the grain size. Hence the vibration velocity was increased and the temperature rise was suppressed due to an increase in the mechanical quality factor (Q_m).

Nb-doped PZT is expressed as the following formula:

$$(Pb_{1\text{-}y/2}\square_{y/2})(Zr_xTi_{1\text{-}x\text{-}y}Nb_y)O_3. \tag{2}$$

Thus, no oxygen vacancy is expected to exist in the Nb-doped PZT, and the space charge effect may not occur in this material. The increment of the mechanical quality factor may not be caused by space charge effect of oxygen vacancy, but by the pinning effect via the grain boundary.

CONCLUSION

Heat generation can be dramatically suppressed in Nb-doped PZTs with a decrease in grain size. The temperature rise of the fine grain ceramic (0.9 μm) is

40 % lower than that of the coarse grain ceramic (3.0 μm) under the maximum vibration velocity of 0.30 m/s.

This phenomenon is caused by the Q_m improvement phenomenologically, further by the domain pinning via the grain boundaries microscopically.

REFERENCES

[1]S. Takahashi and S. Hirose, Jpn. J. Appl. Phys., **32** 2422-25 (1993)

[2]S. Takahashi, Y. Sasaki, S. Hirose, and K. Uchino, Mat. Res. Soc. Proc., **360** 305-310 (1995)

[3]S. Hirose and S. Takahashi, IEICE, **J80-A** [10], 1621-36 (1997) (in Japanese)

[4]C. A. Randall, N. Kim, J. Kucera, W. Cao, and T. R. Shrout, J. Am. Ceram. Soc., **81** [3] 677-87 (1998)

[5]A. H. Webster and T. B. Weston, J. Can. Ceram. Soc., **37** 51-54 (1968)

[6]T. Yamamoto, Am. Ceram. Soc. Bull., **71** [6] 978-85 (1992)

[7]G. H. Haertling, Am. Ceram. Soc. Bull., **43** 875-79 (1964)

[8]N. Okada, K. Ishikawa, K. Murakami, T. Nomura, M. Hagino, N. Nishino and U. Kihara, Jpn. J. Appl. Phys., **31** 3041-44 (1992)

[9]N. Okada, K. Ishikawa, T. Nomura, K. Murakami, S. Fukuoka, N. Nishino and U. Kihara, J. Jpn. Appl. Phys., **30** 2267-70 (1990)

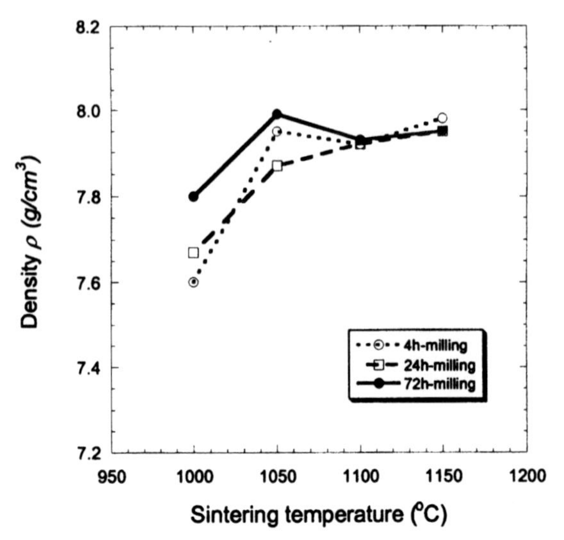

Fig. 1 Changes in density (ρ) in sintering temperature with the milling time.

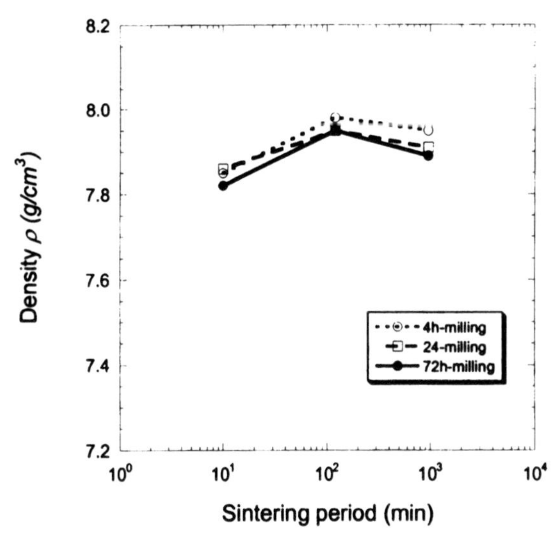

Fig. 2 Changes in density (ρ) in sintering period with the milling time.

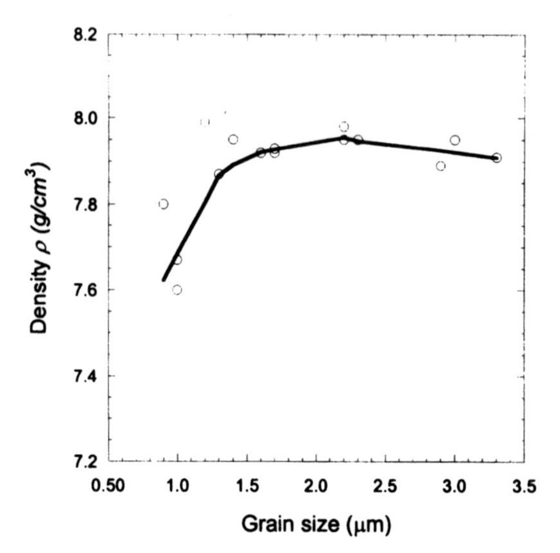

Fig. 3 Grain size dependence of density.

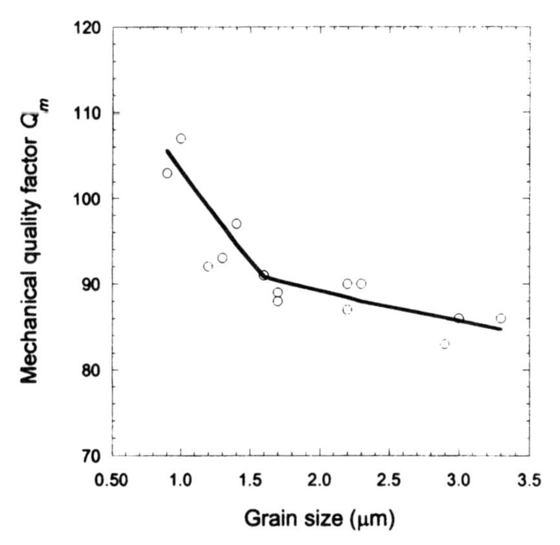

Fig. 4 Grain size dependence of the mechanical quality factor (Qm).

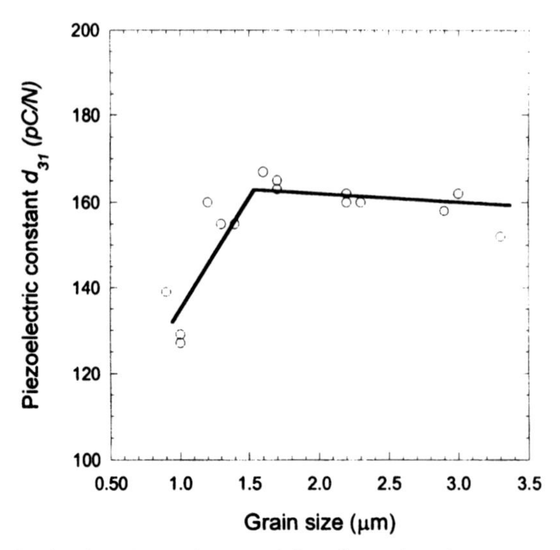

Fig. 5 Grain size dependence of the piezoelectric constant (d_{31}).

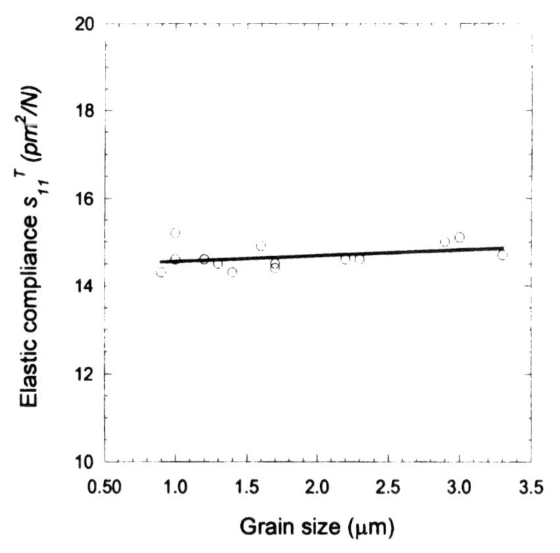

Fig. 6 Grain size dependence of the elastic compliance (s_{11}^T).

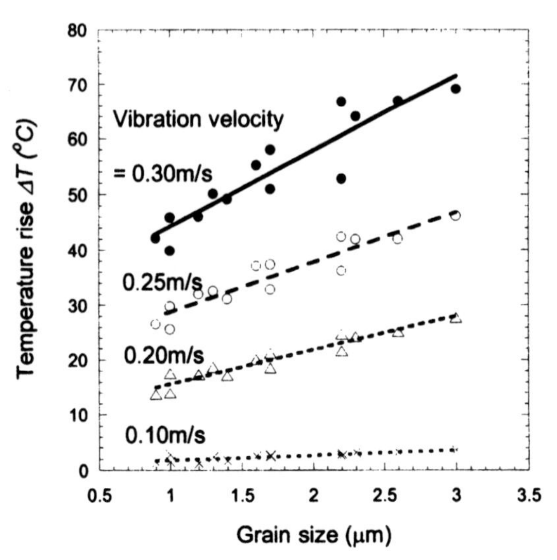

Fig. 7 Temperature rise as a function of grain size compared with vibration velocity.

HIGH POWER PIEZOELECTRICS OF (1-x)$Pb(Zn_{1/3}Nb_{2/3})O_3$–x$PbTiO_3$ SINGLE CRYSTALS

Shashank Priya[1], Uma Belegundu, Alfredo Vázquez Carazo and Kenji Uchino
International Center for Actuators and Transducers
Materials Research Laboratory, Pennsylvania State University, University Park, PA 16802.

Abstract

A novel resonance method for measuring the complex elastic and electromechanical properties of the piezoelectric materials has been developed. In this method, the piezoelectric material is excited by applying electric field and the vibration velocity of the transverse resonance mode is measured as a function of frequency. This method is applied here to determine the elastic and electromechanical properties of the $Pb(Zn_{1/3}Nb_{2/3})O_3$ – $PbTiO_3$ (PZNT) single crystals under high power operational conditions. It has been found that electromechanical nonlinearities are due to elastic ones under high power drive. The elastic and electromechanical properties were found to vary linearly with the square of the strain amplitude. Mechanical quality factor decreases rapidly with increasing electric field indicating thermal instability in the material under high power resonant drive.

Introduction

In recent years, the relaxor based single crystals have emerged as a group of promising materials for acoustic-imaging applications[1]. In these applications, the material is driven under high electric field reaching high vibration velocities. At higher fields, the electromechanical properties are known to exhibit non-linearities, resulting in significant differences in material properties relative to standard data reported in literature and commercial material sources.

Under high power resonant drive several effects are known to occur such as shift of the resonance frequency f_r, an appearance of a jump and/or a hysteresis in the admittance spectrum during a frequency sweep, a decrease of Q_m and heat generation (due to the lower Q_m) that not only degrades the electromechanical properties, but may also result in thermal stability problems for transducer engineers. The linear constitutive relations of the piezoelectrics does not hold under high power as the nonlinearities appear in the elastic and electromechanical terms. The determination of the nonlinear components is essential for designing of those devices in which an accurate working frequency is required. It also enables to select the best device parameters and minimize the instabilities.

In this paper, we report the high power characteristics of $Pb(Zn_{1/3}Nb_{2/3})O_3 - PbTiO_3$ (PZNT) single crystals using a novel high power resonant measurement method. In this technique, the specimen is excited piezoelectrically and detected displacively, an desired feature for separating the linear and nonlinear components of the electromechanical properties. It has been found that this technique is most suitable for high power characterization of piezoelectric materials.

Experimental Procedure

Materials

The single crystal of PZNT of the morphotropic phase boundary (MPB) composition (91/9) were synthesized by using a conventional flux method The single crystal was oriented along <001> direction and a sample of the size 10.2 x 3.4 x 0.70 mm^3 was obtained. The sample was electroded by gold sputtering and poled by cooling under the field of 10kV/cm from 220°C in silicone oil.

Measurement Technique

Figure 1 shows the schematic diagram of the experimental setup. Resonance curves were measured by piezoelectrically-driving retangular shaped specimen into the fundamental length extensional vibration mode. The displacement of the specimen was detected by measuring the vibration velocity using a Polytec vibrometer model number OF-3001. The specimen is mounted in a wedge shaped specimen holder and held at the nodal points of the fundamental vibration, so as to minimize the effects of sample mounting upon the resonance response. Specimens were driven by an ac electrical field in the longitudinal direction. A signal generated by an NF Electronic Instruments frequency response analyzer was fed into an NF Electronic Instruments 4025 high speed power amplifier. The output signal from the high-power amplifier was fed into a Tektronix TDS 420A oscilloscope, in order to accurately measure the magnitude of the applied signal.

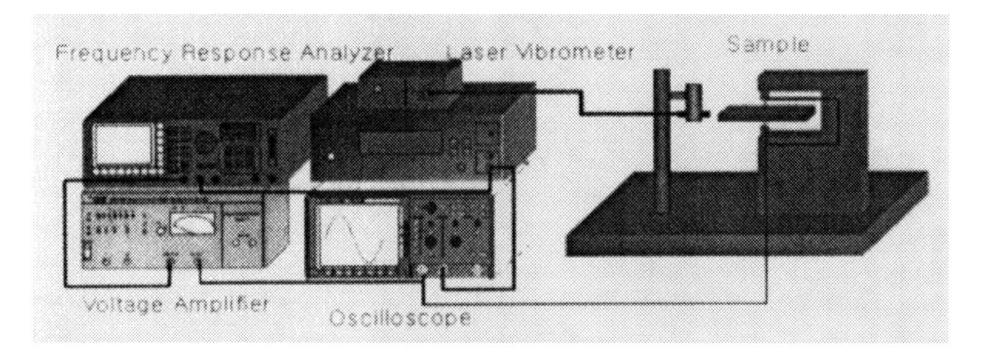

Fig. 1 Schematic diagram of the experimental set up.

The signal from the laser vibrometer was then feedbacked into the NF Frequency Response Analyzer. Measurements were performed by varying the frequency of the output signal from the analyzer into the high-speed amplifier. The output signal from the Polytec Laser Vibrometer was then fedbacked into the NF Frequency Response Analzyer through the rear panel and also to the digital oscilloscope. The resonance response of the specimens was then obtained by measuring the displacement as a function of f. Data acquisition was computer controlled, via a GPIB interface.

Results and Discussion

Figure 2 shows the RMS vibration velocity (v_{rms}) as a function of f for the PZNT single crystal. Data are shown for various E_{ac}'s. In Fig. 2, the specimen velocity can be seen to exhibit a resonance curve typical of that of a linear harmonic oscillator. However, with increasing E_{ac}, strong nonlinearities can be seen to become evident in the resonance spectra. With increasing

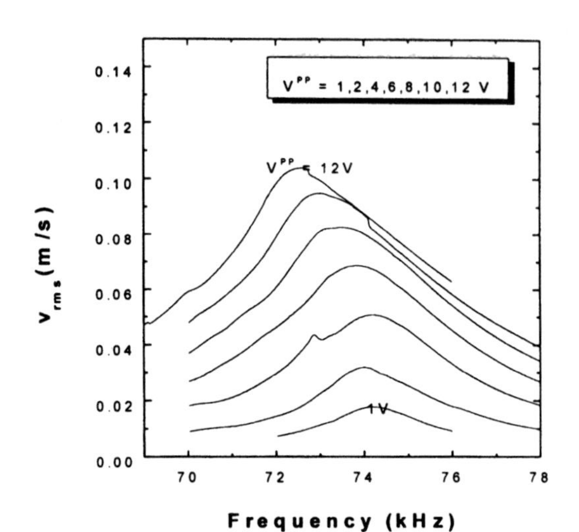

Fig. 2. RMS vibration velocity as a function of frequency at different electric field for PZNT single crystal.

excitation, the peak in the resonance curve can be seen to shift to lower f. The frequency of the maximum in the resonance peak can be seen to decrease from ~74.2 kHz to ~72.64 kHz with increasing E_{ac} between 0.50 and 6.3 V/mm.

According to Nayfeh [2], the shift in the resonance frequency with increasing excitation can be represented by an empirical response relationship:

$$f_r(eff) = f_r^{lin} + \frac{3}{8}\left(\frac{\alpha\varepsilon^2}{f_r^{lin}}\right) \pm \sqrt{\left(\frac{K}{f_r^{lin}\varepsilon}\right)^2 - \delta^2} \quad . \tag{1}$$

where $f_r(eff)$ is the effective resonance frequency defined where the maximum displacement occurs, f_r^{lin} is the amplitude independent linear resonance frequency, α is the non-linear elastic constant, δ the linear anelastic constant (damping or $1/Q_m$), K the external excitation, and ε the RMS strain amplitude of the vibration of the sample.

Figure 3 shows a plot of $f_r(eff)$ as a function of ε^2. In this figure, a linear dependence can readily be seen. A linear fitting of the data to equation yielded a value of f_r^{lin} equal to 74.4 kHz and a value for α/ f_r^{lin} of -7.94×10^8 kHz. These results show that the shift in the effective resonance frequency with increasing E_{ac} is due to negative third order elastic constant. The linear (s_{11}) and nonlinear (s_{1111}) elastic compliances can be determined from this data, as given in Eq. (2)

$$s_{11}(eff) = s_{11} + s_{1111}\varepsilon^2 \quad . \tag{2}$$

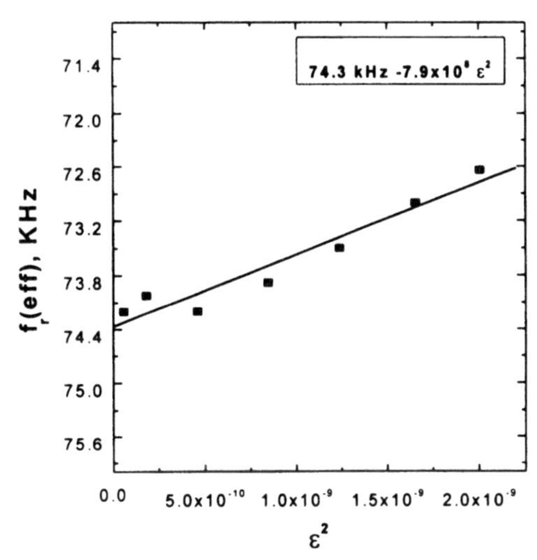

Fig. 3 $f_r(eff)$ as a function of ε^2.

From the measured experimental data linear elastic compliance of 5.60×10^{-11} m^2/N and nonlinear elastic compliance of 1.24×10^{-3} m^2/N was obtained for the PZNT single crystal. The density of the sample was 7.8×10^3 kg/m^3. Figures 4 shows the admittance spectra for the PZNT single crystal.

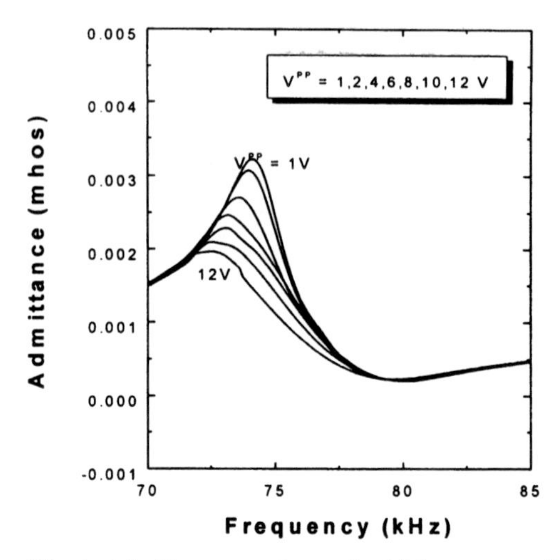

Fig. 4. Admittance spectra under high-power excitation.

Significant changes in these spectra can be seen with increasing E_{ac}. The spectra shifted to lower f and decreased in magnitude with increasing E_{ac}. The relative permittivity (K_3) can be obtained from the measurement of the admittance at low frequencies.

Figure 5 shows the dielectric constant as a function of E_{rms} for PZNT single crystal. In this figure, the value of K_3 can be seen to be independent of E_{rms}. This indicates that the dielectric behavior is nearly constant during the resonance measurements for all values of E_{rms}

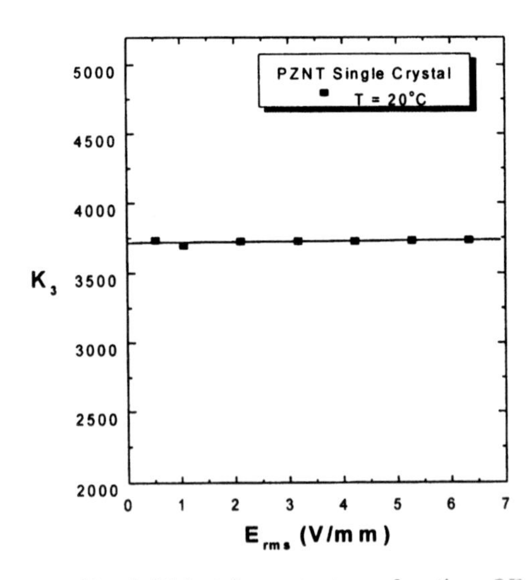

Fig. 5. Dielectric constant as a function of E_{rms}.

investigated. Accordingly, nonlinearities in the electromechanical properties must be related to those in the elastic behavior. The effective electromechanical coupling factor (k_{31}(eff))can be computed from f_r(eff) and the antiresonance frequency (f_a). A relationship between k_{31}(eff) and the linear and nonlinear elastic compliances is given as [3]:

$$k_{31}(\text{eff}) \sim k_{31}^{\text{lin}} + \frac{\pi^2}{8k_{31}^{\text{lin}}} L \rho^{1/2} f_a [\frac{s_{1111}}{s_{11}^{1/2}} \varepsilon^2] \tag{4}$$

where L is the length of the sample and ρ is the density of the sample. Equation (4) predicts a linear relationship between k_{31}(eff) and ε^2. The linear fit yielded the linear electromechanical coupling factor of 0.39. These results show that the nonlinearities in the electromechanical coupling coefficient under resonant drive are due to those in the elastic compliance. The effective transverse piezoelectric constant d_{31}(eff) is well-known to be related to the coupling coefficient, the elastic compliance and the dielectric constant. d_{31} can be expressed in terms of the linear and nonlinear elastic compliances as [3] :

$$d_{31}(\text{eff}) \sim k_{31}^{\text{lin}} \sqrt{s_{11} K_3 \varepsilon_0} + \sqrt{K_3 \varepsilon_0} \left(\frac{k_{31}^{\text{lin}} s_{1111}}{2\sqrt{s_{11}}} + \frac{\pi^2}{8k_{31}^{\text{lin}}} L \rho^{1/2} f_a s_{1111} \right) \varepsilon^2 \tag{5}$$

Equation (5) predicts a linear relationship between d_{31}(eff) and ε^2. The linear fit yielded the linear piezoelectric constant value of 542.4 pC/N.

Figure 6 shows the variation of the mechanical quality factor (Q_m) with E_{ac}. It can be seen that there is significant decrement in the magnitude of Q_m with increasing excitation indicating thermal instability in the material. This data is required for evaluation of the material performance under realistic operational conditions.

The magnitude of α/ f_r^{lin} for soft and hard PZT material was found to be $2.18\text{x}10^6$ kHZ and $5.37\text{x}10^4$ kHz respectively [3]. Comparison of these values with that of PZNT single crystal shows that single crystal is much more elastically nonlinear than the PZT ceramics in the E_{ac} range investigated. This large difference occurs not only due to large change in f_r(eff) with E_{ac} but also due to the fact that the amplitude of the displacement is significantly less compared to PZT ceramics.

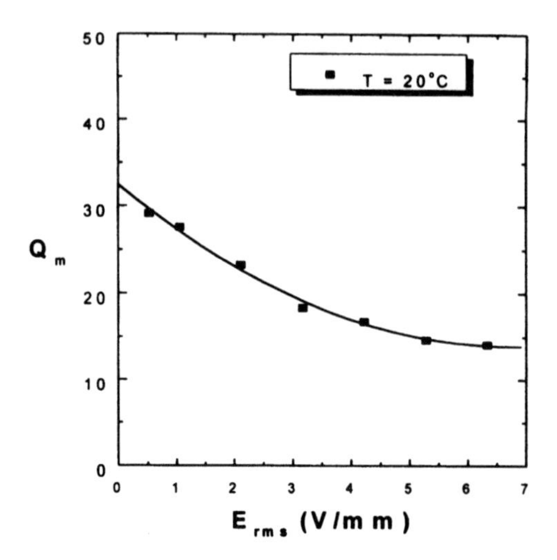

Fig. 6 Variation of Q_m as a function of E_{ac}.

Conclusions

The nonlinear electromechanical properties of PZN-PT single crystals have been determined using a novel high power resonant measurement method. Under high power excitation, the nonlinearities in the electromechanical parameters have been shown to reflect the nonlinearity in the elastic constant. Dielectric nonlinearity was not found to contribute significantly. The determination of nonlinarities in the elastic and electromechanical properties is necessary for designing the devices working under high power conditions and requiring exact operating frequency.

Acknowledgement

This work was sponsored by office of naval research through contact number N00014-99-1-0754.

References

1. K. Uchino, "Recent Trend of Piezoelectric Actuator Developments", *Proc. of Intl. Symp. Micromechatronics and Human Science* 3-9 (1999).
2. A.H. Nayfeh and D. Mook: "*Nonlinear Oscillations*", John Wiley & Sons, New York, NY (1979).
3. S. Priya, D. Viehland, A.V. Carazo, J. Ryu and K. Uchino, "High Power Resonant Measurements of Piezoelectric Mateials: Importance of Elastic Nonlinearities", *J. Appl. P*hys. (2001) submitted.

RESIDUAL STRESS IN HIGH CAPACITANCE BME-MLCCS

Yukie Nakano, Daisuke Iwanaga, Takako Hibi, Mari Miyauchi and Takeshi Nomura
Materials Research Center, TDK Corporation,
570-2, Matsugashita, Minami-Hatori, Narita-shi, Chiba-ken, Japan 2868588

ABSTRACT

The present paper focuses on the residual stress in high capacitance BME(base metal electrode)-MLCCs. Recently developments in the thickness reduction and multiplication of dielectric layers have been accelerated in order to achieve higher capacitance value. With the increase of the dielectric layer number, the internal stress in the capacitors became too high to be ignored. This stress is supposed to cause a problematic reliability of the capacitor. It is considered that the stress is induced by the Ni electrode, globing during firing process. It is also considered that the globing behavior of the Ni-electrode is highly affected by the firing behavior of dielectric material that is co-fired with the Ni-electrode.

The high capacitance 1206-type BME-MLCCs, which have a capacitance value of 100 μ F and Y5V characteristic, have been successfully developed by optimizing the firing program and the improvement of Ni-electrode to control the stress. It have 400 of dielectric layers which is 2.5 μ m.

INTRODUCTION

The demands for the MLCCs have been increased for the last decade, especially it became twice in the production number for the last two years. This increased demand is greatly owing to the spreading use of a mobile cellular phone, as approximately 200 MLCCs are required to fabricate one cellular phone. The success in the development of the high capacitance and miniaturized MLCCs also accelerated the production growth of the MLCCs.

The replacement of the internal electrode material from a noble metal Pd to a base metal Ni is in the background of the rapid progress of the capacitance value in the MLCCs. For the case that Pd is used as the internal electrode, more layers in a MLCCs results in higher cost of electrode material, which could be disadvantageous to the MLCCs against the other competing capacitors, such as aluminum-electrolytic capacitors, tantalum-electrolytic capacitors, and film capacitors. The use of Ni as an internal electrode therefore brought a great cost merit for the high capacitance MLCCs. Since 1990's the production of Ni-MLCCs has expanded, owing to the success in the development of the highly reliable Ni-MLCCs[1-5], with replacing Pd-MLCCs. Now the market of the Ni-MLCCs is becoming wider and wider owing to the cost advantage accompanied with the high capacitance. The Ni-MLCCs have lots of advantage over the electrolytic capacitors, including no requirements for considering polarity, high capacitance value, low ESR(equivalent series resistance), and miniaturized size , and so on.

In order to achieve higher capacitance, the reduction in thickness and the multiplication of the dielectric layers are inevitably requested. However the MLCCs are fabricated by using different materials i.e. dielectric and internal electrode materials, and therefor these materials must be co-fired together. This co-fired process has exposed a difficulty in thinning and multiplication of the layers. The difference in the physical characteristics of these materials, including densification characteristics, could cause defects appearing in the MLCCs. Even if there were not any defects appeared, it is suspected that the residual stress remained in fired body. Therefore the stress in MLCCs and it's effect on the physical properties have been studied and these will be disclosed in this paper.

EXPERIMENTAL PROCEDURE

$(Ba,Ca)(Ti,Zr)O_3$-based materials and additives were used for the dielectrics. The dielectrics were prepared so as to give three different compositions, i.e. STD(B), STD+sintering aids(A) and STD-SiO_2(C), in order to compare the firing behaviors. Green chips were prepared from these three materials and a Ni paste. Then, the green chips were fired in a reducing atmosphere in order to prevent Ni from being oxidized. After the firing procedure, In-Ga alloy was applied so as to form an external electrode for a succeeding electrical measurement. In order to compare the firing behavior, MLCCs were fabricated to have various numbers of dielectric layers. Adding to these MLCCs, a dielectric-only-body and a Ni-only-body, whose sizes were approximately same as that of the MLCCs, were also prepared for the comparison.

The residual stress in the MLCCs was estimated by using a Vickers hardness tester. For the measurement, the MLCCs were lapped at the external electrode side towards the center and then polished. After these procedures, a diamond shape indenter was struck on a margin area of the lapped MLCCs by using the Vickers hardness tester.The residual stress was also estimated by a micro-region XRD analysis by using a micro-diffractometer (Mac Science, Inc).

An optical microscopic, SEM and TEM observations were carried out to study the microstructure of the MLCCs.

The capacitance aging was measured to evaluate the reliability of the MLCCs. For the measurement, the initial capacitance value was defined as the value after annealing at 150°C for 1hr and subsequent keeping at room temperature for 24hrs. The capacitance aging was measured after applying a voltage at a certain temperature for certain time durations.

RESULTS AND DISCUSSION

Effect of the multiplication on the shrinkage.

The cross sections of the MLCCs with the various layer numbers, which are lapped from external electrode, are shown in Fig.1. With increase the layer numbers, the expansion in the thickness direction and the distortion of the shape can be seen in this figure.

Shrinkage along the thickness (T), width (W) and length (L) were measured for MLCCs having different layer numbers, of which results are shown in Fig.2. As can be realized immediately, the layer number is an important factor controlling the shrinkage. According to the multiplication, the shrinkage rate decreased in the thickness direction, while those in the width and length directions increased. The shrinking behaviors of the dielectrics-only-body (without Ni internal electrode) were also measured for the comparison. The shrinkage along the three directions were almost same and no anisotropy was observed for the dielectrics-only-body. The anisotropy of shrinkage of MLCCs is attributed to the origin of the stress.

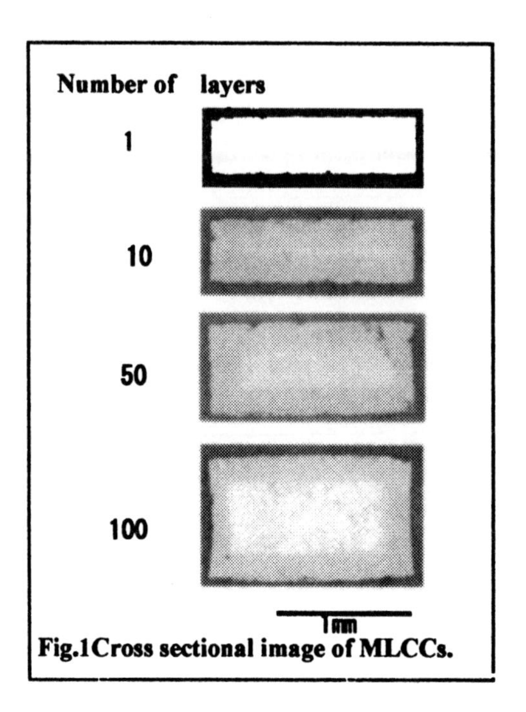

Fig.1Cross sectional image of MLCCs.

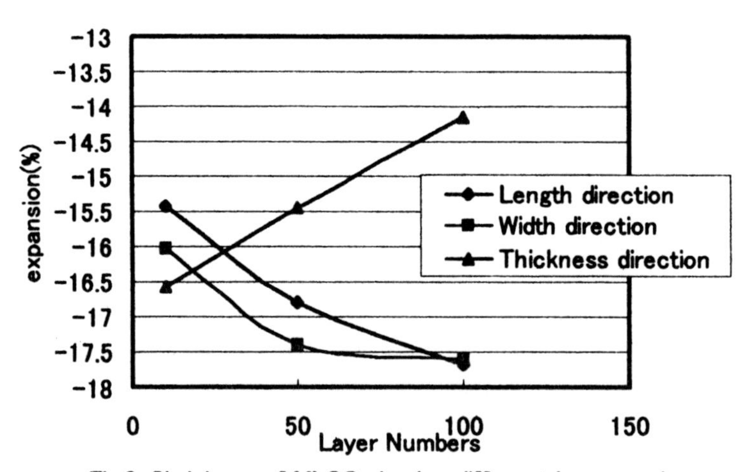

Fig.2. Shrinkage of MLCCs having different layer numbers.

Crack behavior evaluated by Micro-indentation method

A Vickers hardness tester was used to evaluate the residual stress in MLCCs. They were lapped from external electrode, and polished. After that an indenter was struck on a margin area of that. It was observed that length of the cracks occurring from the corners of the impression were so different, asymmetrical. And it varied depending on the position where the indenter was struck. At the top or bottom margin, the cracks were long in the width direction and short in the thickness direction, as can be seen in Fig.3. However, at the side margin, these were long in the thickness direction and short in the width direction. It is supposed that the difference in the crack length by the direction was resulted by the tensile stress in the MLCCs, which should be perpendicular to the longer cracks.

The residual stress can be estimated from the impression size(a) and the crack length(c) by using IF method (equation 1,2).

$K_{IC}=0.026(E^{1/2}P^{1/2}a)/c^{3/2}$ (eq.1)

$K_{IC}= K_{IC}^{0}=+2(c/\pi)^{1/2}\sigma_1$ (eq.2)

E: elastic modulus (Pa)

P: load(N)

σ_1: internal stress (Pa)

On the estimation of the residual stress, the cracks length must be symmetry as precondition. But the cracks length were asymmetrical as the result, therefore the stress could not be estimated in this method.

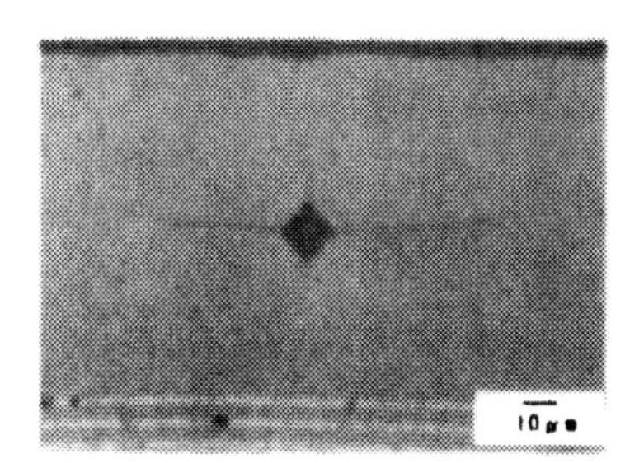

Fig.3 Anisotropy of the crack length from impression at the top margin on lapping MLCC.

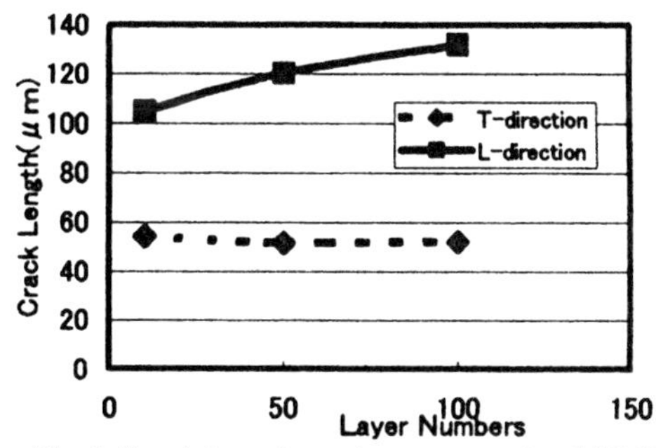

Fig..4 Crack length at the top margin.of MLCCs.

The anisotropy of the crack length by the impression at the top margin was investigated for MLCCs having various numbers of dielectric layers. Fig.4 shows the crack length as a function of the layer number. It can be said that the anisotropy increased with the layer number increasing. This can be explained from the hypothesis that the increased layer number caused the enhanced stress. In order to elucidate this hypothesis more clearly, the residual stress in the MLCCs having 300 dielectric layers was directly measured by using a micro X-ray diffractometer. The result is schematically shown in Fig.5. At the top margin tensile stress were measured in the thickness direction, and compression in the width direction, on the contrary in the side margin, tensile stress in the width direction and compressional stress in the thickness direction were measured. And in the center, the stress was almost zero. This result agrees with the result obtained by the micro-indentation method mentioned above. From these results, it is concluded that there are some stresses in MLCCs, whose size and direction are different depending on the position in MLCCs.

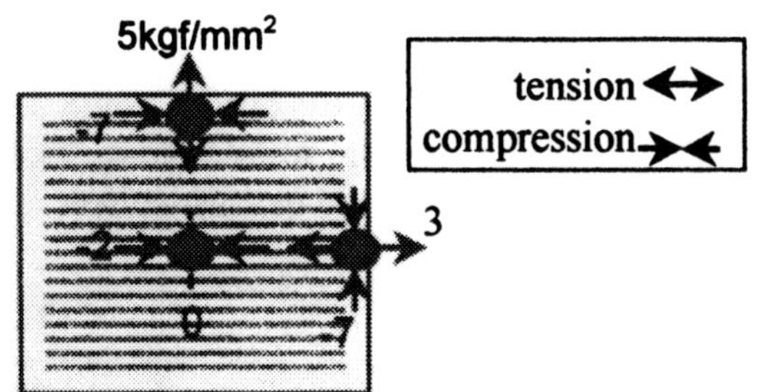

Fig.5 Stress analysis by micro-X-ray diffraction.

Effect of the stress on the reliability

The capacitance aging was measured under an application of a dc voltage at 40°C for MLCCs having various numbers of the dielectric layers. The result showed that the aging characteristic was affected by the layer number, as can be seen in Fig.6. In the case that the margin area was removed by the lapping process after firing, the aging deterioration decreased as shown in Fig.7. The MLCCs, for which the more amount of the margin was removed, showed the more capacitance decreased. As described above, the stress was thought to be caused during the firing process. It is therefore supposed that the stress in the fired body was released by the lapping process where the margin area was removed.

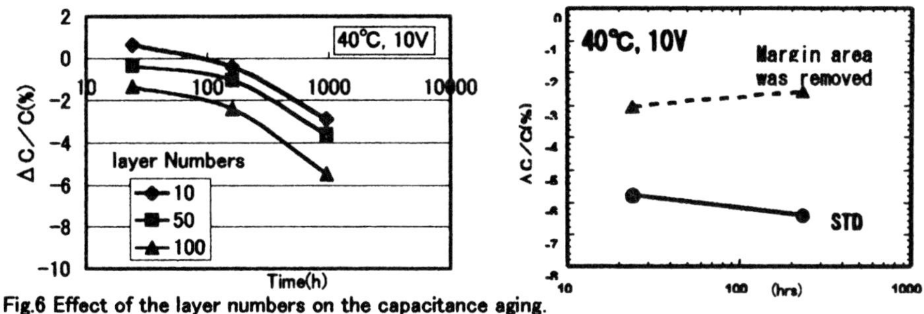

Fig.6 Effect of the layer numbers on the capacitance aging.

Fig.7 Capacitance aging before and after the removal of the margin area.

Thermal shrinkage profiles and microstructure

The thermal expansion(shrinkage) profiles against the firing temperature were obtained for MLCCs having different layer numbers fired in the reducing atmosphere. The measurements were also carried out for the Ni-only-body and dielectrics-only-body for the comparison. The results are shown in Fig.8. For the case of the Ni-only-body, the shrinkage started at a temperature significantly lower than those of the dielectrics-only-body. For the case of the MLCCs, which comprise dielectrics and Ni layers, the shrinkage temperature was almost same with dielectric-only-body. It is supposed that, in this case, the Ni internal electrodes have some stress in MLCCs from these results.

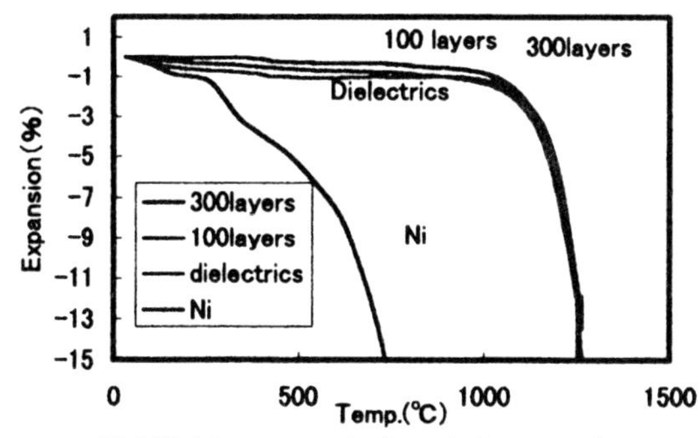

Fig8 Shrinkage curves in the reducing atmosphere.

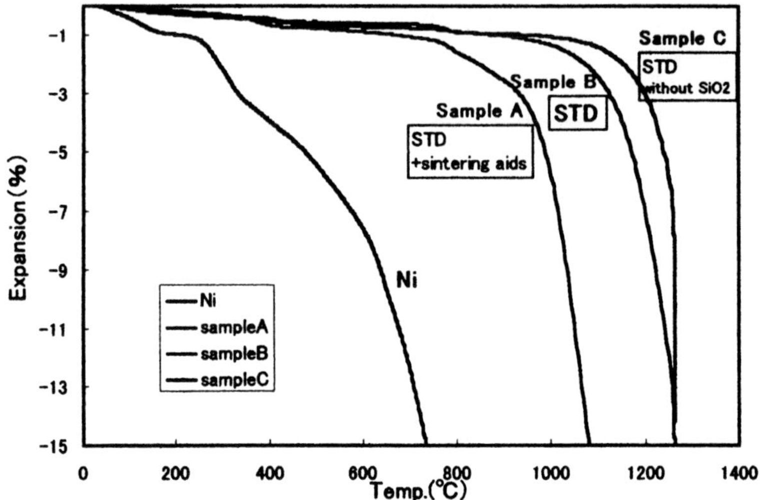

Fig.9 Shrinkage curves of MLCCs with different dielectric composition in the reducing atmosphere.

The microstructural images of the MLCCs fired at different temperatures are also shown in the middle row in Fig.10 (STD). With the increase of the firing temperatures, the Ni internal electrodes became discontinuous and globular shaped. The expansion of the MLCCs is attributed to the Ni electrode globing.

Effect of the sintering temperature of dielectrics on the Ni

In order to adjust the shrinking temperatures of the Ni and dielectrics to fit each other, the sintering aids were added to the dielectrics. The, MLCCs were fabricated by using these dielectrics and Ni-electrode, which will be referred as Sample-A. MLCCs were also fabricated from dielectrics, to which no SiO_2 was added, in order to investigate the effect of the firing behavior of the dielectrics on the microstructure of the Ni electrode. These samples will be referred as Sample-C. All the samples, including the STD (Sample-B), were fabricated by using the same Ni paste to have 100 dielectric layers. The shrinkage curves are shown in Fig.9. It is understand that shrinkage temperature are different from this figure.

The microstructural images are shown in Fig.10 for these samples fired at various temperatures. For the case of the Sample-A, even the low firing temperature, i.e. 1100 and 1180°C, gave the discontinuous and globular microstructure of the Ni. The Sample-B also showed a similar microstructure of the Ni in the firing temperature range between 1180 and 1290°C. However, the Sample-C showed the different firing behavior of the Ni. The Ni internal electrode was more continuous and had less globular shape, even though the firing temperature was relatively high, i.e. 1240 and 1290°C. From these results, it can be concluded that the dielectrics used is more important factor for the firing behavior of Ni rather than the firing temperature. Among several factors of the dielectrics, the different microstructure of the Ni electrode is attributed to the different shrinkage behavior of the dielectrics. The Ni globularity could not be suppressed even if the starting temperature of the shrinking of the dielectrics was lowered.

The reaction between the dielectrics and the Ni or the oxidation of the Ni in the firing process were also suspected. Fig.11 and 12 show TEM images of the interface between the dielectrics and Ni of STD-MLCCs. These images show no distinct interfacial layers at the interface and indicate good coherency between the dielectric and electrode layers.

High capacitance MLCCs.

1206-type MLCCs were successfully developed, which have a capacitance value of 100μF and Y5V characteristic. The cross section image of the MLCC is shown in Fig.13. Between the Ni-electrodes, observed is only one layer consisting of crystal grains of the dielectrics and it have high reliability.

CONCLUSIONS

1. The internal stresses in MLCCS tend to be enhanced by the multiplication of dielectric layers, resulting in the accelerated degradation of the capacitance.
2. The internal stress in MLCCs is attributed to the Ni globing during the firing process. The Ni globing does not seem to depend on the firing temperature but does seem to be highly affected by the firing behavior of the dielectrics.
3. 1206-type MLCCs were successfully developed, which have a capacitance value of 100μF and Y5V characteristic.

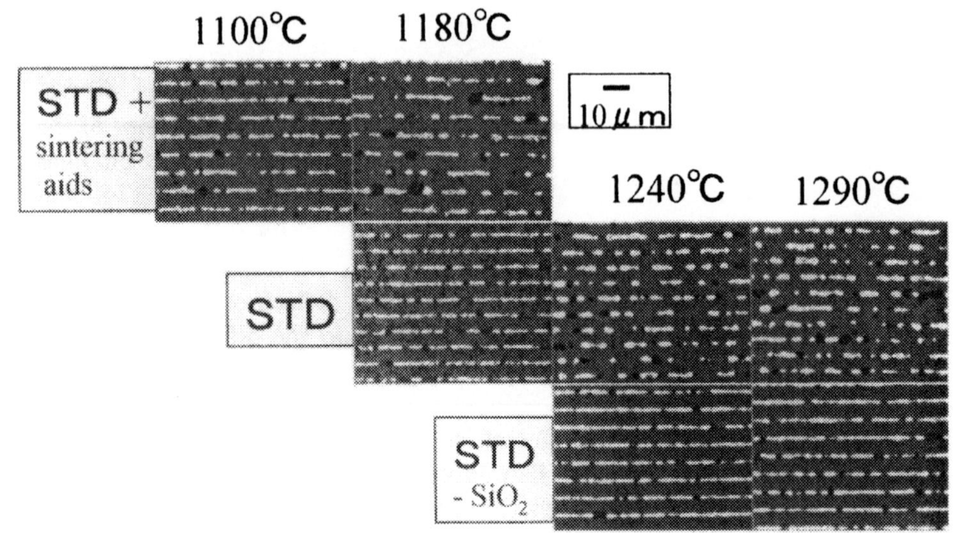

Fig.10 Cross section of MLCCs fired at different temperature.

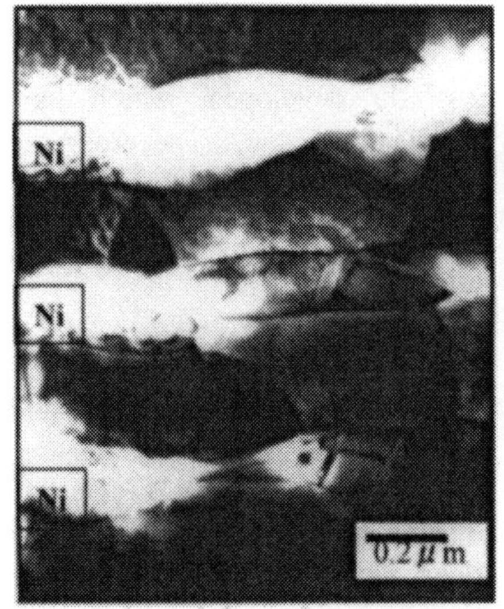

Fig.11 TEM image of cross sections of MLCCs (JEM-200FX, E=200kV)

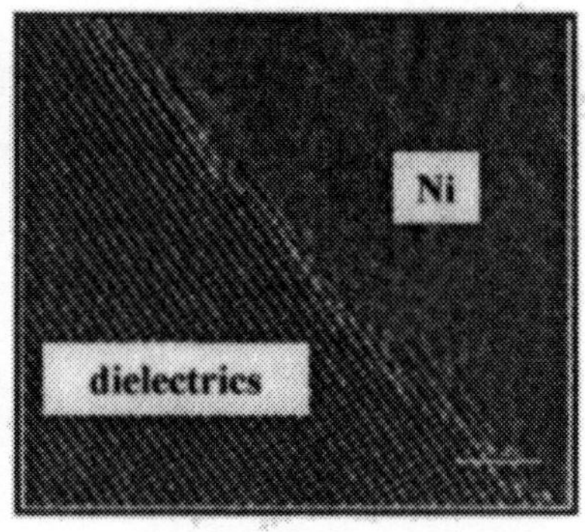

Fig12. HRTEM image of the interface between dielectrics and Ni . (JEM-3010, E=300kV)

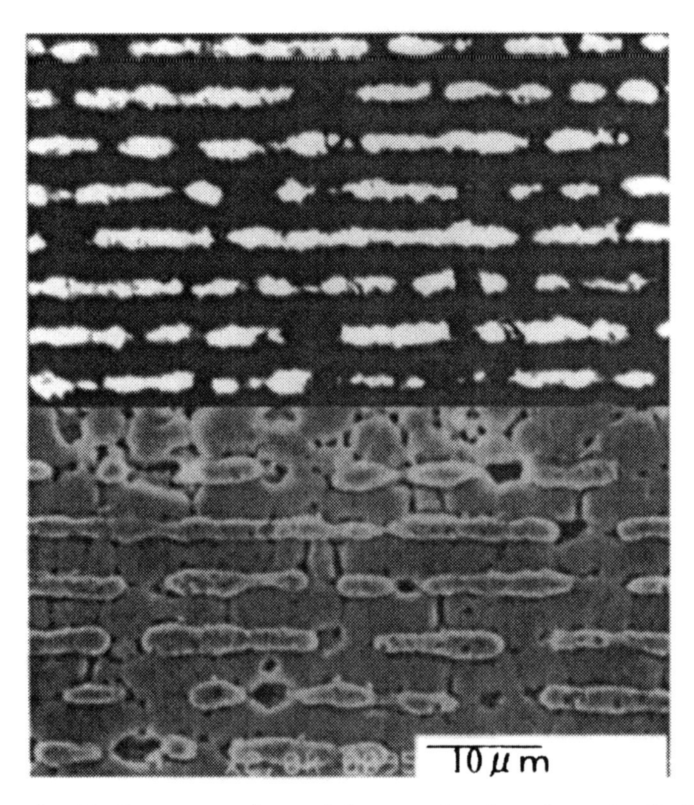

Fig.13 Cross section of the newly developed MLCC having 100 μF in 1206size with Y5V characteristic.

REFERENCES

1) T.Nomura,S.Sumita,Y.Nakano,K.Nishiyama, *Proc.5thUS-l.japan Seminar on Dielectric and Piezoelectric Ceramics*, (1990) 29-32.
2)Y.Yamamatsu et al.,*Jounal of power source*, 60 (1996)199-203.
3)R.Waser,T.Baiatu and K.Hardtl, *J.Am.Ceram.Soc.*, 3 (1990) 1645-1653.
4)Y.Takagi,Y.Kawaguchi,Y.Ueno,Y.Yoneda, *CARTS98:18th Capacitor and resistor Technology Syposium*,(1998)197-202.
5)N.Fujikawa, *NipponCeramiics Kykai Denshizairyou-bukai*, (1990) 86
6)H.Kabasawa,H.Saitoh,S.Tosaka,NEEDS&SEEDS,10(1994)15-18
7)W.Carlson,T.Rutt,M.Wild,Ferroelectric Letters,21(1996)1-9
8) K.Okazaki, *Ceramic Bulletin*, 63,*9* (1984)1150-1152.

PROCESSING OF Pb-Ba-Zr-Ti BASED DIELECTRICS FOR HIGH POWER CAPACITOR APPLICATIONS

Roy J. Rayne, Todd L. Jessen, and Barry A. Bender
Code 6351
U.S. Naval Research Laboratory
Washington, DC 20375-5343

Manfred Kahn and Mark T. Chase
Potomac Research Int.
Fairfax, VA 22030

ABSTRACT

Commercially available raw materials and conventional processing techniques were used to produce a dielectric material that may be suitable for use in a high voltage ceramic capacitor. For this study, a lead barium zirconium titanate base composition was investigated. It was determined that 95% of the precursor zirconia powder needed to be less than one micron in order for the process to yield a single phase material.

INTRODUCTION

For the development of an all electric ship the US Navy needs high voltage ceramic capacitors that meet X7R specifications. The capacitors must be small enough for ship board operations and they must be cost effective to manufacture. Such high voltage ceramic capacitors are not commercially available. We have identified a family of compositions in the lead barium zirconium titanate family which meets this need. This study focused on the development of a reproducible process that uses 'industrially friendly' methods so that the process could be easily transitioned to a capacitor manufacturer. Commercially available raw materials and conventional processing techniques were used to produce a material that is suitable for use in a high voltage ceramic capacitor.

EXPERIMENTAL PROCEDURE

There have been several studies on the dielectric properties of $(Pb,Ba)(Zr,Ti)O_3$[1-4]. The composition for this study, $(Pb_{0.65}Ba_{0.35})(Zr_{0.7}Ti_{0.3})O_3$, was selected based on data reported by Furukawa[3,4]. The determining factors in the powder selection were, in order of importance, particle size, purity and cost. The raw materials used in this study were PbO (purity 99.9%, particle size – 325 mesh), $BaCO_3$ (purity 99.8%, particle size 1 μm), ZrO_2 (purity 99.5%, Hf <100ppm, particle size 1 μm APS) and TiO_2 (purity 99.5, particle size 1-2 μm APS) and were obtained from Alfa Aesar, Ward Hill , MA..

Stoichiometric amounts of precursor powders were blended into a purified water solution containing a dispersant (Tamol 901, Rohm and Haas, Philadelphia, Pa.) and a surfactant (Triton CF-10, Rohm and Haas, Philadelphia, Pa). The resultant slurry was then attrition milled[3] using the following conditions. The slurry solids loading was 20% by volume. The milling media used was zirconia and the media to slurry ratio was 2:1 by volume. The slurry was attrition milled for 2h. It was then washed into a preheated low profile glass dish and dried at about 90°C.

The dried powder was calcined in an alumina dish at 850°C for 2h. After calcination the powder was attrition milled and dried using the same milling and drying conditions as above. The dried powder was uniaxially pressed into a pellet. The pellet was buried in setter powder and sintered at 1280°C for 2h in a covered alumina crucible to provide a lead rich environment. The setter powder contained $PbZrO_3$ as a source of excess lead. Multiple pellets were sintered at the same time and used for subsequent characterization.

Material characterization was done using powders and pellets after each processing step. The typical pellet size was 10.7 mm in diameter and 1.6 mm thick. X-ray diffraction (XRG 3100, Philips Electronic Instruments Inc, Mahwah, NJ) was used to monitor the phase evolution using both powder and pellets. Microstructural characterization was done on the cross section of pellets (both fractured and polished surfaces) using scanning electron microscopy (SEM). Dielectric measurements were done using a Hewlett Packard 4284A LCR and an in-house designed high voltage test fixture on a pellet as a disk capacitor. The ends of the pellet were coated with palladium and then gold as an electrode.

RESULTS AND DISCUSSION

The as-received raw powders were characterized by SEM (see Fig. 1) and only the as-received ZrO_2 powder was analyzed by a particle size analyzer (CAPA 700, Horiba Instruments Inc, Irvine, CA). The particle size of the PbO and $BaCO_3$ powders is not as critical because during calcination PbO is very volatile and $BaCO_3$ decomposes. It is more important that the particle size of TiO_2 and ZrO_2 be as small as practically possible. The platelet morphology of TiO_2 powder made it difficult to obtain reliable results using the CAPA 700 sedimentation technique. Consequently, SEM micrographs were used to confirm that the particle size of the TiO_2 was as advertised to be about 1-2 μm (see Fig. 1C). The particle size of the ZrO_2 was measured using the CAPA 700 and the results showed a median particle size of 0.44 μm (see Fig 2A).

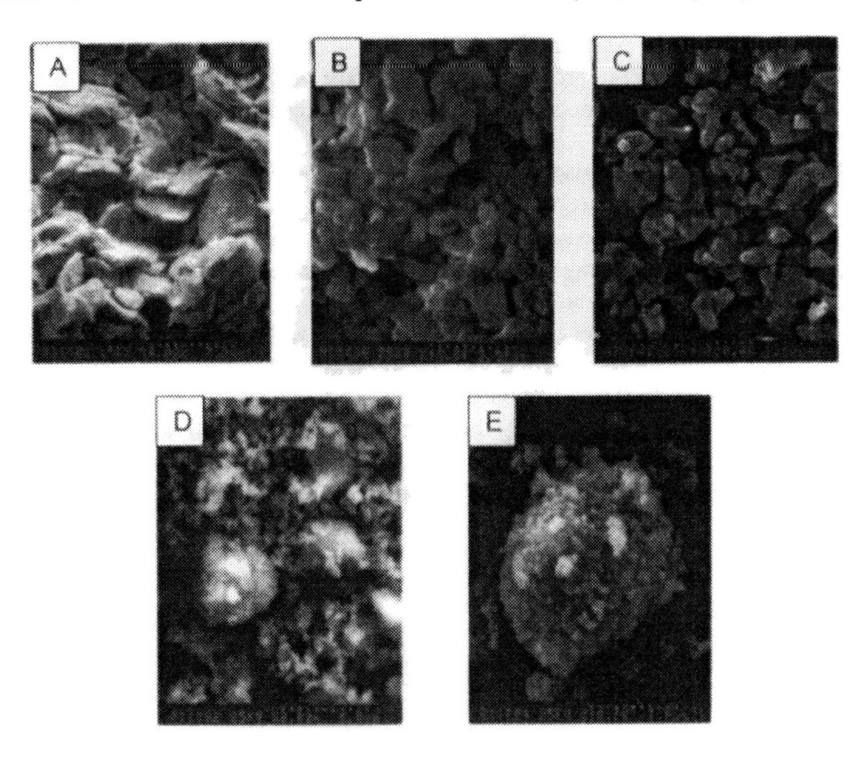

Fig. 1 SEM micrographs of the as-received raw powders (A) PbO, (B) $BaCO_3$, (C) TiO_2, (D) and (E) ZrO_2.

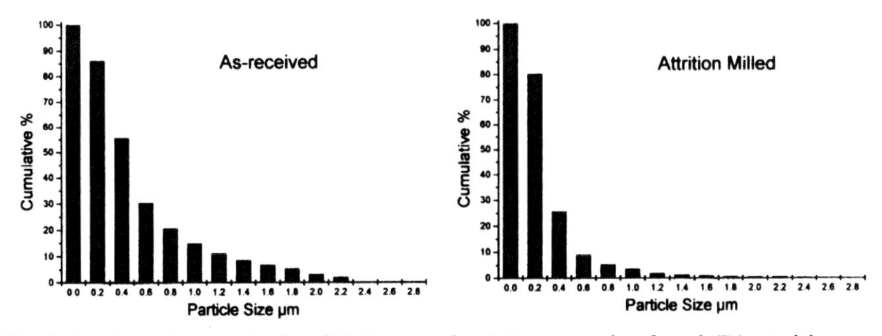

Fig. 2 Particle size analysis of ZrO_2 powder (A) as-received and (B) attrition milled for 2h.

During the attrition milling of the precursor powders minimal particle size reduction was expected because PbO and $BaCO_3$ are softer than ZrO_2 and TiO_2. Therefore the purpose of milling the precursor powders was to obtain a homogeneously mixed powder. So, a dispersant was added to the purified water solution to ensure that the precursor powders were well dispersed. A surfactant was also added to the purified water solution to make sure that the solution wet the precursor powders.

The purpose of the calcination step was to decompose the $BaCO_3$ and partially react the precursor powders while minimizing the lead loss. The calcination temperature must be high enough to decompose the $BaCO_3$ yet low enough to minimize the lead loss. X-ray diffraction (XRD) characterization of the calcined powder indicated that it was not fully reacted (Fig. 3).

Attrition milling of the calcined powder was done in order to break up agglomerates formed during calcination and to make sure that the powder used to make the pellet for sintering was well mixed. The conditions used for attrition milling the calcined powder were the same conditions used for milling the precursor powder. Microstructural characterization of the milled powders showed that they were well mixed with minimal agglomeration.

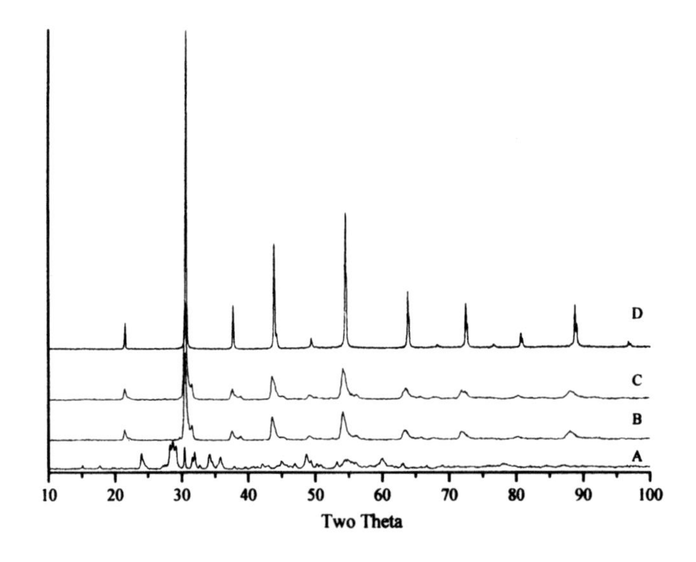

Fig. 3 XRD patterns showing the phase evolution during processing (A) attrition milled precursor powders, (B) calcined powder, (C) attrition milled calcined powder, and (D) sintered pellet.

XRD analysis of a sintered pellet showed that within the limits of detection the pellet was single phase (Fig. 3D). However, SEM analysis of the polished surface of a sintered pellet showed that there was a small amount of a ZrO_2 second phase (Fig. 4A). The cause of the ZrO_2 second phase was believed to be the particle size of the as-received ZrO_2 powder was too large. In an effort to reduce the particle size of the as-received ZrO_2 powder it was attrition milled using the same milling parameters as previously discussed with the exception of the slurry solids loading which was 30% by volume. The particle size distribution of the attrition milled ZrO_2 powder was compared to the as-received ZrO_2 powder (Fig. 2). The significant difference between the powders was the percentage of particles that are below 1μm, 85% for the as-received powder versus 95% for the attrition milled powder. Subsequently, when the attrition milled ZrO_2 powder was used instead of the as-received ZrO_2 powder the ZrO_2 second phase was eliminated (Fig. 4B).

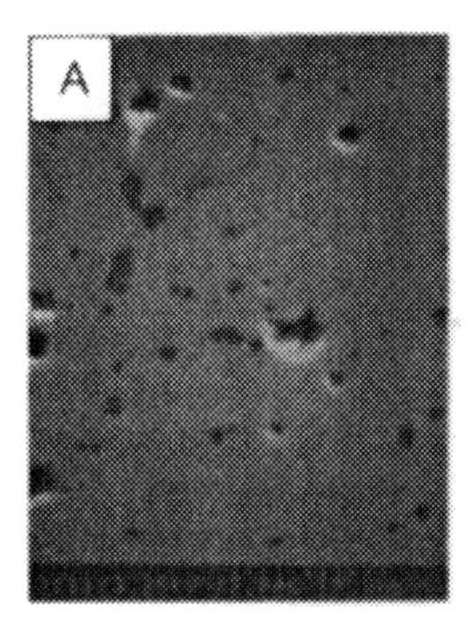

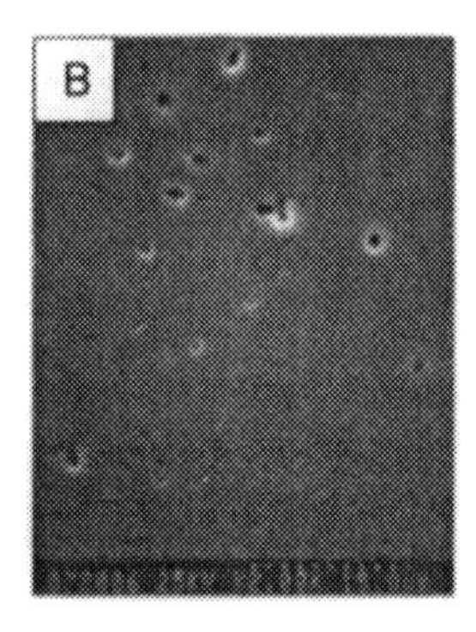

Fig. 4 SEM micrograph of the polished surface of a sintered pellet (A) using as-received ZrO_2 powder as a precursor and (B) using attrition-milled ZrO_2 powder as a precursor.

Electrical properties were measured on numerous pellets at room temperature, with a 600 V/mm dc bias voltage and a frequency of 1000 Hz. Typical results of these measurements were a dielectric constant of 5500 and a loss factor of 1.9%. Measurements taken to determine the breakdown strength showed it to be greater than 5000 V/mm. The dielectric constant was stable over a temperature range -55°C to 125°C (Fig. 5). These measurements indicate that this dielectric material should meet the electrical requirements for high voltage (> 1000 V) applications.

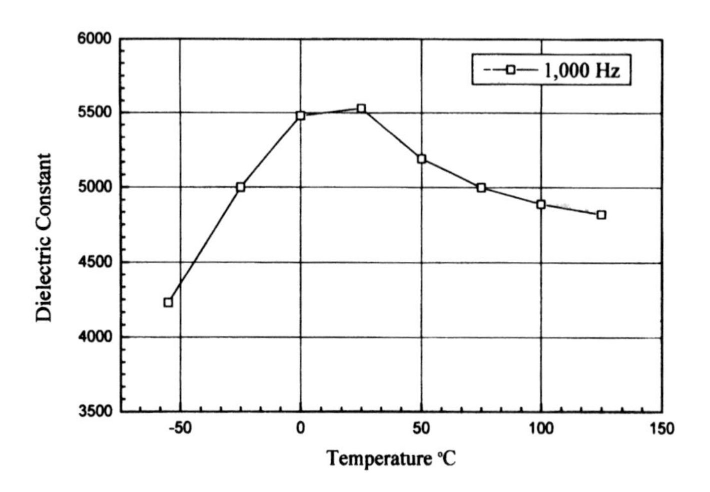

Fig. 5 Dielectric Constant vs. Temperature measured with a 2000 V/mm dc bias.

CONCLUSION

Commercially available raw materials and conventional processing techniques were used to produce a dielectric material. Experimental results showed that in order to produce a single phase dielectric material that 95% of the particles of the precursor ZrO_2 powder had to be less than 1 μm. The dielectric constant and loss factor were measured and it was concluded that this material is suitable for use in a high voltage ceramic capacitor. Efforts are underway to make a multi-layer capacitor using this material and measure its electrical properties to determine the worthiness of this material.

REFERENCES

1. E. Moreira, J. de Mello, J. Povoa, D. Garcia, and J. Eiras, "Dielectric Properties of $(Pb,Ba)(Zr,Ti)O_3$ Ceramics," *1996 IEEE Ultrasonics Symposium,* **1** 527-30 (1996).
2. Z. Ujma, M. Adamczyk, and J. Handerek, "Relaxor Properties of $(Pb_{0.75}Ba_{0.25})(Zr_{0.70}Ti_{0.30})O_3$ Ceramics," *J. Er. Ceram. Soc.,* **18** 2201-07 (1998)
3. O. Furukawa, H. Kanai, and Y. Yamashita, "A New Relaxor Dielectric for High Voltage Multilayer Ceramic Capacitors with Large Capacitance," *Jpn. J. Appl. Phys.*, **32** 1708-11 (1993).
4. H. Kanai, O. Furukawa, H. Abe, and Y. Yamashita, "Dielectric Properties of $(Pb_{1-x}X_x)(Zr_{0.7}Ti_{0.3})O_3$ (X = Ca, Sr, Ba) Ceramics," *J. Am. Ceram. Soc.,* **77** [10] 2620-24 (1994).
5. S. Padden and J. Reed, "Grinding Kinetics and Media Wear during Attrition Milling," *Amer. Ceram. Bull.*, **72** [3] 101-12 (1993)

ADDITIVE INTERACTIONS IN AQUEOUS $BaTiO_3$ SUSPENSION

Chia-Chen Li and Jau-Ho Jean
Department of Materials Science and Engineering
National Tsing Hua University
Hsinchu, Taiwan, ROC

ABSTRACT

The interaction between dissolved Ba^{+2} and dissociated ammonium salt of poly(acrylic acid) ($PAA-NH_4$) in aqueous suspension has been studied. It is found that the dissolved Ba^{+2} causes flocculation of dissociated $PAA-NH_4$, thus degrading its dispersing effectiveness in aqueous $BaTiO_3$ suspensions. The concentration of $PAA-NH_4$ required to stabilize aqueous $BaTiO_3$ suspensions increases with increasing Ba^{+2} concentration at a given pH. A stability map, which is determined by rheology study, is constructed to describe the amount of $PAA-NH_4$ required to have well-dispersed aqueous $BaTiO_3$ suspensions as a function of Ba^{+2} concentrations at different pH values.

INTRODUCTION

In previous papers [1-3], colloidal stability of aqueous $BaTiO_3$ powder suspensions with various solids loading and amounts of ammonium salts of poly(acrylic acid) ($PAA-NH_4$) and poly(methacrylic acid) ($PMAA-NH_4$) at different pH values was investigated. Several critical factors such as solid loading, pH, degree of polyelectrolyte dissociation, molecular weight of polyelectrolytes, quantity of polyelectrolyte added and surface chemistry of powders were identified. It has also been reported [4,5] that barium titanate is not thermodynamically stable in the acidic aqueous solutions. Barium ions are leached out of the powder which results in a titanium rich surface layer. It was also found that the amount of Ba^{+2} leached increases with decreasing pH. This increases the pH of resulting suspension and changes the surface chemistry of $BaTiO_3$ [5,6]. Since the leached barium ions are positively charged and the dissociated $PAA-NH_4$ and $PMAA-NH_4$ are negatively charged, the interaction between them in the aqueous suspensions can be expected and is also the major objective of this study. Ba^{+2} concentration in the suspensions was controlled by adding anhydrous $BaCl_2$. The chemical interaction of free barium ion and dissociated $PAA-NH_4$ was examined by ^{13}C NMR spectroscopy. Effects of $BaCl_2$ on the effective hydrodynamic radius of dissociated $PAA-NH_4$ at various pH values were determined by dynamic light scattering. Colloidal stability of aqueous $BaTiO_3$ suspensions at different pH values with various concentrations of $PAA-NH_4$ and $BaCl_2$ was determined by rheological measurement.

EXPERIMENTAL PROCEDURE

The ceramic powder used in this study was a high-purity $BaTiO_3$ (TICON-HBP, TAM, Niagara Falls, NY) with a Ba/Ti ratio of 0.991. The powder had a median size of 1.0 μm and a specific surface area of 3.05 m^2/g. The polyelectrolyte used was a commercially available ammonium salt of poly(acrylic acid) ($PAA-NH_4$) (Darvan-821A, R.T. Vanderbilt, CT) with an average molecular weight of 6 000 g/mol. Ba^{+2} concentration in the suspensions was controlled by adding anhydrous $BaCl_2$. Deionized and distilled water was used, and the pH was adjusted by HCl and KOH.

The turbidity of aqueous solutions with added $BaCl_2$ and dissociated $PAA-NH_4$ was measured by UV-visible spectroscopy (U-3501, Hitachi) at a wavelength of 600 nm. Chemical interaction between barium ions and dissociated $PAA-NH_4$ was analyzed by ^{13}C NMR spectroscopy (AM-400, Bruker). Methyl carbon of sodium 2,2-dimethyl-2-silapentane-5-sulfonate was used as an external reference for carbon. The hydrodynamic radius of ionized $PAA-NH_4$ was measured by dynamic light scattering. Aqueous suspensions of 17 vol% $BaTiO_3$ with different concentrations of $PAA-NH_4$ were prepared at pH = 7-12. The adsorption amount of $PAA-NH_4$ on the $BaTiO_3$ was measured by the method of potentiometric titration, which details of this technique was described previously [1-3]. The techniques of electrokinetic mobility (EM) (ZetaPlus, Brookhaven Instruments, USA) was used to measure the zeta potential of 5 vol% aqueous barium carbonate suspensions. Rheological behavior of 17 vol% $BaTiO_3$ powder suspensions with different $PAA-NH_4$ concentrations at pH = 7-12 were determined using a concentric cylinder viscometer (Model RV-100/CV-200, Haake, Germany). Viscosity at a shear rate of 100 s^{-1} was used for comparison.

RESULTS AND DISCUSSION

Interactions between Ba^{+2} and $PAA-NH_4$

Various concentrations of $BaCl_2$ were added to aqueous solutions of $PAA-NH_4$ of 0.001 mol/L, and the turbidity was measured by an UV-visible spectrometer. Figure 1 shows the results of transmission for the aqueous solutions with 0.001 mol/L $PAA-NH_4$ and various concentrations of $BaCl_2$. It is found that the transmission remains relatively unchanged when the concentration of $BaCl_2$ is less than 0.01 mol/L, decreases dramatically in the range of 0.02-0.2 mol/L, and becomes completely opaque at a concentration of $BaCl_2$ greater than 0.3 mol/L.

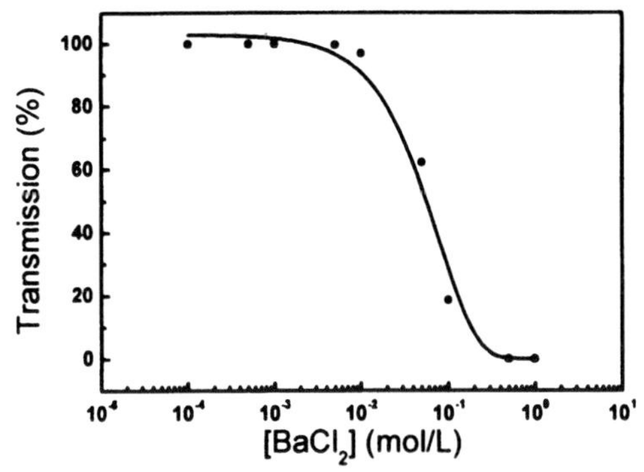

Figure 1 Transmission of aqueous PAA-NH_4 solution as a function of $BaCl_2$ concentration.

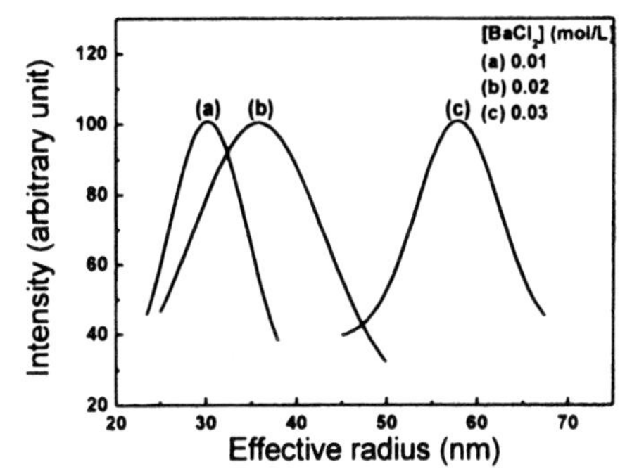

Figure 2 Effective hydrodynamic radius of dissociated PAA-NH_4 as a function of $BaCl_2$ concentration at pH = 7.5.

To find out if chemical reaction takes place between Ba^{+2} and dissociated PAA^-, ^{13}C NMR spectroscopy was used. No significant change in the spectra between pure PAA-NH_4 and white residue obtained from opaque suspensions with dissociated PAA^- and Ba^{+2} was found, indicating no new compounds formed. The above results conclude that the interaction between Ba^{+2} and dissociated PAA^- is not chemical rather than physical in nature. One of possibilities is that the electrostatic attraction between positively charged Ba^{+2} and negatively charged PAA^- reduces the repulsion between polymer chains of PAA^- [7], causing the polymer to coil up. It is also believed that the electrostatic attraction minimizes the repulsive force between dissociated PAA^-, resulting in a higher degree of flocculation. This can be confirmed by the results of hydrodynamic radius measurement given below. Figure 2 shows the results of hydrodynamic radius of dissociated PAA^- as a function of $BaCl_2$ concentration at pH = 7.5. The hydrodynamic radius distribution becomes broader and moves to a larger effective size with increasing concentration of $BaCl_2$. Similar results are also observed at pH = 9.2 and 12.0, and the data of median effective radius are listed in Table I.

Table I. Median effective radius ($\langle S^2 \rangle^{1/2}$) and monolayer adsorption (C_m) of $PAA\text{-}NH_4$ on $BaTiO_3$ as a function of pH value and $BaCl_2$ concentration

	pH 7.5		pH 9.2		pH 12.0	
$[BaCl_2]$ (mol/L)	$\langle S^2 \rangle^{1/2}$ (nm)	C_m (mg/gm BT)	$\langle S^2 \rangle^{1/2}$ (nm)	C_m (mg/gm BT)	$\langle S^2 \rangle^{1/2}$ (nm)	C_m (mg/gm BT)
0	4.5	0.58	6.3	0.42	9.5	0.19
0.01	30.0	1.84	49.0	1.77	59.5	1.47
0.02	35.0	2.09	57.5	2.08	67.5	1.69
0.03	57.5	2.80	57.5	2.60	68.5	2.29

Adsorption of $PAA\text{-}NH_4$ on $BaTiO_3$ with $BaCl_2$

Typical results to show the effects of $BaCl_2$ on the specific adsorption of $PAA\text{-}NH_4$ on $BaTiO_3$ powder at pH = 7.5, 9.2, and 12.0 are shown in Fig. 3. For all conditions investigated, each adsorption curve shows a characteristic plateau, which corresponds to the amount of $PAA\text{-}NH_4$ required for monolayer coverage.

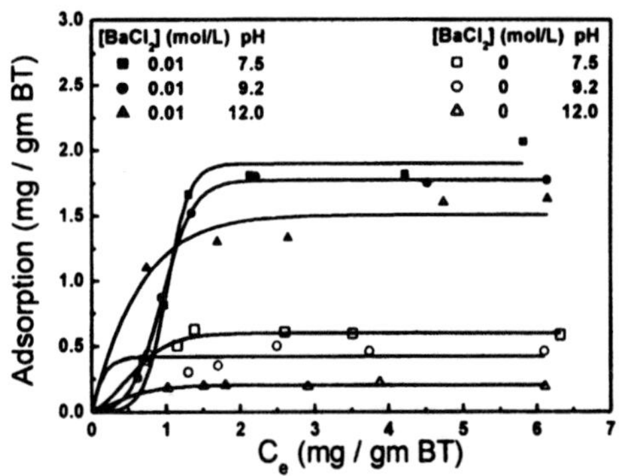

Figure 3 Adsorption isotherms of $PAA\text{-}NH_4$ on $BaTiO_3$ surface as a function of pH value and $BaCl_2$ concentration.

It is found that the saturated adsorbance of $PAA\text{-}NH_4$ decreases with increasing pH, in agreement with those reported [1-3]. Moreover, a higher, saturated adsorbance of $PAA\text{-}NH_4$ at a higher concentration of $BaCl_2$ is observed. To determine the monolayer adsorbance of PAA-NH_4 quantitatively, the data in Fig. 3 are analyzed using the Langmuir monolayer adsorption equation [8]. The calculated data of monolayer adsorbance (C_m) as a function of pH are listed in Table I. It is found that the monolayer adsorption of PAA-NH_4 decreases with increasing pH for all conditions investigated. At a given pH, moreover, a larger monolayer adsorption of PAA-NH_4 is observed at a higher concentration of $BaCl_2$.

Effect of adsoprtion of PAA-NH_4 on zeta potential of $BaTiO_3$ with $BaCl_2$

To investigate the effect of PAA-NH_4 adsorption on the surface chemistry of $BaTiO_3$ in the presence of $BaCl_2$, the zeta potentials are measured. Supernatants from the $BaTiO_3$ powder suspensions with 0.01 mol/L $BaCl_2$ and various concentrations of PAA-NH_4 at pH = 7.5, 9.2 and 12.0 are used, and the results are

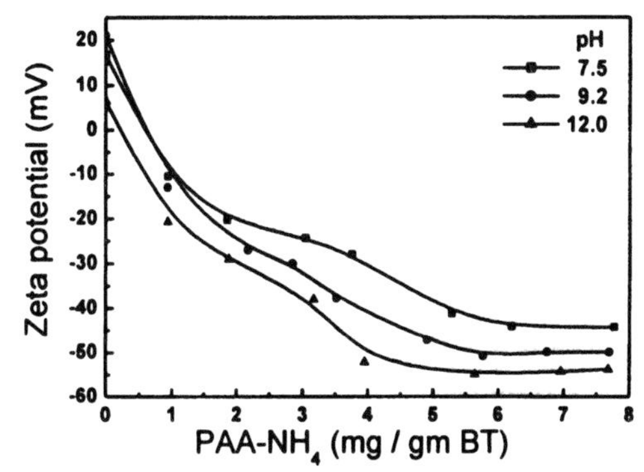

Figure 4 Zeta potential of $BaTiO_3$ as a function of initial concentration of PAA-NH_4 at pH = 7-12 and [$BaCl_2$] = 0.01 M.

shown in Fig. 4. Since the specific adsorption of Ba^{+2} takes place for the suspensions without PAA-NH_4 [5], the zeta potentials are positive. For all pH values investigated, the zeta potential becomes negative with increasing concentration of PAA-NH_4. This is due to the fact that the negatively dissociated PAA^- polyelectrolytes are adsorbed onto $BaTiO_3$ surface, reversing the charge from positive to negative on the particle surface. With PAA-NH_4 concentration increasing, the zeta potential shows a continuous decrease initially, followed by a constant value. A smaller constant zeta potential is also observed at a lower pH. Similar results are also observed at $BaCl_2$ concentrations of 0.02 and 0.03 mol/L investigated.

Rheological behavior

The colloidal stability of aqueous $BaTiO_3$ suspensions is also assessed by its rheological behavior. To identify the effect of the free barium ion on the rheological behavior, viscosity of the suspensions with various amounts of PAA-NH_4 at different pH values and $BaCl_2$ concentrations is measured. For clarity of discussion, only the rheological results at pH = 7.5 and [$BaCl_2$] = 0.01 mol/L are

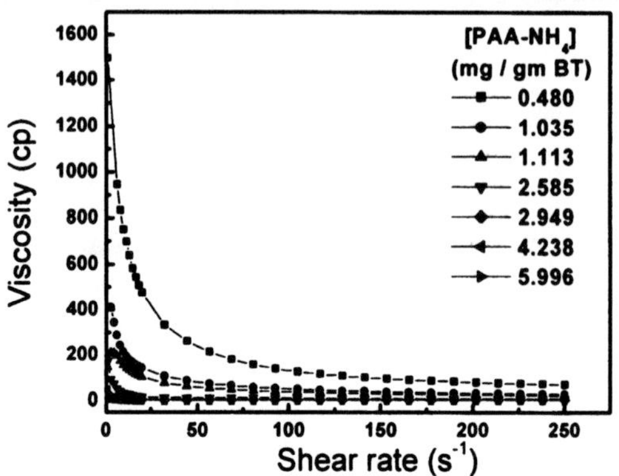

Figure 5 Viscosity as a function of shear rate for 17 vol% $BaTiO_3$ suspensions with various concentrations of PAA-NH_4 at pH = 12.0 and [$BaCl_2$] = 0.01 M.

given in Fig. 5. It is found that a shear-thinning behavior is observed for the suspensions with PAA-NH_4 concentrations less than 2.585 mg/gm BT, suggesting a flocculated powder suspension. However, when the PAA-NH_4 concentration continues to increase up to 2.949 mg/gm BT, a relatively constant viscosity with increasing shear rate is observed, which is indicative of a Newtonian response that corresponds to a well-dispersed powder suspension. Similar phenomena are also observed at other pH values and Ba^{+2} concentrations investigated. For comparison, the viscosity at an arbitrary shear rate, e.g. 100 s^{-1}, was chosen and the data are summarized in Fig. 6. For all pH values investigated, the viscosity results show a significant decrease initially, followed by relatively little dependence on the PAA-NH_4 concentration. A critical PAA-NH_4 concentration, which characterizes the above transition, can be determined from the intercept of the slopes of these two stages.

Stability map for $BaTiO_3$ / PAA-NH_4

Armed with the results of viscosity (Figs. 5 and 6), a stability map that summarizes the critical concentration of PAA-NH_4 polyelectrolyte required to achieve colloidal stability of aqueous $BaTiO_3$ suspensions as a function of pH and $BaCl_2$ concentration is summarized in Fig. 7. Three major areas, which include dissolution of barium ions and stable and unstable regions, are indicated. The area

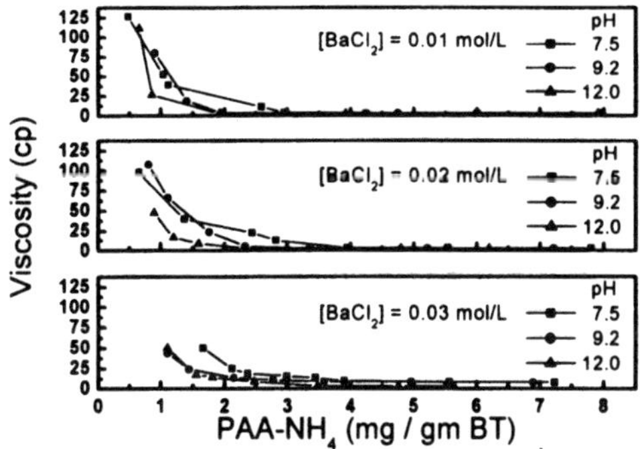

Figure 6 Viscosity at a shear rate of 100 s^{-1} at pH = 7-12 as a function of PAA-NH_4 concentration and (a) 0.01M, (b) 0.02M and (c) 0.03M $BaCl_2$.

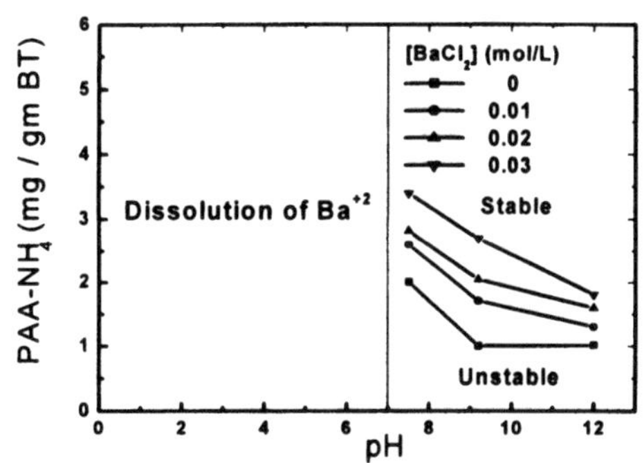

Figure 7 Stability map outlining the required PAA-NH_4 concentration to stabilize aqueous $BaTiO_3$ suspensions at different pH values and $BaCl_2$ concentrations.

of barium ion dissolution, which appears only at pH < 7, is not considered to be suitable for powder processing [1-3]. At pH > 7, the curves that represent the transition between stable and unstable area are identified, according to the results of the required PAA-NH_4 concentration to reach Newtonian rheology (Figs. 5 and 6). At a given concentration of $BaCl_2$, only the area above the curve is colloidally stable and a well-dispersed suspension is obtained. In other words, each curve, which exhibits a decreasing trend with increasing pH, represents the minimum PAA-NH_4 concentration required to achieve colloidal stability of aqueous $BaTiO_3$ suspensions at a given $BaCl_2$ concentration. It is further noted in Fig. 7 that the curves shift upward with increasing $BaCl_2$ concentration, indicating that the required PAA-NH_4 concentration to achieve colloidal stability of aqueous $BaTiO_3$ suspensions increases with increasing $BaCl_2$ concentration.

CONCLUSIONS

The colloidal stability of aqueous $BaTiO_3$ powder suspensions with various amounts of PAA-NH_4 at different pH values and $BaCl_2$ concentrations has been investigated. The presence of free Ba^{+2} causes flocculation of negatively dissociated PAA^-, degrading the dispersion effectiveness of PAA-NH_4. This in turn requires a higher concentration of PAA-NH_4 with increasing free Ba^{+2} to obtain a colloidally stable aqueous $BaTiO_3$ suspension. A stability map, which describes the critical amount of PAA-NH_4 required to obtain colloidally stable aqueous $BaTiO_3$ powder suspensions as a function of pH value and Ba^{+2} concentration, is constructed according to rheological results.

ACKNOWLEDGMENTS

This study was supported by the National Science Council of Republic of China under Grant No: NSC 88-2216-E-007-036.

REFERENCES

[1] J. H. Jean and H. R. Wang, "Dispersion of Aqueous Barium Titanate Suspensions with Ammonium Salt of Poly(methacrylic acid)," *J. Am. Ceram. Soc.*, **81** [6] 1589-99 (1998).

[2] J. H. Jean and H. R. Wang, "Stabilization of Aqueous $BaTiO_3$ Suspensions with Ammonium Salt of Poly(acrylic acid) at Various pH Values," *J. Mater. Res.*, **13** [8] 2245-50 (1998).

[3] J. H. Jean and H. R. Wang, "Effects of Solids Loading, pH, and Polyelectrolyte Addition on the Stabilization of Concentrated Aqueous $BaTiO_3$ Suspensions," *J. Am. Ceram. Soc.*, **83** [2] 277-80 (2000).

[4] H. W. Nesbitt, G. M. Bancroft, W. S. Fyfe, S. N. Karkhanis, and A. Nishijima, "Thermodynamic Stability and Kinetics of Perovskite Dissolution," *Nature*, **289** [29] 358-62 (1981).

[5] M. C. Blanco-Lopez, B. Rand, and F. L. Riley, "The Properties of Aqueous Phase Suspensions of Barium Titanate," *J. Eur. Ceram. Soc.*, **17** [3] 281-87 (1997).

[6] A. Neubrand, R. Lindner, and P. Hoffmann, "Room-Temperature Solubility Behavior of Barium Titanate in Aqueous Media," *J. Am. Ceram. Soc.*, **83** [4] 860-64 (2000).

[7] D. A. Mortimer, "Synthetic Polyelectrolytes–A Review," *Polym. Inter.*, **25** [8] 29-41 (1991).

[8] T. Sato and R. Ruth, *Stabilization of Colloidal Dispersions by Polymer Adsorption*; pp. 1-36. Marcel Dekker, New York, 1982.

AQUEOUS TAPE CASTING OF SURFACE-MODIFIED CORDIERITE GLASS CERAMICS POWDERS

Sen Mei[a], José Maria F. Ferreira[a✉], Juan Yang[a], Rodrigo Martins[b]

[a]Department of Ceramics and Glass Engineering, UIMC, University of Aveiro,3810-193, Aveiro, Portugal

[b]Materials Science Department, Faculty of Science and Technology, New University of Lisbon, 2824-114 Monte de Caparica, Lisbon, Portugal

✉Tel: 351-234-370242; Fax: 351-234-425300; email:jmf@cv.ua.pt

ABSTRACT

In the present work, cordierite and glass powders coated with Al_2O_3 layers were prepared by a precipitation method. Particle size distribution (PSD), SEM, HRTEM, XRD and electrophoretic measurements were used to characterize the as-prepared powders after calcination at 500°C. The SEM morphology of the coated powders showed coiled worm-like surface, which was amorphous as revealed by HRTEM. The coating Al_2O_3 layer has dominated the surface properties of the coated glass and cordierite powders, improving their processing ability in aqueous media. The dielectric properties of sintered tapes (1100°C, 2h) derived from the coated cordierite and glass powders were measured (dielectric constant of ~5, loss of ~0.001 at 1MHz), which were comparable to the tapes derived from un-coated powders.

INTRODUCTION

Cordierite is a promising low-k material that is used as substrate in the interconnecting and packaging industries, particularly in the fabrication of multichip module (MCM-C),[1-5] owing to its low dielectric constant and compatible thermal expansion with that of Si chip. Some low softening-point borosilicate glasses are usually added to stoichiometric cordierite in order to lower the sintering temperature. Tape casting is a prevalently colloidal technique for fabricating substrates and multilayers, while a number of other new applications for producing complex laminated shapes, thin structural ceramics and microstructures with grain orientation have also been explored.[6-8]

In our previous work, the processing of cordierite-glass powders by aqueous tape casting has been investigated.[7] However, the maximum solids volume fraction was relatively low, which often originated low packing density values and defects in the green tapes. Such features, in turn, deteriorated the sinterability with some (un-removable) large pores between particles remaining in the sintered microstructures. Furthermore, the relatively high volume fraction of liquid phase in the suspension make difficult to control the thickness and flatness of the green tapes along the drying process due to the significant shrinkage values experienced.[8]

It was already shown that the dispersion ability of Si_3N_4 powders could be improved by coating the Si_3N_4 particles with a thin layer of the sintering additives.[9] The surface layer endowed the composite particles with the surface character of the coating components (yttria and alumina). The coated layer could also improve the processing ability of Si_3N_4 particles in water in comparison with the pure Si_3N_4 powder or the mechanical mixture of this powder with the sintering additives.[10] Luther *et al.*[11] also investigated the dispersion properties of the Si_3N_4 powder coated with Al_2O_3 layer, and found that the electrophoretic behaviors of the coated powders resemble that of the surface layer of Al_2O_3. Shih *et al.*[12] and Lidén *et al.*[13] also observed that the Si_3N_4 powder with a boehmite coating and the Si_3N_4 powder with adsorbed colloidal fine Y_2O_3 particles exhibited better processing ability than the pure Si_3N_4 powder. Such enhancements encouraged us to follow a similar approach to improve the processing of cordierite-glass system.

In the present work, cordierite and glass powders were coated with an Al_2O_3 layer by a precipitation method in order to increase the solids volume fraction in the aqueous tape casting slurries. The selection of Al_2O_3 as the coating surface layer was motivated by the fact that this oxide compounds both cordierite and glass, and would not greatly change the dielectric properties of the composite. The amount of Al_2O_3 should be high enough to modify the surface proprieties of the particles without increasing significantly the dielectric constant of the composite. Therefore, a ratio of 10-wt% of Al_2O_3 relative to the sum of glass and cordierite components was used. The surface layer was deposited by a precipitation process, in which the formation of coating Al_2O_3 layer was controlled by the slow release of OH^- from the hydrolysis reaction of urea.

EXPERIMENTAL PROCEDURE

Preparation of the coated powders

Cordierite powders were synthesized in the lab by using alumina powders (A16 SG, Alcoa Chemicals, USA), $Mg(NO_3)_2 \cdot 6H_2O$ (Merck KgaA, Germany) and silica powders (Aldrich chemical company, Inc., Spain) according to the stoichiometric ratio.[14] The synthesized cordierite and a commercially available glass powder (Schott glass package, Germany) were coated with alumina by a precipitation method. The flow chart of the process is shown in Figure 1. Aluminium nitrate nonahydrate ($Al(NO_3)_3{\cdot}9H_2O$) were used as a source of Al_2O_3. An amount of 25g of cordierite was first ultrasonically dispersed into 500ml distilled water. Urea (A. R. grade, Riedel-deHaen) was added as precipitating agent. When the pH value of the suspension was around 9, $Al(NO_3)_3{\cdot}9H_2O$ was added according to the weight ratio Al_2O_3:cordierite = 10:90. The suspension was then heated at 90°C for 4 hours under magnetic stirring. The resulting coated cordierite powder was then washed with distilled water to remove the remaining NO_3^- and urea, followed by washing with absolute ethanol in an attempt to alleviate agglomeration among the powder particles. The obtained paste was ultrasonically re-dispersed into absolute ethanol, and dried at 60°C for 48 hours. The dried powder was finally calcined at 500°C for 1 hour. The same procedure was followed when preparing coated glass powders.

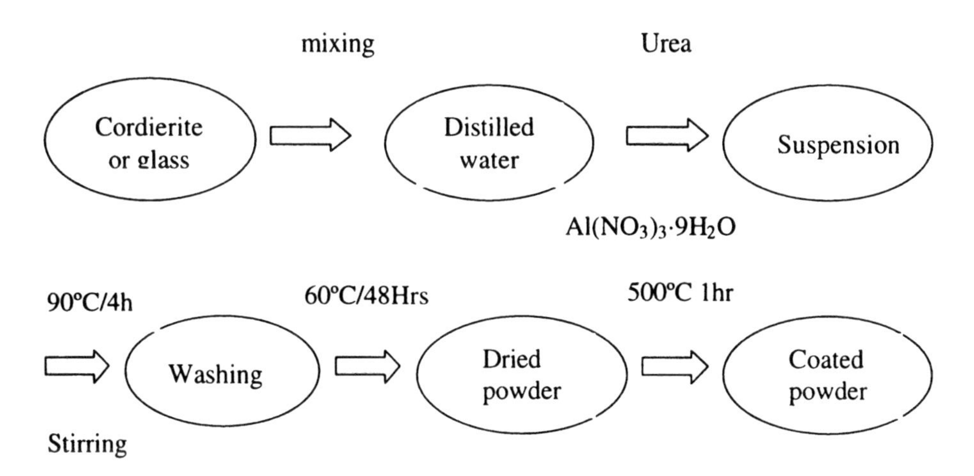

Figure 1. Synthesis the coated cordierite and glass powders

Characterization of the powders

The particle size distributions and mean particle sizes of cordierite and glass powders were measured using a Coulter LS230 instrument (Coulter Electronics Limited, UK). Surface morphology of the coated and un-coated powders were observed by a High Resolution Electron Microscope (HREM) (Hitachi H9000-NA, Japan). Electrophoretic measurements of coated and un-coated powders under different pH values were performed on a zeta potentiometer (Coulter Delsa 440 SX, USA), using a solution of 0.001M KCl as background electrolyte. XRD patterns of the coated powders (calcined at 500°C for 1h) and un-coated powders were characterized with an X-ray diffractometer (D/MAX-C, Rigaku, Japan), using Cu Kα radiation. The microstructures of coated and un-coated powders were observed by a scanning electron microscope (S-4100, Hitachi, Japan).

Preparation of the suspensions and green tapes

Suspension with different solids contents were prepared by mechanically mixing equal weight proportions of the coated cordierite and glass powders into distilled water in the presence of 1.0-wt% (relative to the solids) of Dolapix CE 64 (Zschimmer & Schwarz, Germany) as dispersing agent. After being stirred for 30 min for uniform distribution of the components, the as-obtained slurries were deagglomerated in polyethylene bottles using ZrO_2 balls for 6 hours. The binder (Duramax B-1080, Rohm and Haas, USA) was then added to the as-prepared suspensions. After that, a de-airing and conditioning step was performed for further 6 hours by rolling the slips in the milling container without balls. The green tapes were prepared by casting the as-prepared suspensions onto a plastic film (Polypylene, Western Wallis, USA) using a laboratory tape caster (Elmetherm, Oradour Sur Vayres, France). A gap of 300 μm under the blade and a fixed casting speed of (0.6cm s^{-1}) were selected for tape casting processing. The process was carried out at the room temperature (≈ 20°C) and humidity (≈ 45-50% RH). The sintered tapes were obtained by heating the green tapes at 1100°C for 2hrs. The microstructures of green tapes and sintered tapes derived from coated and un-coated powders were also observed by a scanning electron microscope (S-4100, Hitachi, Japan).

RESULTS AND DISCUSSION

Characterization of the powders

Figures 2a and 2b show, respectively, the TEM morphologies and HREM patterns of the coated cordierite powders after calcination at 500°C for 1h. A core-shell particle structure is observed in the coated powders as shown in Figure 2a. This structure could be further confirmed by HREM imaging where a distinct arrangement between the regular diffraction lattices of polycrystalline cordierite and the random diffraction patterns of the amorphous alumina surface layer could be observed. The thickness of the amorphous alumina layer is around 10 nm. The amorphous nature of the coating layer could be also deduced from the XRD patterns (Figure 3) of the coated and un-coated cordierite powders, which do not show significant differences. Precipitation also occurred at the surface of the glass particles. However, since in this case both the core particles and surface layers are amorphous, the HREM technique was not helpful in detecting any distinct boundary between the core particles and surface layers.

Figure 4a and 4b show the SEM morphologies of the coated cordierite and glass powders, respectively. It is interesting to note that a worm-like layer covers the surfaces of the core particles. On the other hand, the un-coated cordierite and glass powders (Figure 4c and 4d, respectively) show relatively smooth surface compared with the coated ones. The reasons for the formation of the worm-like layer are not clear at the present. Further investigation will be necessary to clarify this point.

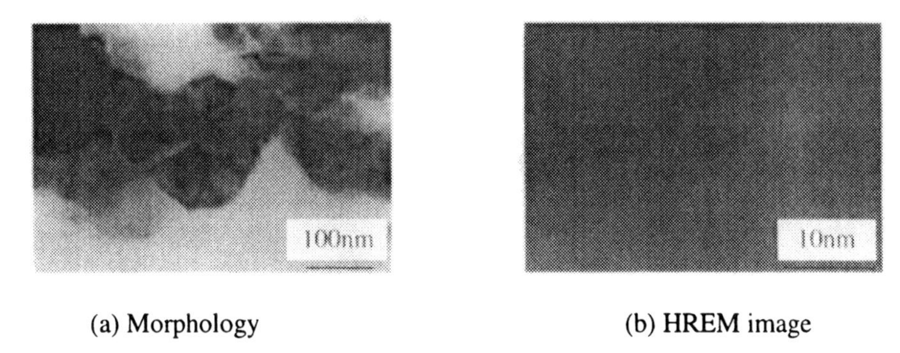

(a) Morphology (b) HREM image

Figure 2. TEM images of coated cordierite particles.

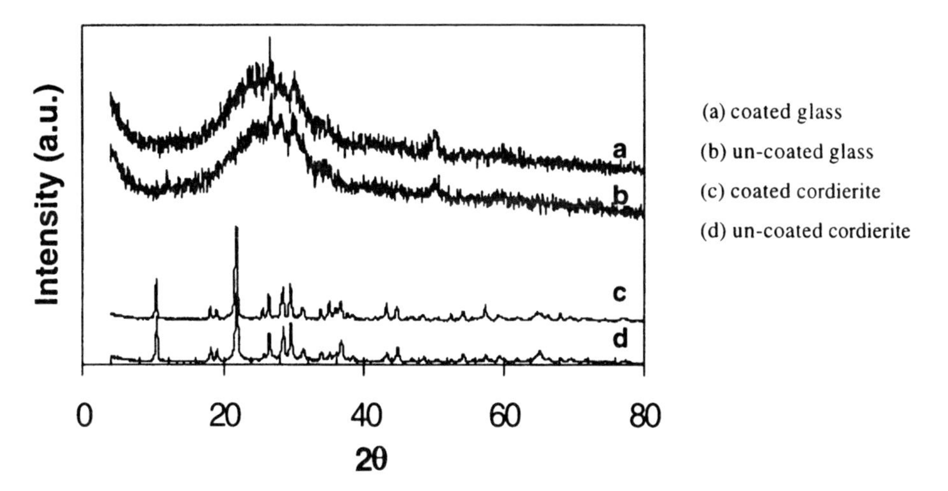

Figure 3. XRD results of the coated or un-coated powders

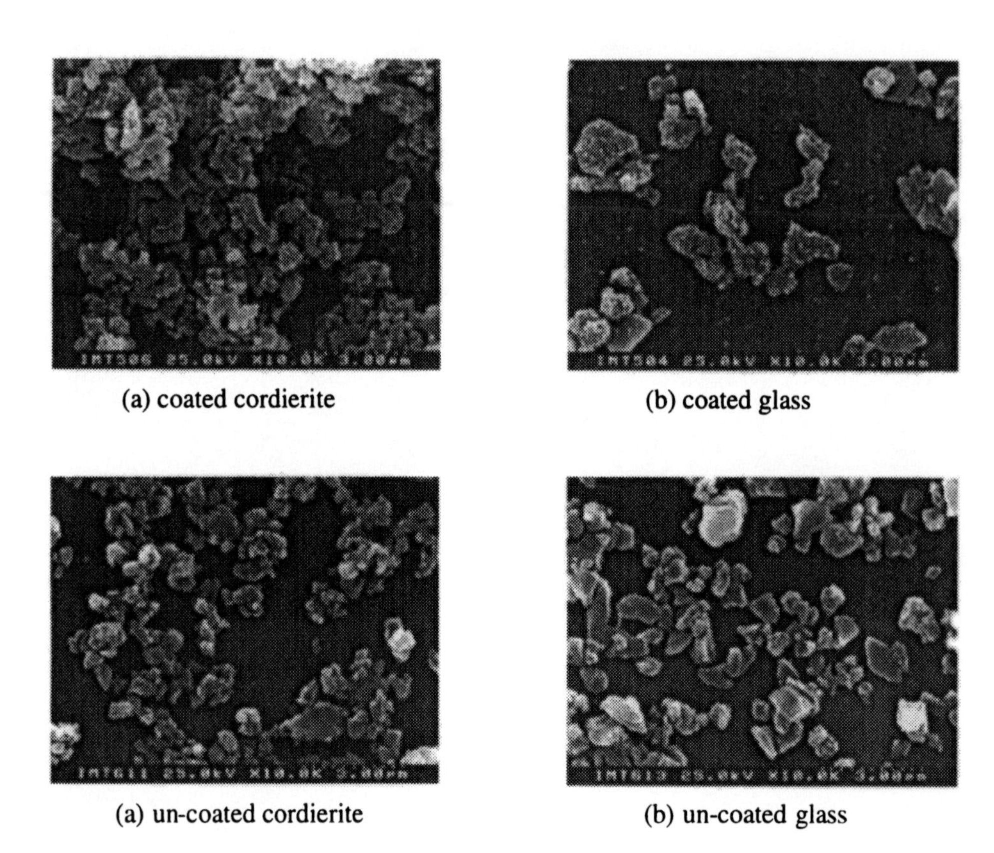

(a) coated cordierite (b) coated glass

(a) un-coated cordierite (b) un-coated glass

Figure 4. SEM morphologies of the coated and un-coated powders

Obviously, the coating layer leads to larger particles. Figures 5a and 5b show the particle size distributions of the coated and un-coated cordierite and glass powders, respectively. Average particle sizes of 1.92 μm and 0.82 μm were measured for the coated and un-coated cordierite powders, respectively, whereas 2.59 μm and 1.85 μm were the registered values for the glass powders, respectively. However, the significant differences in the average particle sizes measured for the coated and un-coated powders can not be attributed only to the increase of particles' diameter, as the results of TEM suggest. Particle size measurements were carried out at a pH value around 6, which is closed to the isoelectric point (IEP) of the coated powders. Under these conditions, agglomerates could be easily formed and contributed to the larger average particle sizes measured.

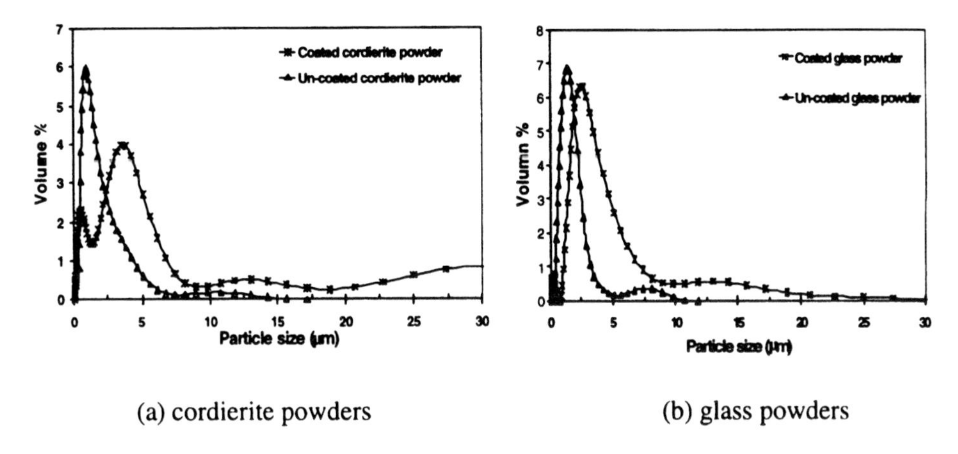

(a) cordierite powders (b) glass powders

Figure 5. Particle size distributions of coated and un-coated powders

Properties of green and sintered tapes

In the present work, the maximum values of solids loading that enabled to obtain crack-free green tapes were 70-wt% and 80-wt%, for un-coated and coated powders, respectively. The SEM morphologies of the green and sintered tapes prepared from suspensions containing 70-wt% of un-coated powders and 80 wt-% of coated ones are shown in Figures 6a to 6d, respectively. The tapes prepared from coated system show denser microstructures and a homogeneous distribution of the glass and cordierite phases, whereas agglomerates and large holes could be observed in the tapes prepared from the un-coated powders.

Finally, just as a note, the measured values of dielectric constants are almost the same for the substrates prepared from both coated and un-coated powders (≈5, at 1MHz). However, the substrates fabricated from the coated powders exhibited a slightly lower dielectric loss (0.001, 1 MHz), compared with that from the un-coated powders (0.002, 1 MHz).

Based on the above results, the surface modification of the cordierite and glass powders by coating them with a thin Al_2O_3 layer improves the electrophoretic properties of the particles and rheological behavior of the suspensions. The enhanced dispersion ability of the coated powders enables to more easily process the composite particles in aqueous media and produce denser-packed green and sintered substrates by tape casting.

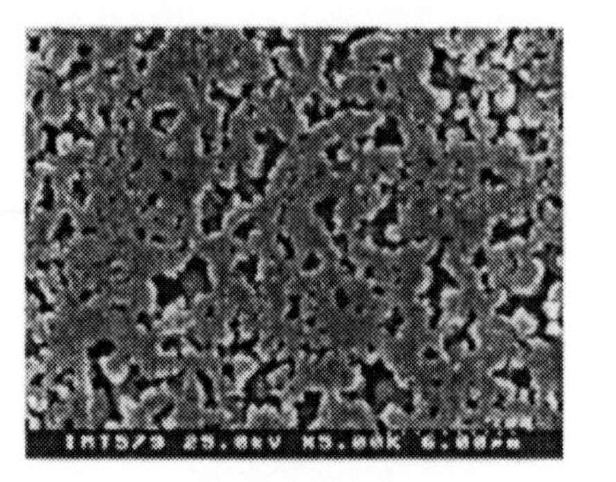

(a) green tape prepared from a 70-wt% un-coated powders suspension

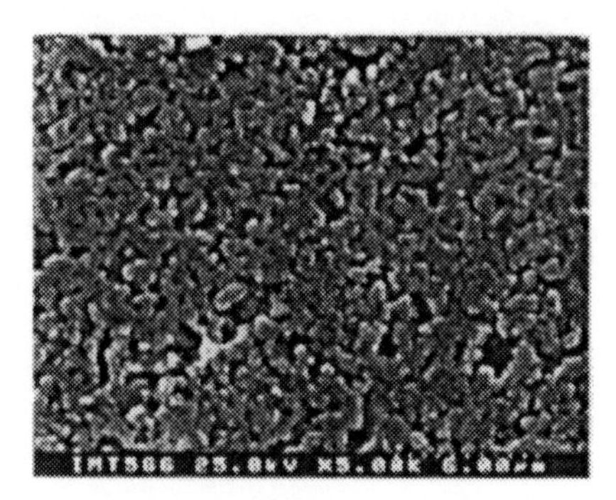

(b) green tape prepared from a 80-wt% coated powders suspension

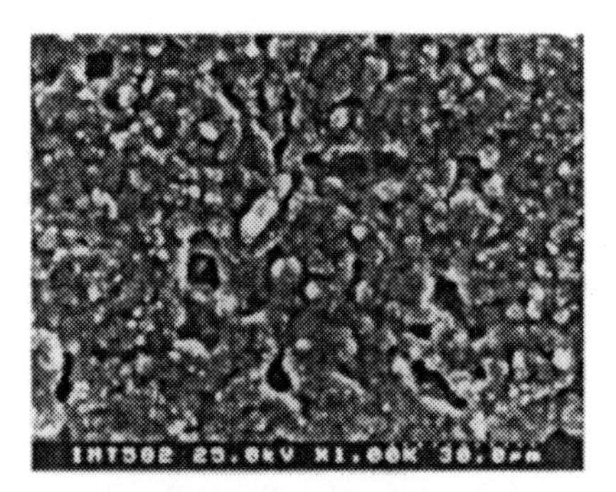

(c) sintered tape prepared from a 70wt% un-coated powders

(d) sintered tape prepared from a 80-wt% coated powders

Figure 6. The microstructural images of the green tapes and sintered tapes

CONCLUSIONS

Cordierite and glass powders were coated with an alumina precursor by using a heterogeneous precipitation method in the presence of urea. The surface layer endowed the coated powders the surface character of alumina, making the composite particles more easily to disperse in water. Accordingly, suspensions with higher solids volume fraction could be prepared from coated powders leading to denser and more homogeneous green and sintered tapes, compared with un-coated powders. These enhanced microstructural features lead to improve dielectric properties of the sintered substrates.

ACKNOWLEDGEMENTS

The first and second authors are grateful to Fundação Para Ciência e a Tecnologia of Portugal for the grants in the frame of PRAXIS program.

REFERENCES

1. R. R. Tummala, "Ceramic and Glass-Ceramic Packaging in the 1990s," *J. Am. Ceram. Soc.*, **74** [5], 895-908 (1991).

2. J. U. Knickerbocker, G. B. Leung, W. R. Miller, S. P. Young, S.A. Sands, and R. F. Indyk, "IBM System/390 Air-Cooled Alumina Thermal Conduction Module," *IBM J. Res. Devel.*, **35** [3], 330-340 (1991).

3. J. F. MacDowell, G. H. Beall, "Low K Glass-Ceramics for Microelectronic Packaging," *Ceramic Transactions*, Vol. 15, The American Ceramic Society, Inc, 259-277, 1990.

4. J. U. Knickerbocker, "Overview of the Glass-Ceramic/Copper Substrate —A High-Performance Multilayer Package for the 1990s," *Am. Ceram. Soc. Bull.*, **71** [9], 1393-1401 (1992).

5. S. H. Knickerbocker, A. H. Kumar, and L. W. Herron, "Cordierite Glass-Ceramics for Multilayer Ceramic Packaging," *Am. Ceram. Soc. Bull.*, **72** [1], 90-95 (1992).

6. T. P. Hyatt, "Electronics: Tape Casting, Roll Compacting," *Am. Ceram. Soc. Bull.*, **74** [10], 56-59 (1995).

7. S. Mei, J. Yang, J. M. F. Ferreira, "Optimization of Parameters for Aqueous Tape-Casting of Cordierite-Based Glass Ceramics by Taguchi Method," submitted to *Mater. Sci. & Eng.*

8. M. Descamps, M. Mascart, B. Thierry, and D. Leger, "How to Control Cracking of Tape-Cast Sheet," *Am. Ceram. Soc. Bull.*, **74** [3], 89-92 (1995).

9. J. Yang, J. M. F. Ferreira, W. Weng, "Dispersion Properties of Silicon Nitride Powder Coated with Yttrium and Alumina Precursors", *J. Coll. Interf. Sci.* **206,** 274-280 (1998).

10. J. Yang, F. J. Oliveira, R. F. Silva and J. M. F. Ferreira, 'Pressureless Sinterability of Slip Cast Silicon Nitride Bodies Prepared from Sol-Gel Coated Powders," *J. Europ. Ceram. Soc.*, **19**, 433-439 (1999).

11. E. P. Luther, F. F. Lange, and D. S. Person, ""Alumina" Surface Modification of Silicon Nitride for Colloidal Processing," *J. Am. Ceram. Soc.* **78** [8], 2009-2014 (1995).

12. W. H. Shih, L. L. Pwu, and A. A. Tseng, "Boehmite Coating as a Consolidation aand Forming Aid in Aqueous Silicon Nitride Processing," *J. Am. Ceram. Soc.* **78** [8], 1252-60 (1995).

13. E. Liden, M. Persson, E. Carlstrom and R. Carlsson, "Electrostatic Adsorption of a Colloidal Sintering Agent on Silicon Nitride Particles," *J. Am. Ceram. Soc.,* **74** [6], 1335-39 (1991).

14. S. Mei, J. Yang, J. M. F. Ferreira, "Cordierite-Based Glass-Ceramics Processed by Slip Casting," *J. Euro. Ceram. Soc.,* **21** [2], 185-193 (2001).

15. G. A. Parks, "The Isoelectric Points of Solid Oxides, Solid Hydroxides, and Aqueous Hydroxy Complex Systems", *Chem. Rev.*, **65**,177-198 (1965).

16. G. Tari, "Advances in Colloidal Processing of Alumina", Ph.D dissertation, University of Aveiro (1999).

17. G. Tarì, J. M. F. Ferreira, and O. Lyckfeldt, "Influence of the Stabilizing Mechanism and Solid Loading on Slip Casting of Alumina", *J. Europ. Ceram. Soc.*, **18** (1998) 479-386.

18. S. Mei, J. Yang, and J. M. F. Ferreira, "Effect of Dispersant Concentration on Slip Casting of Cordierite-Based Glass-Ceramics" Submitted to *J. Coll. Interf. Sci.*

19. Hayashi, "Surface Chemistry of Ceramic Shaping Processes", Annual Report for Overseas Readers, Japan Fine Ceramics Association (1991).

EMBEDDING A PASSIVE MATERIAL LAYER IN LOW TEMPERATURE CO-FIRED PACKING

E. R. Twiname
Richard E. Mistler, Inc.
1430 Bristol Pike
Morrisville, PA 19067

G.L. Messing
The Pennsylvania State Univ.
Materials Research Lab.
State College, PA 16802

C.A. Randall
The Pennsylvania State Univ.
Materials Research Lab.
State College, PA 16802

ABSTRACT

Integral substrates, the inclusion of passive device material layers within multilayer ceramic substrates or packages, are predicted to be the next evolutionary step in size reduction for packaged electronic components. This work investigates co-sintering and magnetic property issues encountered during the fabrication of integral substrates composed of low temperature co-fired ceramic (LTCC) insulators and nickel copper zinc ferrite inductor material. Co-sintering incompatibilities including glass redistribution, shrinkage rate, and sintering onset temperature were quantified. LTCC and ferrite densification temperatures differed by over 100°C. The depth of glass redistribution from the LTCC into the ferrite layer was proportional to $t^{1/2}$ as predicted by the Rideal-Washburn equation.

Adjustment of ferrite densification by the addition of lead silicate glass sintering aids was explored, as was the effect of the glass additions on ferrite magnetic properties. Glass additions of 3, 5, and 8 vol.% were used to modify ferrite densification. Additions of 8 vol.% of low T_g glasses were found adequate for the production of LTCC/ferrite integral substrates. Ferrites with sintering aid glass additions had magnetic permeabilities of 60 to 80 H/mm and quality factors of over 100.

INTRODUCTION

The high levels of interest and resulting technological advances in low temperature co-fired ceramics (LTCC) have resulted in the development of insulating, resistive, capacitive and inductive materials that densify in the same peak temperature range, 800°C - 950°C. These low sintering temperatures permit the use of high conductivity metallization such as copper, silver and nickel. The demand for lighter, cheaper and faster electronic devices continues to fuel demand for smaller electronic packages. This pressure for miniaturization, combined with the ability to co-sinter passive device materials, has spurred the development of array and network passive components that use package surface mount area more efficiently. Significant increases in passive component densities have been achieved through area array packaging technology as well as the joining of discrete passive components such as a resistor and capacitor, capacitor and inductor, or capacitor and varistor. It has also been shown that significant package size reduction may be achieved by embedding an entire layer of passive material within the multilayer substrate. This "integral substrate" approach to passive integration is predicted to be the most cost effective design approach for passive component densities in excess of three per square centimeter of a single component type.

Inductors, formed into multilayer surface mount components, are used in microelectronic packaging applications such as frequency filtering. While recent developments in planar inductor design address the effects of circuit configuration on inductor performance, detailed evaluation of ferrite inductor material integration into LTCC packaging is noticeably scarce. A number of low frequency device opportunities exist for LTCC packages with integral inductors such as power electronics[1]. The current work investigates co-sintering phenomena involved with the fabrication of integral inductor substrates using commercial LTCC systems as well as the effect of sintering aid additions on ferrite densification and magnetic properties.

The co-sintering of two or more dissimilar materials to yield a monolithic composite requires a balance of static and dynamic mechanical, chemical and thermodynamic phenomena to obtain a composite structure without defects and without adversely affecting properties. The avoidance of unwanted mixing[2] or chemical reactions and the matching of densification behaviors to avoid stress generation[3,4,5] is therefore necessary and has been investigated for a number of packaging applications.

Strength of a composite body is increased by the elimination of porosity. Materials to be co-sintered must therefore reach the desired density through the same heating profile. Porosity is undesirable in integral inductor substrates since it will degrade both mechanical and magnetic properties. Chemical reactions during co-sintering must also be avoided since new phase formation removes some of the desired phases and alters the stoichiometry of the surrounding area, affecting mechanical, electrical or magnetic properties. Interdiffusion, in a similar manner, alters the local distribution of ions. Diffusion, however, cannot be avoided when joining dissimilar materials due to the chemical potential gradient across the interface. Interdiffusion, therefore, must be minimized or tolerated to some extent since the thermodynamic driving force for diffusion cannot be removed. Interdiffusion is a concern with high glass content LTCC materials since it is typically seen that ions diffuse more rapidly through amorphous materials than through crystalline phases due to the more open structure.

EXPERIMENTAL PROCEDURE

Ferrite A - $(NiO)_{0.20}(CuO)_{0.16}(ZnO)_{0.16}(Fe_2O_3)_{0.48}$ and ferrite B - $(NiO)_{0.23}(CuO)_{0.15}(ZnO)_{0.14}(Fe_2O_3)_{0.48}$ were prepared by wet mixing of component oxides in a plastic container with water and YSZ media, drying and calcining at 700°C for 4 hours. Hard aggregates were removed and ferrite surface area increased to 9.5-10 m^2/g by attrition milling in water using YSZ media in a polymer lined cup.

Two commercially available LTCC tapes were used, a ceramic filled glass (CFG) and a devitryfying glass (DVG). Investigation using both a devitrifying glass and a ceramic filled glass allowed flexibility in material-system design. The CFG is a lead-borosilicate glass frit (T_g ~ 820°C) combined with 1-5 μm alpha alumina powder whereas the DVG material is a calcium borosilicate glass frit that crystallizes into CaB_2O_4 (calcium borate), $CaSiO_3$ (wollastonite) and $Ca_3Si_2O_7$ (rankinite) upon sintering leaving some residual amorphous phase. Three sintering aid glasses, designated SCB-1, SCB-2 and SCB-3 (SEMCOM Company, Toledo, OH) with measured T_g's of 520°, 390° and 390°C respectively (DSC 2920, 1600 DTA attachment, TA Instruments), were used to modify ferrite sintering temperature.

Ferrite tapes were cast using a methyl ethyl ketone/ethanol solvent system, blown menhaden fish oil dispersant, polyvinyl butyral binder and butyl benzyl phthalate plasticizer. Pellets (6.35 mm OD) and toroids (12.1 mm OD, 5.35 mm ID) were uniaxially pressed using a steric acid lubricant. Pellets were pressed from LTCC materials after organic removal at 500°C for one hour.

Crystalline phases were determined by X-ray powder diffraction (Pad V X-ray diffractometer, Cu K_α source, Scintag, Inc.). Specific surface area was measured using BET surface area analysis (Gemini BET, Micromeritics). Microstructure and atomic composition was analyzed using a scanning electron microscope with EDS attachment (SEM: S-3500N, Hitachi, Ltd., EDS: IMIX Digital Spectrometer,

PGT, Inc.). Since green formation processes have only minor effects on sintering temperature, pellets were used instead of tape cast layers for shrinkage rate analysis in a thermo-mechanical analyzer (TMA) (TMA-50, Shimadzu Corp.). Organic removal and pellet sintering was done in a box kiln while substrates were sintered, after organic removal, in a continuous infrared tunnel kiln (Model LA-306, Radiant Tech. Corp.) to achieve higher heating rates.
Fired densities were measured by immersion in kerosene. Magnetic inductance and quality factor were measured from wire wound toroids using an impedance analyzer (4194A Impedance / Gain Phase Analyzer, 16047D fixture, Hewlett Packard). Toroids were wound with 6.75 or 10.75 turns of 610 μm (0.024 inch) diameter enameled copper wire. Lead length from the toroid was kept at 38 mm (1.5 inch).

RESULTS AND DISCUSSION

Material Interactions:

Theoretical densities of the prepared ferrites were estimated from the six major spinel peaks between 15° and 90° 2Θ. Ferrite theoretical density was calculated to be 5.4 g/cm^3 for both ferrite compositions, which agrees well with measured values. X-ray diffraction analysis of co-sintered laminates was compared to that of individual materials to identify new phases (Fig. 1). Laminates used for this analysis were sintered at 950°C, above the recommended LTCC sintering temperatures, to promote phase formation and were ground prior to analysis. Since every diffraction peak from the laminate corresponds to a phase in the component materials, it was concluded that no new phases are formed between the LTCC and ferrite materials up to a sintering temperature of 950°C.

The extent of ferrite cation diffusion into the LTCC during co-sintering is important since addition of transition metal ions may affect the insulating and dielectric properties of the LTCC host and loss of these ions affect magnetic properties of the ferrite. Nakano et. al.[7] also report that the presence of Bi_2O_3, WO_2 or PbO in NiCuZn ferrite has deleterious effects on silver electrodes during co-sintering. SEM/EDS analysis performed on DVG / ferrite laminates showed a change in LTCC microstructure near the ferrite interface (Fig. 2a) with an observable diffusion region and increased porosity as compared to the bulk LTCC. Selected area EDS analysis of a DVG / ferrite laminate (Fig. 2b) shows the presence of iron, copper and zinc in the DVG layer. Selected area EDS of a CFG/ferrite laminate (Fig. 3) did not detect ferrite cations in the CFG host and there was no observable change in CFG microstructure. It is noticeable, however, that glass wicked from the CFG into the ferrite layer during co-sintering (Fig. 4). Glass redistribution can be seen as a region of high ferrite density since the CFG glass acted as a sintering aid for the ferrite. Redistribution from the DVG layer into the ferrite had the opposite effect, separating the ferrite particles and inhibiting sintering, perhaps due to the concurrent devitrification process. Due to the similarity between CFG and DVG densification temperatures (Fig. 5), greater diffusion of ferrite cations into the DVG, and inhibition of ferrite sintering by DVG redistribution, further work was done on the CFG alone.

Glass Redistribution:

Glass penetration from an LTCC host material into a layer of adjacent material is a co-sintering challenge to overcome for the embedding of any material into integral substrates. This redistribution of glass has also been seen as a detriment to conductor integrity of LTCC multilayer packages[2] even though the conductor material already contained a glass phase.

Glass redistribution was observed in CFG - ferrite laminates fired at temperatures above 800°C (Fig. 6). At temperatures above 900°C, densification of the ferrite layer obscures the penetration region. During co-sintering, porosity within the packed CFG bed is filled with the glass as the LTCC layer densifies, primarily by viscous flow[4,5]. When porosity is eliminated from the LTCC body, the adjacent pore network of the ferrite interface can also be filled with glass due to capillary forces. This penetration into the ferrite layer does not take place until around 800°C when the CFG porosity is eliminated. Since porosity becomes non-continuous at densities above approximately 92%, glass redistribution was not expected to be a significant problem for ferrites that reached this density by 800°C.

Densification in the ferrite interfacial region due to CFG glass redistribution acted to constrain the ferrite layer during ferrite densification. Effectively placing the interior of the ferrite layer in tension during further densification, CFG glass redistribution resulted in separation of the ferrite layer normal to the interfaces (Fig. 7). This separation is consistent with other laminate layers sintered in tension[6].

The most appropriate model for this capillary flow behavior is the Rideal-Washburn Equation, which defines penetration length (λ) as:

$$\lambda = \left(\frac{r \cdot \gamma_{LV} \cdot \cos\theta \cdot t}{2\eta} \right)^{\frac{1}{2}} \tag{1}$$

where r is average pore radius, γ_{LV} is glass surface tension, θ is the wetting angle, t is time and η is glass viscosity. Despite discrepancies between experimental phenomena and assumptions made in the derivation of the Rideal-Washburn Equation, such as non-uniform pore size and densification of the ferrite layer due to redistribution, average glass penetration depths measured from laminates co-sintered at 825°C and 850°C were found to follow the $t^{1/2}$ dependence predicted (Fig. 8).

Co-sintering:

TMA analysis showed the axial shrinkage behavior of the CFG and ferrite materials to be separated by over 100°C (Fig. 5) with CFG shrinkage ceasing by approximately 800°C while ferrite samples continue to densify at temperatures over 900°C. Since the intent of this research is to explore co-sintering of NiCuZn ferrites with a commercially available LTCC, adjustments to shrinkage behavior centered exclusively on the ferrite body. Since previously reported work[7] shows that densification of a similar ferrite composition is not complete at 900°C for a ferrite surface area of 20 m^2/g, particle size reduction alone was not considered an effective tool for lowering ferrite sintering temperature in this application.

The addition of sintering aid glasses to the ferrite material was investigated since a change of densification mechanism from solid-state to liquid phase sintering would lower the temperatures at which densification and shrinkage occurred as seen in the CFG redistribution region. The use of glass-ferrite composite materials has also been reported to dampen the effect of external stress on magnetic properties[8,9] by creating a net tensile stress state within the ferrite grains due to differences in thermal expansion between glass and ferrite during cooling. This same effect has been shown to increase ferrite strength due to the resulting compressive stress in the glass[10].

Additions of 3,5 and 8 vol.% glass were made to the ferrite batches. TMA analysis of the glass added ferrites (Fig.'s 9, 10) showed that additions of SCB-2 or SCB-3 lowered shrinkage temperatures according to the amount added while additions of SCB-1 or CFG had an opposite effect. CFG additions increased the ferrite sintering temperatures and introduced a shoulder in the shrinkage plot indicating a second region of sintering shrinkage. Since the redistribution of glass from the CFG layer promoted sintering in earlier experiments, the increase in sintering temperature and the introduction of a second sintering region with intentional CFG addition are attributed to the alumina particles inherent to the CFG material.

TMA Conversion Technique:

Baseline corrected temperature-displacement plots from TMA analysis were transformed into temperature-density plots by assuming a linear relationship between axial shrinkage and relative density. A series of pellets was prepared from each composition and sintered in a box furnace at temperatures within the respective material densification regions. Axial shrinkage and apparent density were measured for these pellets and the various shrinkage-density points plotted. Least squares regression was performed for each data group to obtain the shrinkage - density conversion equation. The 0% shrinkage point, or y-intercept, was defined prior to regression as equal to the average relative density of the pellet group. Reported densities were used as ρ_{Th} for sintering aid glasses while the maximum measured density of 3.26 g/cm^3 was used for the CFG. ρ_{Th} for composite materials was calculated by weighted average. Regression yielded a line equation in the form of:

$$(\%\rho_{Th}) = (\text{Slope})(\%\ \text{Axial Shrinkage}) + (\text{Green}\ \%\rho_{Th}) \qquad (2)$$

Axial shrinkage data, acquired from the TMA, was converted to densification curves using the appropriate conversion equation for each material (Fig. 11). CFG densification data was gathered by standard techniques since the CFG material displayed creep in the TMA. Creep was not seen in the ferrite or ferrite/glass composite samples.

Resulting densification curves show that, at a temperature of approximately 800°C, the CFG host material was 98.3 % dense while the ferrite was less than 80% dense (Fig. 12), allowing glass redistribution from the CFG into the continuous pore network of the ferrite layer. De-densification of the CFG material above 800°C, was not investigated. Additions of 8 vol.% SCB-2 or SCB-3 yield ferrite bodies over 92% dense prior to 800°C. Glass redistribution was avoided by the removal of continuous ferrite porosity above 92% ρ_{Th}.

While shrinkage behavior was not matched precisely, and total shrinkage strain not addressed, addition of 8 vol. % SCB-2 or 8 vol. % SCB-3 to the ferrite layer aided the production of crack-free, co-sintered CFG / ferrite laminates (Fig. 13). The absence of ferrite layer separation normal to the interfaces supports the previous assertion that glass redistribution has been avoided.

MAGNETIC PROPERTIES

Ferrite B [$(NiO_{0.23}CuO_{0.15}ZnO_{0.14})(Fe_2O_3)_{0.48}$] displayed a similar quality factor (Q) and lower permeability than ferrite A [$(NiO_{0.20}CuO_{0.16}ZnO_{0.16})(Fe_2O_3)_{0.48}$]. A plateau exists for Q in the 5 MHz frequency region for both formulations. The 1 MHz permeabilities (μ') were higher and Q values lower than literature values[11] for parts sintered to 93 % density, at 875°C for 4 hours. Permeabilities of 91 H/mm and 78 H/mm were measured at 1 MHz for ferrite A and ferrite B, respectively, as compared to literature values of 75 and 60 H/mm. Quality factors of approximately 110 were measured for both ferrites. Ferrite A grain sizes were estimated using SEM analysis and found to be 1-2 μm, 2-5 μm and 4-5 μm for sintering temperatures of 875°, 925° and 950°C respectively.

Permeability (μ') and loss (μ'') were measured as a function of ferrite density to determine the effect of porosity on magnetic properties. Ferrite density was adjusted by adjusting specific surface area and sintering profiles. Magnetic properties were measured on the 5 MHz plateau region. Five-megahertz permeability was found to gradually increase with increasing density to a maximum of 213 H/mm at 99% density while loss was seen to remain <1 H/mm up to approximately 92% of theoretical density above which it increased to a maximum of 5.42 H/mm at 99 % density (Fig.'s 14, 15). Similar behavior was seen for frequencies below (1 MHz), within (5MHz) and above (10MHz) the plateau region. The increase in μ' is attributed to the decrease in porosity as well as increasing grain size. Annealing of ferrite samples at 700°C for 6 hrs. in air was not seen to significantly affect magnetic properties. Change of cation valence is therefore not considered an issue for this study.

The degradation of magnetic properties has been attributed to copper on the grain boundaries in NiCuZn ferrites by other authors[7,12]. Copper residing on ferrite grain boundaries causes compressive stress that, through magnetostriction, degrades the magnetic properties of the ferrite. The increase in particle-particle contact during final stage sintering may be a factor in the high losses seen in the high-density ferrite parts. This loss mechanism was not investigated by experimentation.

The use of sintering aids resulted in magnetic losses between 0.5 and 0.7 H/mm up to densities of 98.4% of theoretical (Fig. 15). Permeability values for glass added ferrites remained between 57 and 82 H/mm (Fig. 14), similar to the low-density, pure ferrite samples reported above. These values are also similar to the 75-100 H/mm values shown in literature[11].

Though both μ' and μ'' were decreased by the addition of sintering aid glasses, the relative decreases were not equal. Permeability decreased by a factor of three, from 183.6 H/mm in the 94.5% dense pure ferrite to 62.1 H/mm in the 97% dense, 5vol. % SCB-3 added sample, while loss was decreased by a factor of eight, from 4.56 H/mm to 0.565 H/mm for the same samples. The difference in the effect of glass additions on permeability and loss resulted in a net increase in quality factor from 37 in the 93 % dense pure ferrite to 113 in the 8 vol% SCB-3 added ferrite.

The creation of a net tensile stress on the ferrite grains, discussed previously, may overshadow detrimental effects from phenomena such as CuO_{1-x} on the ferrite grain boundaries. Sintering aid glass additions resulted in ferrite magnetic properties similar to those found with <92% dense ferrite parts. Increases in both permeability and loss previously seen with increasing ferrite density were not observed in ferrite parts containing glass.

CONCLUSIONS

Nickel copper zinc ferrites were prepared from component oxides and used to investigate co-sintering and magnetic property issues involved with embedding inductor layers into LTCC substrates. Two ferrite compositions, detailed above, were found to have theoretical densities of approximately 5.4 g/cm^3. Co-sintering of the ferrites with commercially available LTCC materials, CFG and DVG, revealed co-sintering incompatibilities in the areas of: densification behavior, ferrite cation diffusion, and glass redistribution into the ferrite layer. Diffusion of ferrite constituent cations was found to be less in the CFG than in the DVG.

Glass redistribution from the CFG into the ferrite promoted ferrite densification at the interfaces and resulted in crack-like separations within the ferrite layer normal to the CFG/ferrite interfaces. Penetration depth of the CFG glass at 825°C and 850°C displayed the square root dependence on time predicted by the Rideal-Washburn equation despite non-conformity to assumptions made in the derivation of that equation.

Lead-silicate glass sintering aids were found to be effective tools for matching sintering behavior, lowering ferrite sintering temperatures to more closely match that of the CFG. The highest glass addition used, eight volume percent, was not adequate to exactly match ferrite and CFG sintering temperatures but was adequate to avoid glass redistribution during co-sintering. Ferrite sintering aid additions of either 8v% SCB-2 or 8v% SCB-3 were adequate to produce CFG / ferrite laminates without separations in the ferrite layers.

REFERENCES

[1]M.C. Smit, J.A. Ferreira, J.D. van Wyk, M. Ehnasi, "Technology for manufacture of integrated planar LC structures for power electronic applications,' Proceedings of the 5th European Conference on Power Electronics and Aplications (EPE'93), Vol. 2, Brighton, pp. 173-178.

[2]W.S. Hackenberger, "Reduction of Sintering Damage and Differential Densification in a Low Temperature Cofired Electronic Substrate using Rate Controlled Sintering," Ph.D. Thesis, The Pennsylvania State University, University Park, PA, January 1997.

[3]P.Z. Cai, D.J. Green, and G.L. Messing, "Constrained Densification of Alumina/Zirconia Hybrid Laminates, II: Viscoelastic Stress Computation," J. Am Ceram. Soc. **80**[8] (1997) pp. 1940-1948.

[4]G.Q. Lu, R.C. Sutterlin, and T.K. Gupta, "Effect of Mismatched Sintering Kinetics on Camber in a Low-Temperature Cofired Ceramic Package," J. Am. Ceram. Soc., **76**[8] (1993) pp.1907-1914.

[5]J. Jean, C. Chang, "Cofiring Kinetics and Mechanisms of an Ag-Metallized Ceramic-Filled Glass Electronic Package," J. Am. Ceram. Soc., **80**[12] (1997) pp.3084-3092.

[6]P.Z. Cai, D.J. Green, and G.L. Messing, "Constrained Densification of Alumina/Zirconia Hybrid Laminates, I: Experimental Observations of Processing Defects," J. Am. Ceram. Soc. **80**[8] (1997) pp. 1929-1939.

[7]A. Nakano, and T. Nomura, "Multilayer Chip Inductors," Multilayer Electronic Ceramic Devices, Ceram. Trans. Vol. 97, Edited by J. Jean et. al., Proc 100th Ann. Meeting of the Am. Ceram. Soc., May 1999, pp.285-304.

[8]T. Yamaguchi, and M. Shinagawa, "Effect of Glass Addition and Quenching on the Relation Between Inductance and External Compressive Stress in Ni-Cu-Zn Ferrite-Glass Composites," J. Mat. Sci., **30** (1995) pp. 504-508.

[9]M. Kumagai, and Y. Ikeda, "Stress Insensitive Ferrite for Microinductors," Advances in Ferrites, Proc. 5th Int. Conf. on Ferrites, January 1989, Vol. 1, Trans Tech Publications, pp. 625-630.

10 S. Murayama, M. Kumagai, and Y. Ikeda, "High-Strength Ni-Cu-Zn Ferrite for Surface Mount Devices," Ferrites, Edited by T. Yamaguchi et. al., Proc. of the 6th Int. Conf. on Ferrites (ICF-6), Tokyo, September 1992, pp. 366-369.

11 T. Nomura, A. Nakano, "New Evolution of Ferrite for Multilayer Chip Components," Ferrites, 6th International. Conf. on Ferrites, Tokyo, September 1992, pp. 1198-1201.

12 M. Fujimoto, "Inner Stress Induced by Cu Metal Precipitation at Grain Boundaries in Low-Temperature-Fired Ni-Cu-Zn Ferrite," J. Am. Ceram. Soc., 77[11] (1994) pp. 2873-2878.

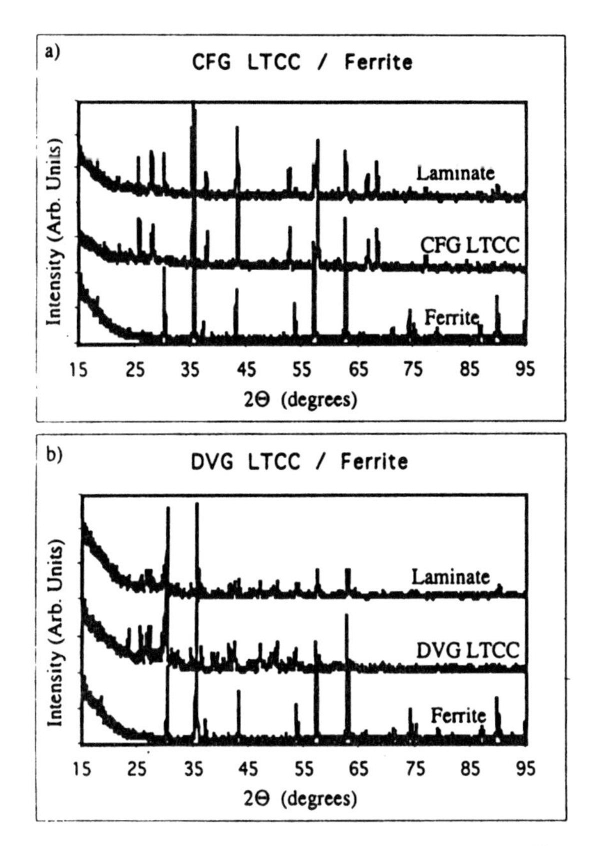

Figure 1: X-ray diffraction revealed no new crystalline phases formed during co-sintering of ferrite and a) ceramic filled glass LTCC or b) devitrifying glass LTCC. All samples sintered at 950°C for four hours.

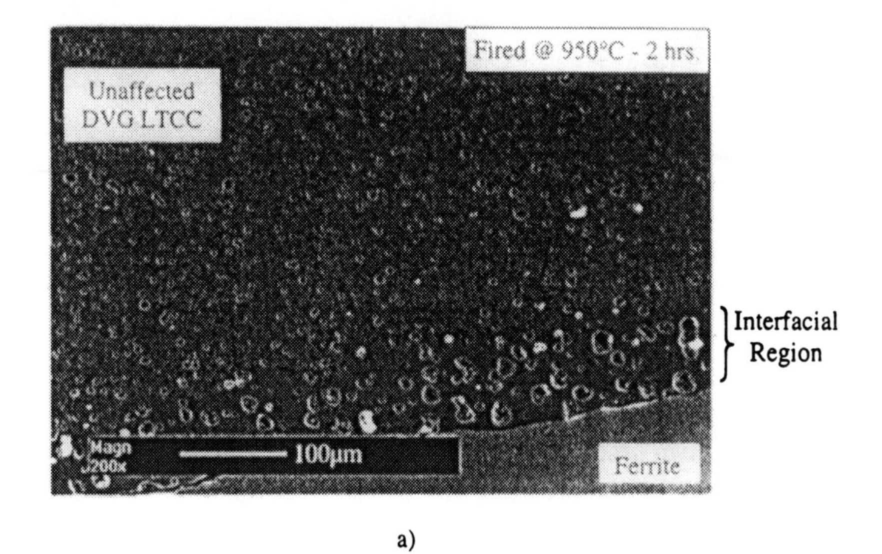

a)

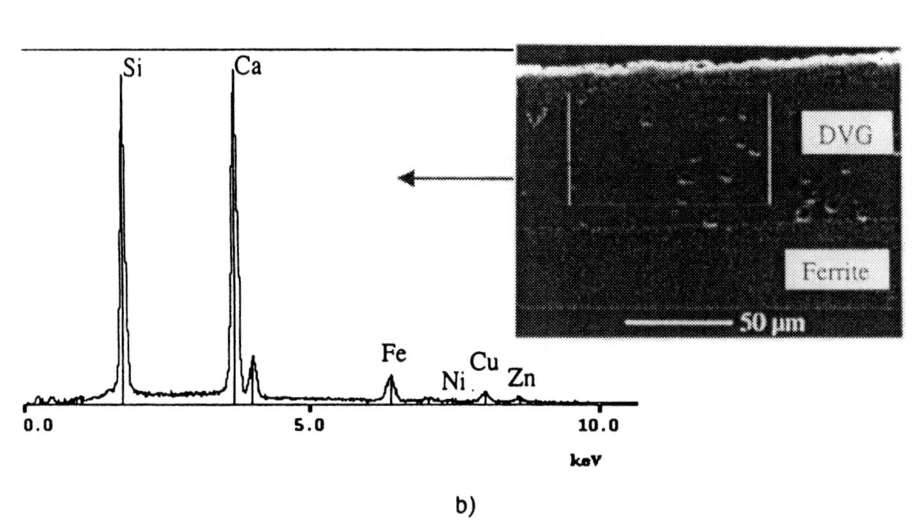

b)

Figure 2: SEM micrographs showing a) an increase in porosity in the DVG interfacial region after sintering at 850°C for 4 hours and b) EDS confirming the presence of ferrite cations in the LTCC material after sintering at 950°C for 4 hours.

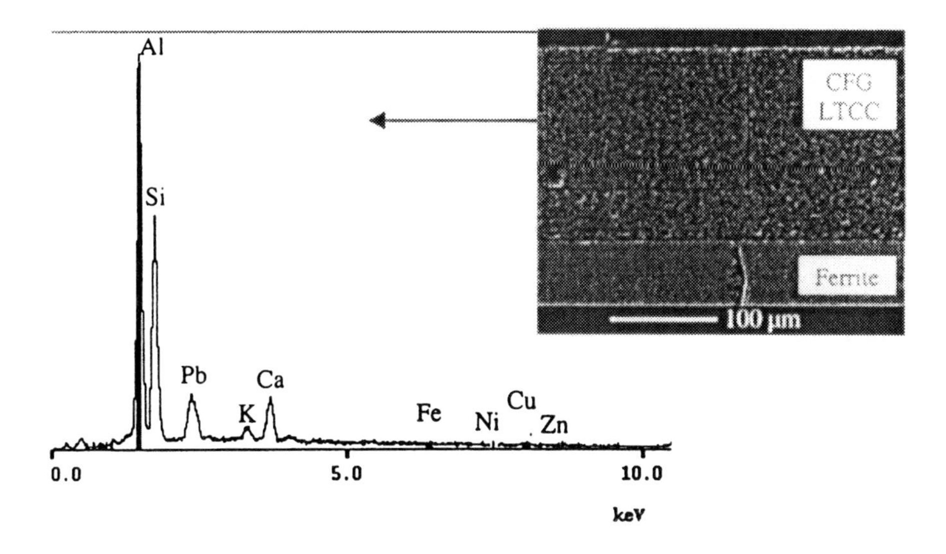

Figure 3: EDS of CFG / ferrite laminate shows the presence of little or no cations diffused from the ferrite into the CFG layer. Laminate fired for 4 hrs. at 950°C.

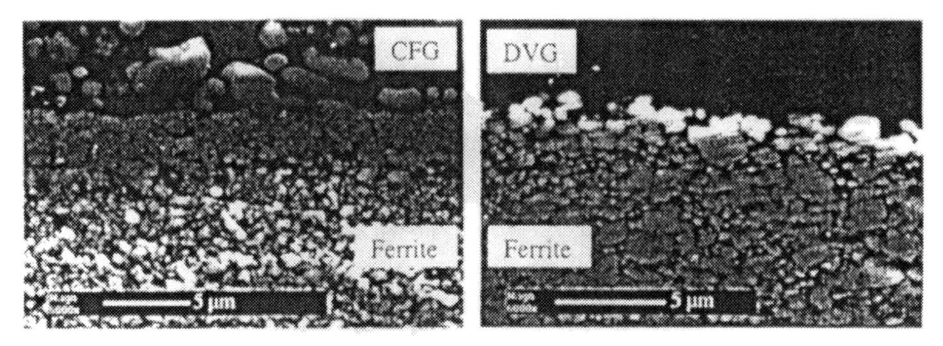

Figure 4: Glass redistribution a) from the CFG promoted sintering of the ferrite, b) from the DVG inhibited ferrite sintering.

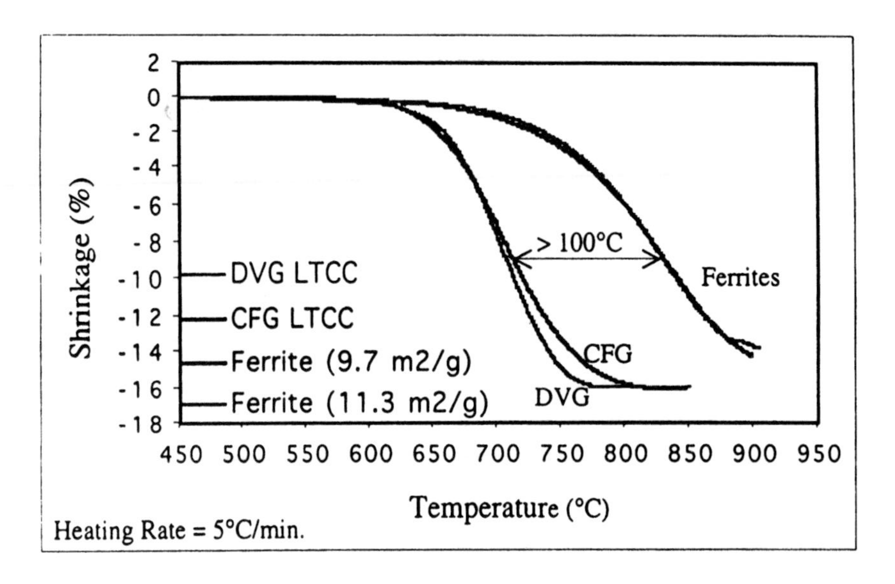

Figure 5: Baseline corrected TMA shrinkage curves for DVG, CFG and ferrites of three surface areas.

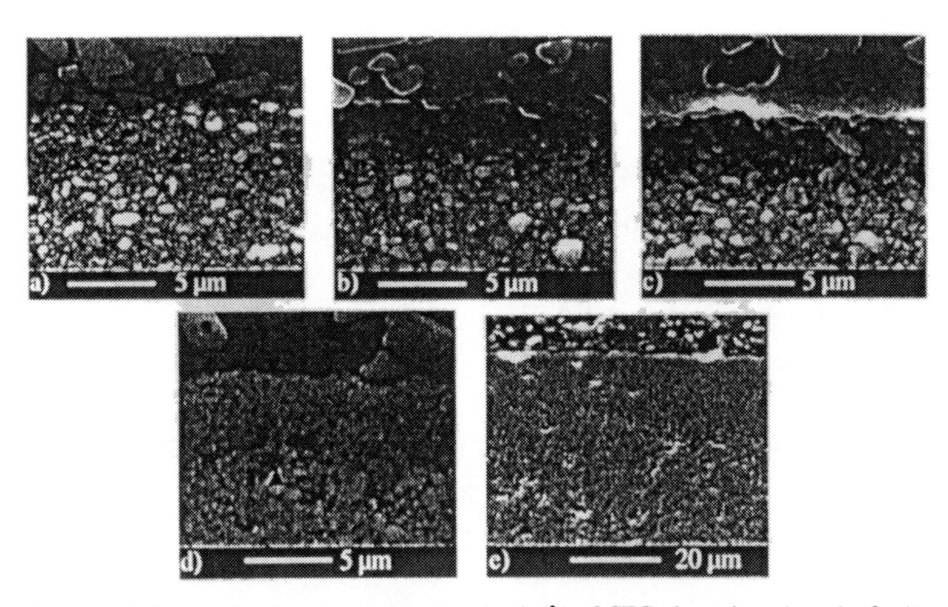

Figure 6: Micrographs showing penetration depth (λ) of CFG glass phase into the ferrite layer during co-sintering for 30 minutes at a) 700°C, b) 750°C, c) 800°C, d) 850°C and e) for 15 min. at 900°C.

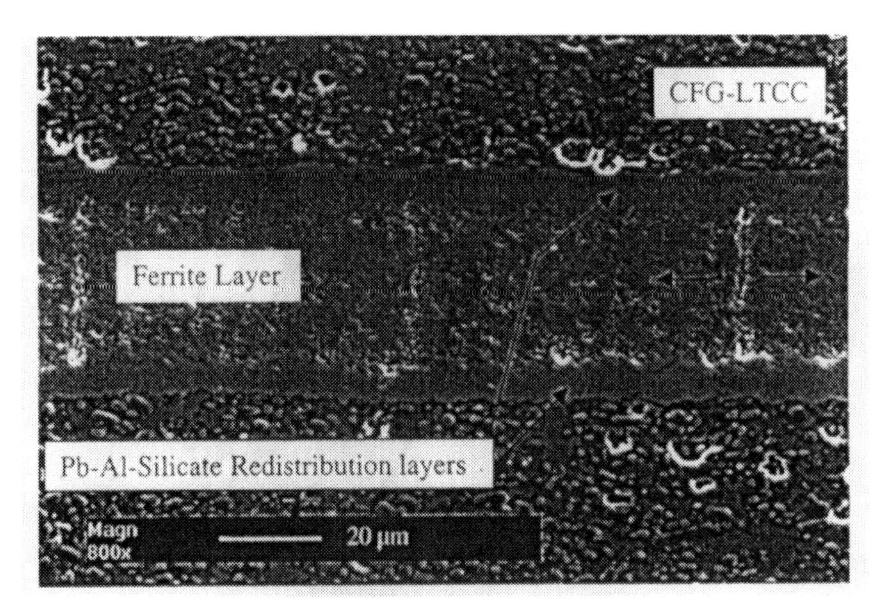

Figure 7: Co-sintered CFG LTCC / ferrite laminate. CFG glass redistribution promotes densification of the ferrite. The ferrite layer separates normal to the CFG interfaces due to tensile forces arising from constrained sintering.

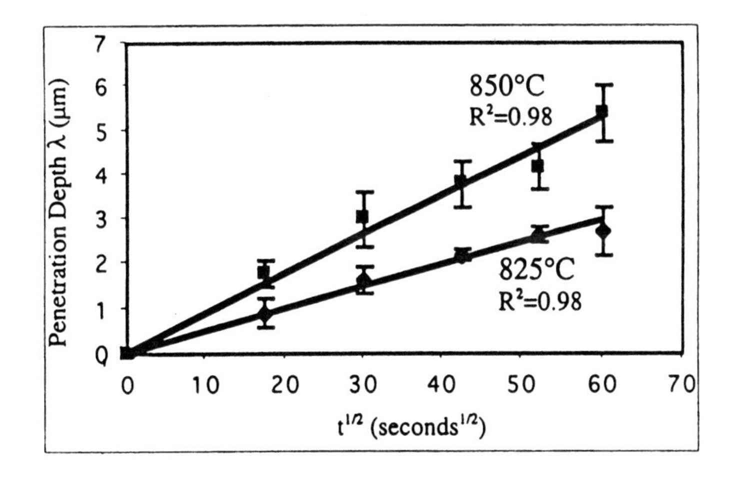

Figure 8: Glass penetration depth follows $t^{1/2}$ dependence predicted by the Rideal Washburn Equation.

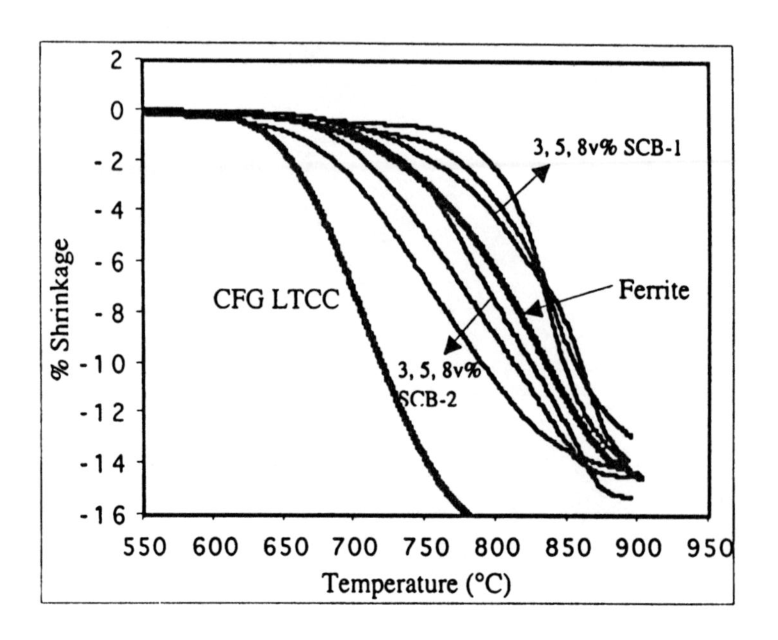

Figure 9: TMA analysis show the effect of 3, 5 or 8 volume percent additions of sintering aid glasses SCB-1 or SCB-2 on ferrite sintering temperature.

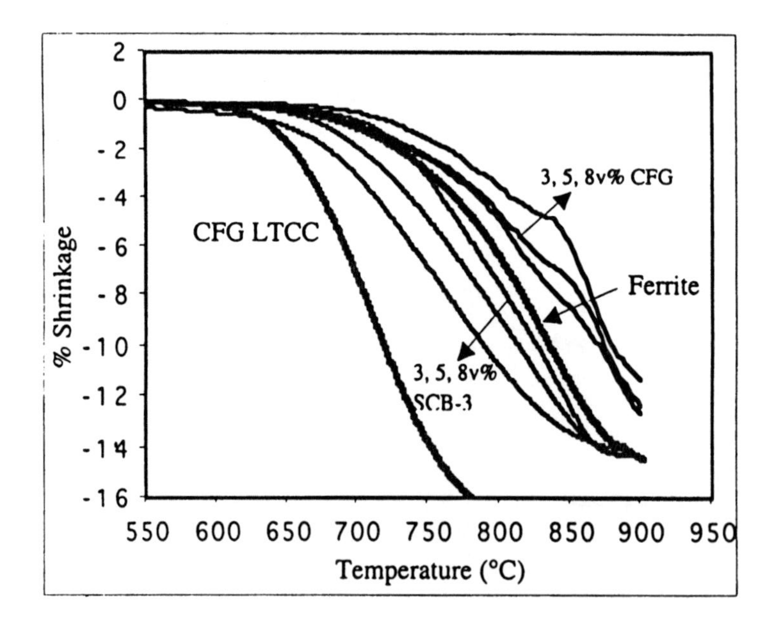

Figure 10: TMA analysis show the effect of 3, 5 or 8 volume percent additions of sintering aid glass SCB-3 or CFG LTCC material on ferrite sintering temperature.

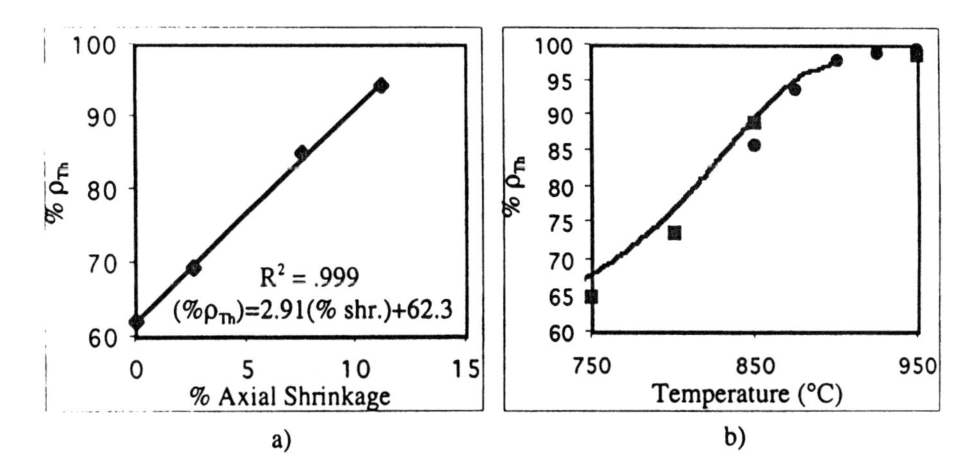

Figure 11: TMA shrinkage data converted to a densification curve through a) the assumed linear relationship between axial shrinkage and density. b) Resulting curve for 11.3 m^2/g ferrite agrees well fired samples of ■ 11.3 m^2/g ferrite, ● 9.2 m^2/g ferrite.

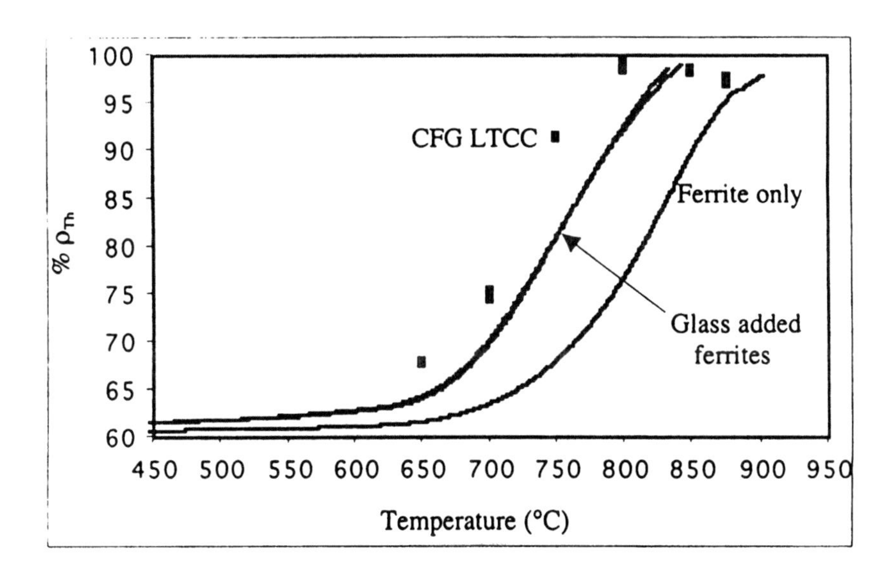

Figure 12: Densification curves for (■) CFG, pure ferrite and ferrite with 8 vol% additions of SCB-2 or SCB-3 sintering aid glasses.
Curves calculated using TMA shrinkage data.

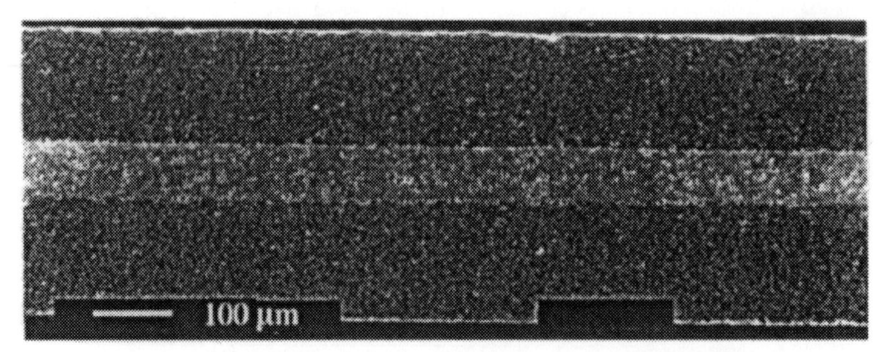

Figure 13: Integral inductor substrate with 8 vol.% glass added $(NiO)_{0.20}(CuO)_{0.16}(ZnO)_{0.16}(Fe_2O_3)_{0.48}$ embedded in CFG LTCC.

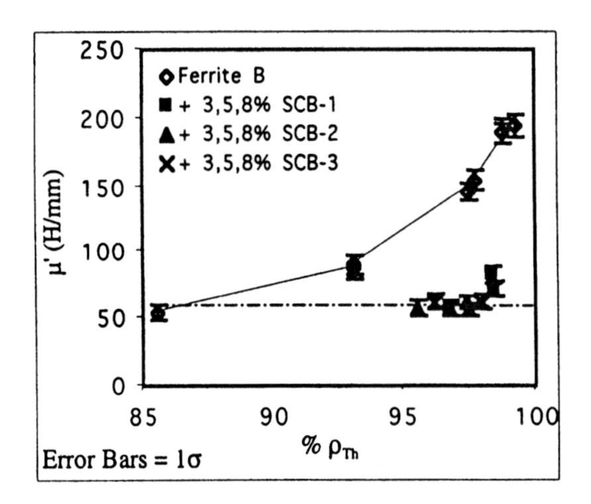

Figure 14: Magnetic permeability (µ') of pure and glass added ferrite B.
Pure ferrites sintered at various temperatures, SCB-1 added samples sintered at 925·C for 4 hours, SCB-2, SCB-3 added samples sintered at 875°C.

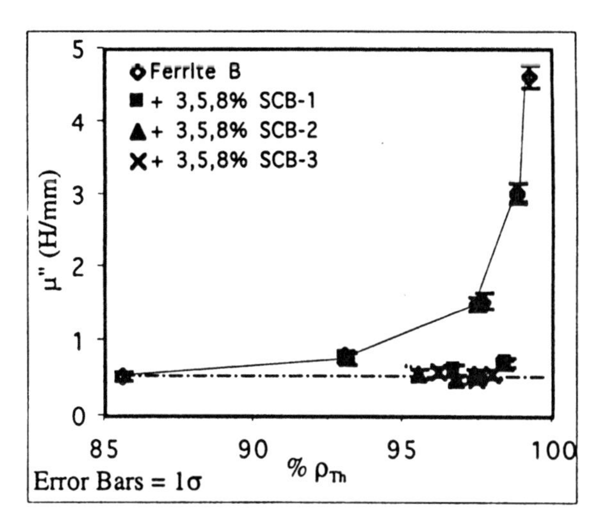

Figure 15: Imaginary part of magnetic permeability (μ") of pure and glass added ferrite B. Samples sintered as in Fig. 14.

RECENT TOPICS IN FERRITE MATERIALS FOR MULTILAYER CHIP COMPONENTS

Atsuyuki Nakano, Hirohiko Ichikawa, Isao Nakahata, Masami Endo
and Takeshi Nomura
Materials Research Center, TDK Corporation
570-2 Minamihadori Matsugashita Narita, Chiba, 286, Japan

ABSTRACT

The multilayer chip ferrite components developed by thick film printing method or sheet method and low temperature sintering NiCuZn ferrite are presently been used for many electrics[1]. The incessant demand for digital higher density circuits in electronics during recent decades has required the continued miniaturization, high frequency and high reliability of the multilayer chip ferrite components. It is well known that hexagonal ferrite can work at high frequency range. We have succeeded in developing the low temperature sintering hexagonal ferrites which can be sintered below Ag melting point temperature. Therefore the multilayer Z type hexagonal ferrite chip inductors which can be use at GHz frequency range are developed. It was found that the inductance of MgCuZn ferrite chips are higher than that of NiCuZn ferrite chips. Because of the higher inductance of MgCuZn ferrites fewer windings can be used in the chips. Lower profile type of chip is available to use the low temperature sintering MgCuZn ferrites. In order to get a low profile chip, the thickness of the ferrite layer must be decreased while maintaining the high reliability. Minute amounts of MnO addition increase the break down voltage and life time (by Highly Accelerated Life Testing) of NiCuZn ferrite chip capacitor.

Fig. 1 Structure of multilayer ferrite chip

INTRODUCTION

Today multilayer ferrite chips are a common component used to attenuate Electro-Magnetic Interference (EMI) as shown in figure 1. Basically the multilayer ferrite chips, as EMI suppression parts will reduce overshoots/undershoots and other high frequency characteristics that cause excessive EMI emissions. The main trend of electric devices is lower voltage for digital circuits, higher clock frequencies and higher density of surface mounting. The requirements of EMI parts are higher "noise" depression capability, "noise" attenuation from MHz to GHz and down sizing/low profit. The size of the chips is decreasing every year, and the chip capacitor industry trends likewise. Presently, the 1005 type has been mainly used for cellular phones and computers. Moreover, the market is requiring smaller and lower profile types. Now we are developing 0603(0.6mm x 0.3mm) chip and 1206 (1.2mm x 0.6mm) array type of chip.

HIGH FREQUENCY

In the case of conventional chips, stray capacitance occurs between internal conductor and terminal electrode as shown in figure 2. The self-resonant frequency is given by equation as below.

$$fr = 2\pi\sqrt{\frac{1}{LC}} \quad (1)$$

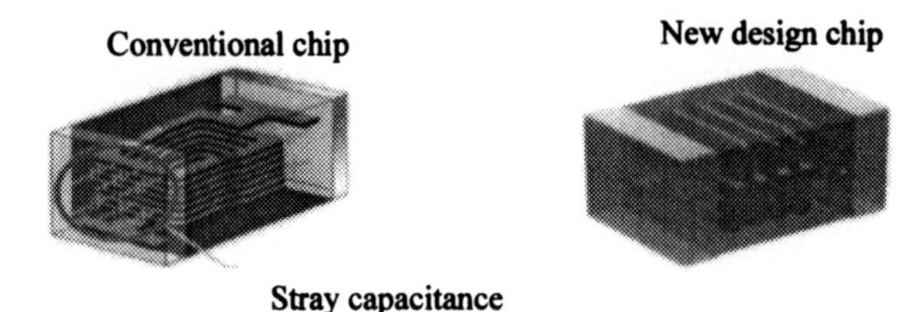

Figure 2 Structure of conventional and new design multilayer ferrite chip.

Where C is the stray capacitance. New design has been studied to decrease the stray capacitance. This stray capacitance decreased by the internal conductors wounding toward to terminal electrode. Figure 3 shows frequency dependence of Impedance of multilayer ferrite chip[2]. The self-resonant frequency of the new type of chip shifts to higher frequencies than that of conventional ferrite chip. It is necessary to design the internal conductor patterns carefully as well as selecting the proper ferrite materials.

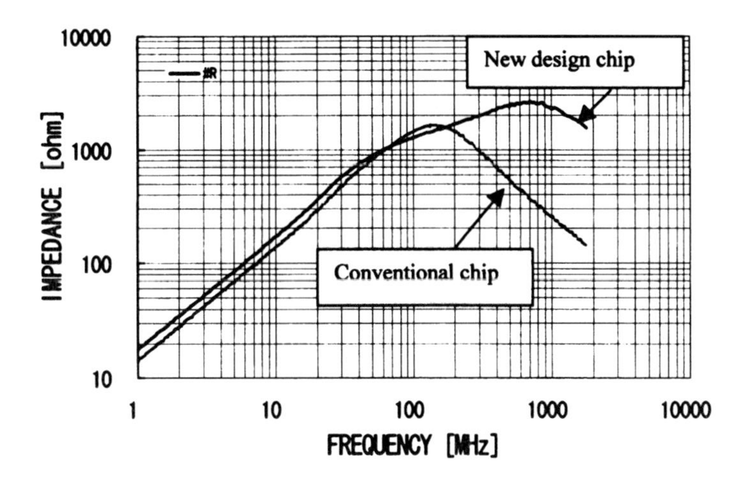

Figure 3 Frequency dependence of Impedance of multilayer ferrite chip.

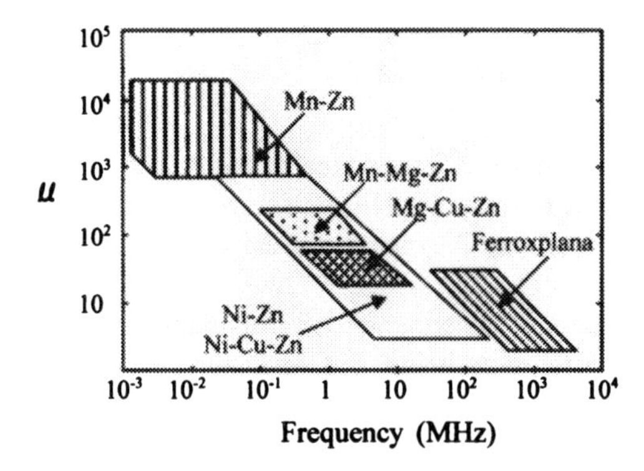

Figure 4 The relationship between frequency and permeability of ferrite systems.

Regarding the development of materials, the relationship between frequency and permeability of ferrite systems is shown in figure 4. At present, NiCuZn ferrites are used for the multilayer chip ferrite components. This is due to the fact that NiCuZn ferrites can cover wide frequency ranges. However, spinel ferrites such as NiCuZn ferrites cannot exceed the permeability of *"Snoek's limit"*. Contrarily, only hexagonal ferrites can be applicable in the high frequency range, beyond the *"Snoek's limit"*, owing to their magnetic anisotropy. However, the ferrites are typically sintered at high temperatures, e.g. 1200-1300℃. Therefore,

in order to co fire with Ag electrode, we had to develop a low temperature sintereable hexagonal ferrite which can be sintered below 900°C[3].
First we focused on Z type hexagonal ferrites because Z type shows high permeability compared with other hexagonal ferrites. We investigated composition, particle size, preparation conditions and addition technique on the sintering behavior of hexagonal ferrites. With Bi_2O_3 and CuO addition, the densification temperature of Z type hexagonal ferrite decreased significantly. Thus we were able to develop low temperature sintering Z type hexagonal ferrite below 900°C.

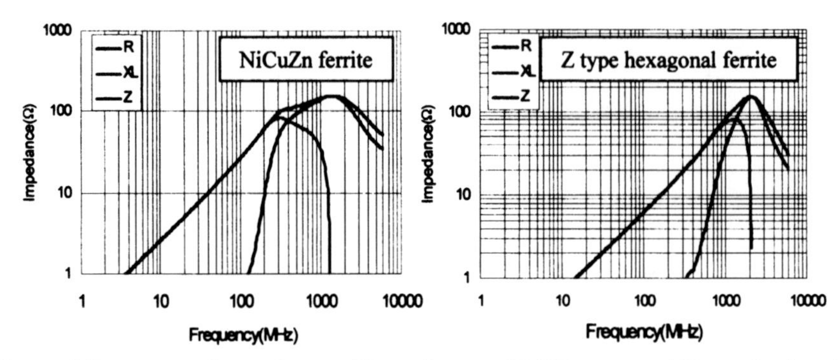

Figure 5 Frequency dependence of impedance of 1608 type multilayer ferrite chip using the low temperature sintering NiCuZn ferrite and the low temperature sintering Z type hexagonal ferrite.

Figure 5 shows frequency dependence of impedance of 1608 type (1.6mm x0.8mm) multilayer ferrite chip using the low temperature sintering NiCuZn ferrite and the low temperature sintering Z type hexagonal ferrite. Comparing impedance between NiCuZn ferrite and Z type hexagonal ferrite, the impedance peak of Z type hexagonal ferrite shows higher frequency than that of NiCuZn ferrite. It is known that the Z type hexagonal ferrite chip works at GHz frequency range as shown in these figures. In order to get higher frequency properties, we tried substitution techniques for Z type hexagonal ferrite ($Ba_3Co_2Fe_{24}O_{41}$). Bivalent metals were substituted with Co of $Ba_3Co_2Fe_{24}O_4$. In the case of substitution with NiO-5%, the impedance peak of the multilayer ferrite chip using the ferrite material shows at 3.5GHz. In addition as to the case of substitution with CuO-25%, the chip obtains 1060-ohm impedance, and the peak shifts higher frequency than that with as shown in figure 6. Therefore, using these materials and with changing design, impedance of 100 to 2000 can be attained, and frequency of impedance peaks can change from 1GHz to 3.5GHz.

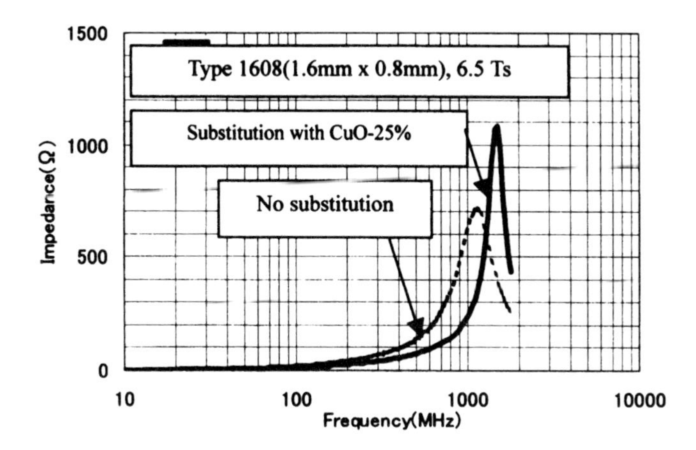

Figure 6 Frequency dependence of impedance of multilayer ferrite chip.

HIGH PERFOMANCE

It is well known that the magnetic property of ferrites is very sensitive to stress[4]. Particularly, the permeability of the magnetic characteristics is sensitive to the stresses and depends on microstructure strongly. Initial permeability, μ i, can be shown

$$\mu\, i = AMs^2 / (aK_1 + b\lambda_s \sigma) \quad (2)$$

M_s:saturation magnetization, K_1:anisotropy constants,

λ_s: saturation magnetostriction, σ :stress, A,b,c: constant number

From this expression, we already investigated as to decreasing or depressing the residual stress at the electrode - ferrite interface[5].

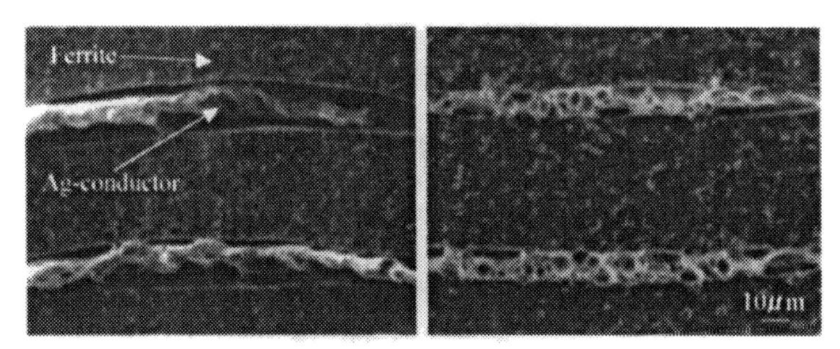

Figure 7 Typical SEM`s of fracture surface of chip inductors.

Figure 7 shows SEM micrograph of a typical chip inductor fracture surface. It is known that the minimizing the contact between the Ag conductor and the ferrite layer can reduce the overall residual stress. Adequate control of the condition of the Ag metallization is a key to achieve high performance in a multilayer ferrite chip. Figure 8 shows TEM images of sintered NiCuZn ferrites of multilayer chip inductor at 890°C fired[6]. It was observed that a great number of interference fringes existed at the grain boundary. On enlarging this area, it was confirmed that low angle tilt boundaries are formed and the d-spacing shrinks. Also, at the center of interference, it was confirmed that CuO and Ag are presented in high concentrations. By in-depth investigations of the microstructure, methods of decreasing the interference fringes were discovered. Therefore, we can control electromagnetic properties of a chip by control of the microstructure[7].

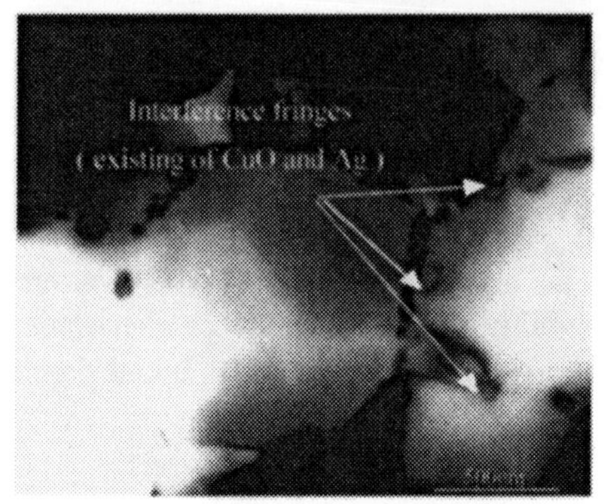

Figure 8 TEM image of sintered multilayer chip inductor.

On the other hand, high performance materials which are MgCuZn ferrites developed. This is probably the first investigation that focuses on magnetostriction in ferrites in order to achieve high performance chip. According to the equation (2), even if the stress is large, a chip would have high performance as magnetostriction is kept low. It is known that magnetostriction of MgCuZn ferrites is lower than that of NiCuZn ferrites. When the ferrite compositions are same, the magnetostriction of MgCuZn ferrites is about 1/4 of the magnetostriction of NiCuZn ferrites. Therefore we studied low temperature sintereable MgCuZn ferrite for multilayer chip ferrite. The first objective was to achieve a sintering temperature below the melting point of the Ag conductor. First of all, we investigated the composition and the particle size of MgCuZn ferrite. However it was difficult to drop the sintering temperature below 900°C only with the adjustment of composition and the particle size. In order to get more shrinkage, the calcination condition was investigated focusing on a homogeneous composition of MgCuZn ferrite. By performing an in-depth study, it was found out that the x-ray peak of Fe_2O_3 disappears over 760 °C. Thus, we developed low temperature sintereable MgCuZn ferrites.

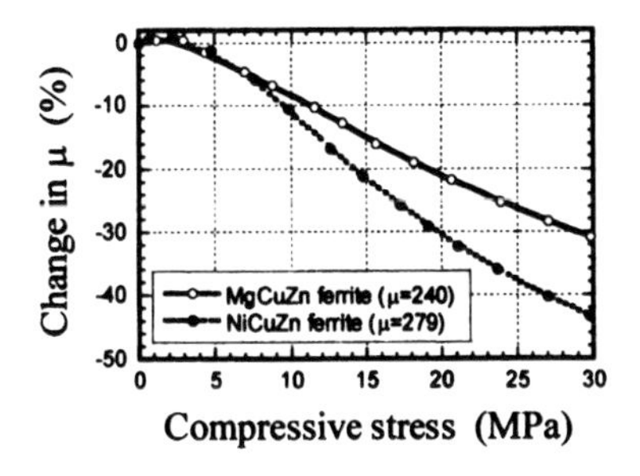

Figure 9 Effect of compressive stress on change in μ.

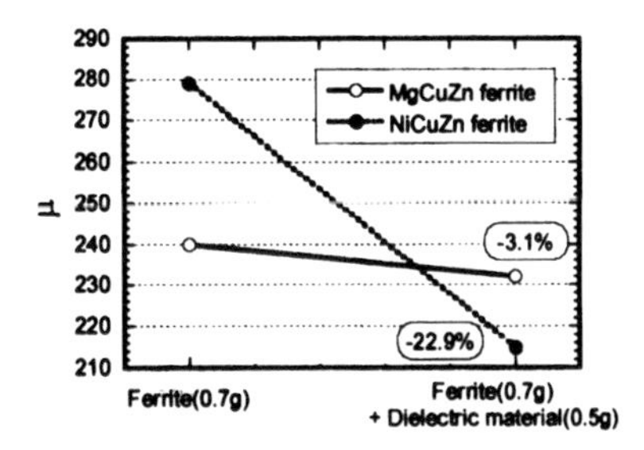

Figure 10 Effect of co-fired with dielectric material on change in μ .

The effect of compressive stress on change in μ of MgCuZn ferrite is shown in figure 9. It is known that compressive stresses cause a decrease in μ of both materials. However, the change in μ of MgCuZn ferrite is smaller than that of NiCuZn ferrite. Figure 10 shows the effect of co-fired with dielectric material on change in μ of MgCuZn ferrite. As for this experiment, when the green body was pressed, dielectric material was pressed with the ferrite likewise, and then the samples were fired. The dielectric material is also the low temperature sintering material which is TiO_2 added CuO 5wt%. It was expected that stress occurred from interface between dielectric material and ferrite material. In the case of co-fired sample of NiCuZn ferrite, the μ shows 22.9% decrease. On the other hand, μ of the MgCuZn ferrite sample decreases only 3%. These results are consistent with our assumption. Therefore, it was thought that MgCuZn ferrite has a high potential for multilayer chip components.

We attempted to make the multilayer chip ferrite by printing method with low temperature sintering MgCuZn and NiCuZn ferrites. Figure 11 shows the relationship between permeability of ferrites and inductance of chips used for the ferrites. Sintering temperature changed the permeability. Regarding the case of the same permeability of MgCuZn ferrites and NiCuZn ferrites, it was found that the inductance of MgCuZn ferrites is higher than that of NiCuZn ferrites. Because of the higher inductance of MgCuZn fewer windings can be used in the chip. Lower profile type of chip is available using the low temperature sintering MgCuZn ferrite.

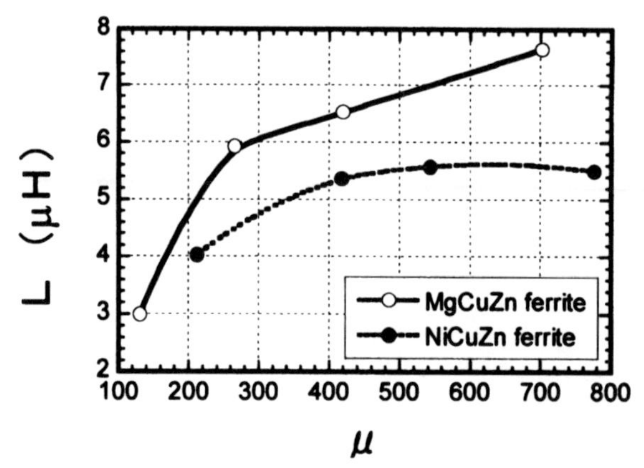

Figure 11 The relationship between permeability of ferrites and inductance of chips used for the ferrites.

HIGH RELIABILITY

The requirement of smaller size and lower profiles for multilayer ferrite chip components is expanded in market. In order to get a low profile chip, the thickness of the ferrite layer must be decreased more and more while maintaining high reliability. Therefore, we began to investigate what factors affected reliability of ferrite layer. First, we tried to study the effect of Mn. As for reliability test, the chip capacitors were prepared for using NiCuZn ferrite. Table 1 shows the effect of MnO on break down voltage and life time (by Highly Accelerated Life Testing) of ferrite chip capacitor. Minute amounts of MnO increase the break down voltage and life time. According to these results, surface of sintered body and cross section around electrode of chip ferrite were investigated the microstructure using EPMA. It was observed that Cu segregation of NiCuZn ferrite decreased with increasing MnO from this investigation as shown in figure 12. At present, we are not sure of the correlation between MnO addition and Cu segregation of ferrite yet. We have to further investigate this phenomenon.

Table 1 the effect of MnO on reliability of ferrite chip capacitor.

Mn content	wt%	0.0711	0.168	0.357	0.544
Break dawn voltage	V/μm	85	109	74	51
Life time	min.	135.7	718.1		0.5

Break dawn voltage condition : Type 3126, 4 layer, 50μm

HALT condition : Type 3126, 4 layer, 50μm, 200℃,100V

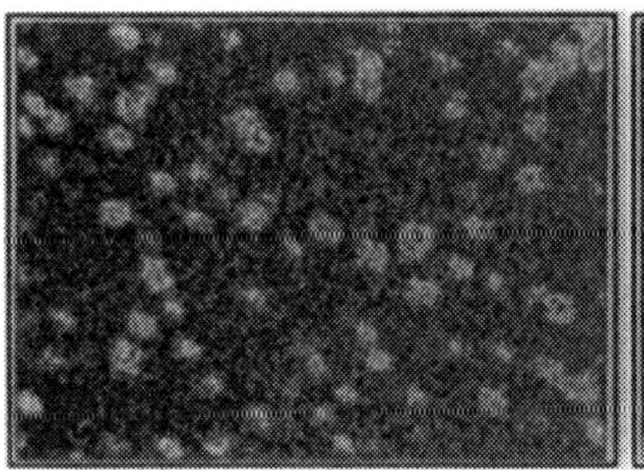

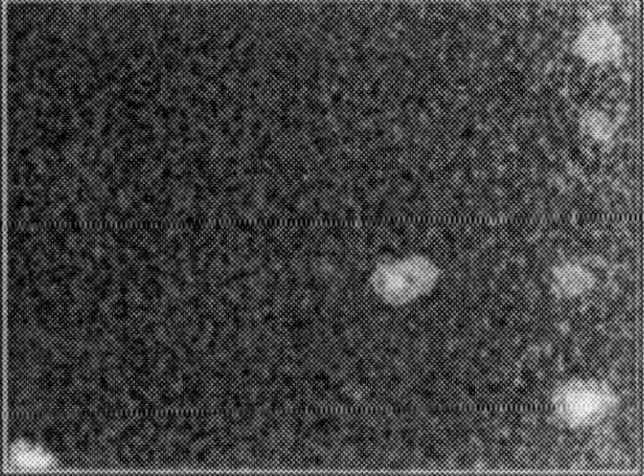

MnO:0.081wt% MnO:0.449wt%

Figure 12 E ffect of MnO addition on Cu segregation of NiCuZn ferrites. (EMPA images)

CONCLUSIONS

Production of multilayer-ferrite chip components has grown remarkably in the past years, and keeping up with the surface-mount-technology boom in the field of electronic components. The ferrite chip components have been progressing by requiring multifunction, using for digital circuits and using at higher frequencies. Further, new materials, that include MgCuZn ferrite and hexagonal ferrite materials are developed at present in order to produce high performance for the multilayer ferrite chip components. We strongly believe that these ferrite materials have potential materials for new requirements.

REFERENCES

[1]T. Nomura, M. Takaya, HYBRIDS 3, 15(1987).

[2]K.Yasuda, Proc.of 8th International Conference on Ferrites(2000) to be published.

[3]M.Endo, A.Nakano, Proc.of 8th International Conference on Ferrites(2000) to be published.

[4]J.Smit and H.P.Wijin, "Ferrite", Phlips Res.Lab.(1959)

[5]Nakano, T. Suzuki, Y. Kanagwa, H. Watanabe and T. Nomura, Proc.of 10th Takei Seminar, P1-8 (1990).

[6]Nakano, H. Momoi, T. Nomura, Proc.of 6th International Conference on Ferrites, P1225-1228 (1992).

[7]Nakano, T. Sato, T. Nomura, Proc.of A.Cer.S, Vol.47 (1995), P241-249.

LEAD FREE MULTILAYER DIELECTRIC SYSTEM FOR TELECOMMUNICATIONS

R. L. Wahlers, S. J. Stein, C. Y. D. Huang, M. R. Heinz
and A. H. Feingold
ElectroScience Laboratories, 416 E. Church Rd., King of Prussia, PA, 19406

ABSTRACT

A totally lead-free, low loss multilayer materials system was developed for telecommunication applications. It includes cofireable silver based conductors, embeddable ferrites for inductors and embeddable dielectrics for capacitors covering a K range from 18 to 265. The LTCC tapes in which these materials can be buried are available in K values of 4, 7.5 & 13, so that one can choose whether they want the speed and isolation advantages associated with low K or the size and cost advantages associated with high K. Meeting the cost target was addressed by using low cost raw materials, parallel processing and low firing temperatures.

INTRODUCTION

There is a common desire among designers of telecommunication systems to provide more functions in an equal or smaller space. They want to do this using proven manufacturing procedures while lowering overall costs. The LTCC based approach is an excellent way to achieve this goal. [1] It is a flexible technology with proven reliability in which one can increase functionality by burying components. In addition, low loss dielectrics and conductors are available providing the desired increased battery life for portable devices.

It was recognized at the onset of this program that even with all the performance advantages associated with LTCC, the device cost was going to be the factor determining its success. With this in mind a number of features were built into the program.

Material costs were reduced by using silver based conductors while manufacturing costs were kept in check by using parallel processing methods to form the modules. In this scheme the printing, hole punching and tape sizing operations are done to the various tapes in parallel. The processed tapes are then inspected before they are brought together for lamination and a final cofiring step. Schemes like this are routinely used to make reliable capacitors of 50 layers or more for less than a penny [2].

There is another advantage LTCC has over PC boards that translates into cost savings. This relates to the variety of materials which can be used in the LTCC approach. The LTCC system includes high K dielectrics and ferrites, that bring about reduced component sizes which usually result in lower overall costs.

This improved LTCC approach is highly reliable. It uses (1) metal ion diffusion inhibiting dielectrics, (2) sintered metal interconnects and (3) has mechanically and electrically stable structures. The high reliability affects cost in that dissatisfied customers and/or returned parts affect the bottom line.

One of the issues which was a subject of considerable debate early on in the program was whether lead containing materials should be used in developing the various components of our system. It is present in one or more of the components in most all of the present LTCC systems. Its presence is understandable as it provides unique contributions to the properties of electro-ceramic devices like capacitors, thermistors and to glasses used in IC's, coating, displays and thick film hybrids. In addition, its combination with tin provides an excellent combination of low cost, low melting temperature and good ductility in solders.

Unfortunately, lead in even minute quantities can be shown to cause brain, nervous system, liver and kidney problems, especially in children. One of the primary concerns is that lead containing products disposed in landfills will be leached into the soil and eventually wind up in water supplies resulting in the poisoning of humans and the Ecosystem.

Electronic products are being upgraded and new ones are coming to market so fast that substantial quantities are indeed going to landfills. It is estimated that every citizen in the European Union generates 23 kg of electronic/electrical trash each year [3], which has the potential for a lot of hazardous material getting into waste streams. Each economic region has a different approach toward solving this problem.

One could argue that the threat posed depends on the types of leaded material and further that the commercial importance of the electro-ceramic and glass should be considered. There is certainly logic to that argument, but it is clear that there will probably be added costs and certainly headaches associated with the use of leaded materials in the future. We decided to avoid the problems by developing a completely lead-free system, i.e., the LTCC tape, the buried components, conductors and solders are all lead-free.

EXPERIMENTAL DESIGN

The designation and description of the non-leaded cofireable dielectrics used in this study are given in Table 1. The LTCC tapes are the glass/ceramic sheets, which form the matrix in which the components are buried and which forms the substrate on which the IC's are placed. These tapes are available with three values for dielectric constant (4, 7.5 & 13). Each has its advantages. We chose the K-13 tape for this study because using it as the matrix tape allows one to make the smallest parts which result in lowest overall cost.

The designations of the lead-free compatible capacitor tapes and pastes and the ferrite tapes which were developed for embedding in the K-13 tape are also listed on this table. The K values are approximate values for the tape (pastes) because they also depend on the firing conditions and electrodes used.

The compatible non-leaded conductors developed for use with the dielectrics listed in the previous table are shown in Table 2. Shown in this table are the buried silvers for conductive traces and vias, buried platinum / silver and

Table I
Non-Leaded, Cofireable Dielectrics

Designation	Description
41110	K-4 LTCC Tape
41020	K-7.5 LTCC Tape
41050	K-13 LTCC Tape
41230	K-18 Capacitor tape
41240	K-50 Capacitor Tape
41250	K-100 Capacitor Tape
41260	K-250 Capacitor Tape
40010	Ferrite Tape
4162	K- 50 Capacitor paste
4163	K- 85 Capacitor paste
4164	K-100 Capacitor paste

Table 2
Non-Leaded, Cofireable Conductor Pastes

Designation	Description
903-CT-1	High conductivity buried Ag
903-CT-1A	Buried Ag, best shrink match
953-1G	Buried Pt/Ag, low cost term.
963-G	Buried Pd/Ag
902-G	Ag via fill
962-G	Ag to Au transition via fill
903-CTA	Solderable top layer Ag
963-G	Solderable top layer Pd/Ag
953-AG	Low cost, leach resist Pt/Ag
803-MG	Top layer, wire bond gold
Solder	95.5 Sn, 3.8 Ag, 0.7 Cu
9904	Top layer photoimage Ag*
8804	Top layer photoimage Au*
*post fireable only	

palladium / silver for component termination. There is also a lead free top layer, wire bondable gold and a Pd/Ag via fill that allows transitioning from the buried silver to the top layer gold without the deleterious (Kirkendall voiding) effects. Leach resistant, solderable metallizations and a lead free solder are listed. The table also lists photoimageable Au and Ag pastes that can be used to create fine lines.

The buried capacitor data shown later in the paper was obtained on tape laminates in which conductors and capacitor dielectrics in either tape or screen printed thick film formats had been placed. The tape stacks were laminated with a uniaxial press using pressures from 1000 to 3000 psi at temperatures of 50 to 70°C. The laminated tape stacks were slowly heated to 450°C and held at this temperature for one hour. They were then heated at various rates to the peak temperature (normally 875°C) and held at that temperature for 15 to 60 minutes.

The dielectric loss versus frequency for the three different LTCC tape matrix materials is shown in Figure 1. The data is from measurements made on 903-CT-1 silver terminated ring resonator samples. The K-4 sample was fired at 850°C with a 12 minute hold at peak while the K-7.5 and K-13 test samples were fired at 875°C with a 15 min hold at peak. The loss curve for FR-4/Cu is shown for comparison since it is a competitive technology.

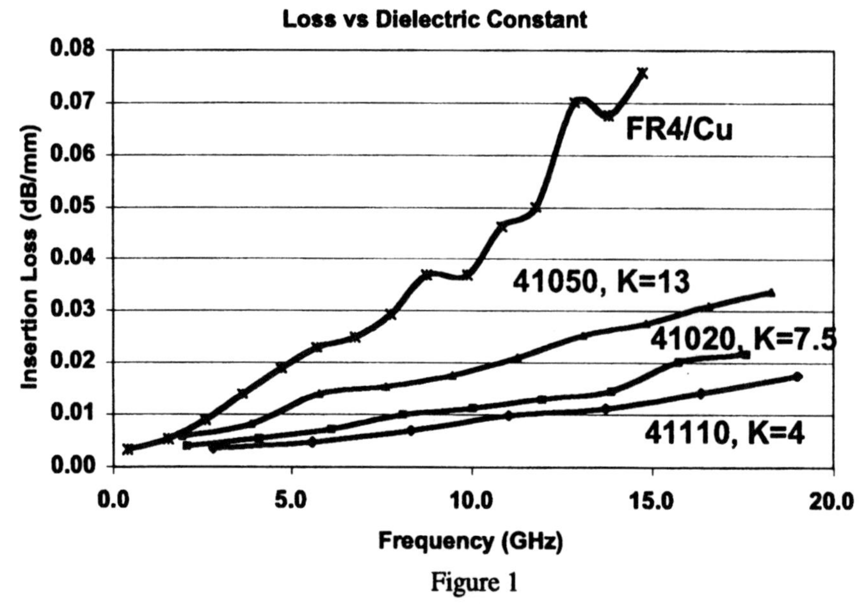

Figure 1

All testing of the K-13 tape indicates it is very robust. Its loss versus frequency curve doesn't change with time at peak firing temperature (at least over the 15 to 60 minute interval) (See Figure 2). Changes in the heating rates after burnout from 2°C to 10°C / min also has no affect on the loss curve (See Figure 3).

Measurements were made of the Dielectric Constant, Dissipation Factor and Insulation Resistance for the buried capacitor test samples at 1 KHz. In addition, the temperature coefficient of capacitance was measured at –55, -30, +85 & 125°C on all these test samples. The ability of the ferrite tape to enhance the inductance of a buried spiral was also tested.

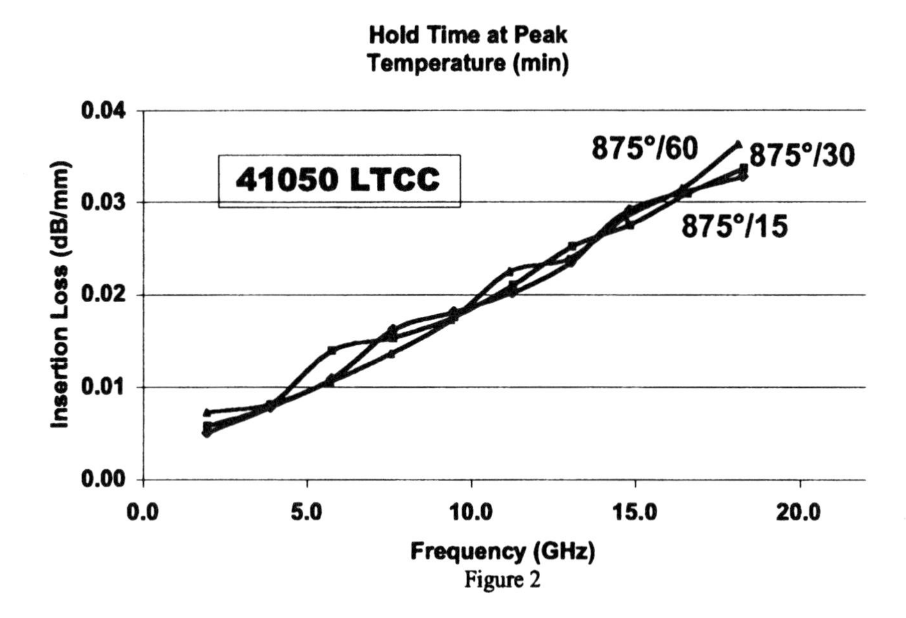

Figure 2

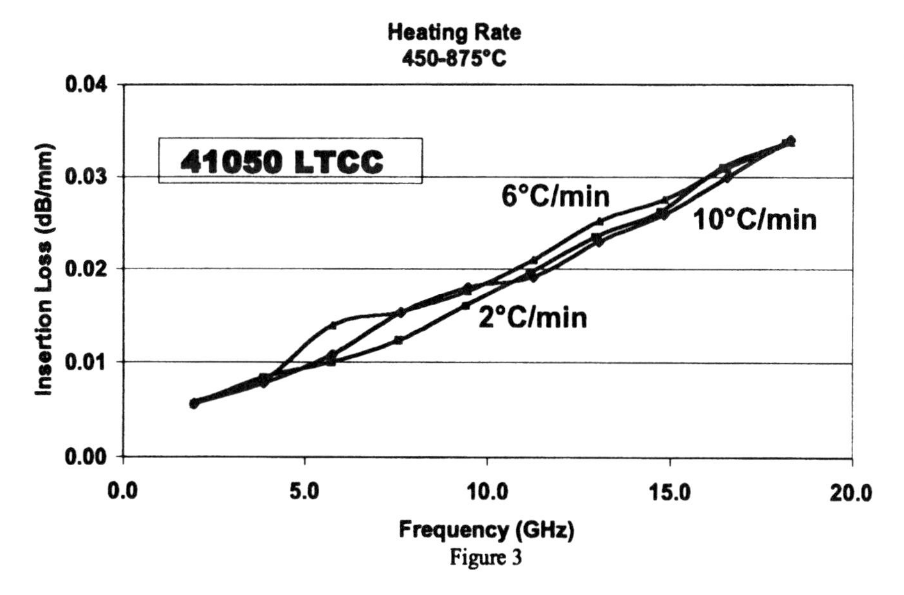

Figure 3

RESULTS

Table 3 shows the results of burying the four different capacitor tapes in the K-13 LTCC tape matrix. These capacitor tapes were formulated to give specific K values but as indicated in the table, the actual values and their environmental stability are influenced by firing conditions and electrode composition.

Table 3

Buried Capacitor Tape	Electrode Designation	Time at 875°C (min)	Dielectric Constant K	Dissipation Factor DF	Insulation Resistance IR	Maximum |%ΔC| -55 to 125°C
41230	953-1G	60	18	0.1%	2×10^{11}	1.4
41240	953-1G	15	60	0.3%	4×10^{11}	5.8
41250	953-1G	15	103	0.6%	2×10^{11}	8.2
41250	903-CT-1	15	98	0.7%	9×10^{10}	5.5
41250	953-1G	60	108	0.6%	1×10^{11}	5.7
41250	963-G	15	110	0.6%	4×10^{11}	5.2
41250	963-G	60	107	0.5%	3×10^{11}	4.6
41260	953-1G	15	225	1.3%	7×10^{9}	15.2
41260	963-G	15	265	1.0%	2×10^{11}	13.2

Comparable data for non-leaded screen-printed capacitor dielectrics developed for use with the K-13 LTCC tape is presented in Table 4. We didn't investigate as many processing and electrode variations on screen printed buried capacitor pastes, so the data available is limited.

The fired laminates were cross-sectioned to get dielectric thickness for calculating K values and to examine the interface bonds between the LTCC matrix and the capacitors. Figure 4 shows backscattered micrographs of capacitor tape (41250) embedded in K-13 tape at magnifications of 100X and 500X. The conductive electrodes are 953-1G Pt/Ag. The samples were fired at 875°C – 15 minutes

Figure 5 shows comparable backscattered micrographs of screen- printed 4164. It also used the 953-1G Pt/Ag and the 875°C - 15 minute firing cycle.

Table 4

Buried Capacitor Paste	Electrode Designation	Time at 875°C (min)	Dielectric Constant K	Dissipation Factor DF	Insulation Resistance IR	Maximum |%ΔC| -55 to 125°C
4162	953-1G	15	54	0.3%	8×10^{11}	0.8
4163	953-1G	15	84	0.8%	8×10^{10}	9.2
4164	953-1G	15	102	2.5%	1×10^{11}	17.5
4164	963-G	60	70	1.1%	1×10^{9}	13.7

Note that in the case of both the tape sheet capacitor and the screen-printed capacitor, the bonds between the tape, capacitor dielectric and conductor is very good. No delamination, bubbles or blisters were detected and no warpage occurred.

In an experiment to evaluate the effectiveness of the ferrite tape, the inductance of a screen-printed silver spiral pattern was measured on samples in three configurations. In the first the silver spiral was printed directly on the 41050 (K-13 LTCC) tape. This was buried between other sheets of 41050, laminated and fired. In the second set the silver spiral pattern was printed on a sheet of 40010 ferrite tape and covered with a second so the printed spiral inductor had one ferrite sheet on each side of it. This two layer structure was then embedded in sheets of 41050 (K-13) tape, laminated and fired. The third set was formed like the second except that two layers of 40010 ferrite tape were placed on either side of the printed spiral. Firing in all three cases was at a peak temperature of 875°C with a hold time at this temperature of 60 minutes.

While printed conductive spirals can be used to create inductors in LTCC based modules, the results shown on Table 5 indicate that 40010 ferrite cover layers can be used to enhance inductance. They can be used to either reduce the area to achieve a given inductance value or to increase the inductance achievable on a given area. Optical micrographs of buried ferrite and capacitor tape

Table 5

Part Description	Inductance (μh)
Silver conductive spiral buried in LTCC	2.0
Silver conductive spiral with 2 layers ferrite tape *(one each side) buried in LTCC	3.5
Silver conductive spiral with 4 layers ferrite tape *(two each side) buried in LTCC	6.5

*each layer about 60μm thick

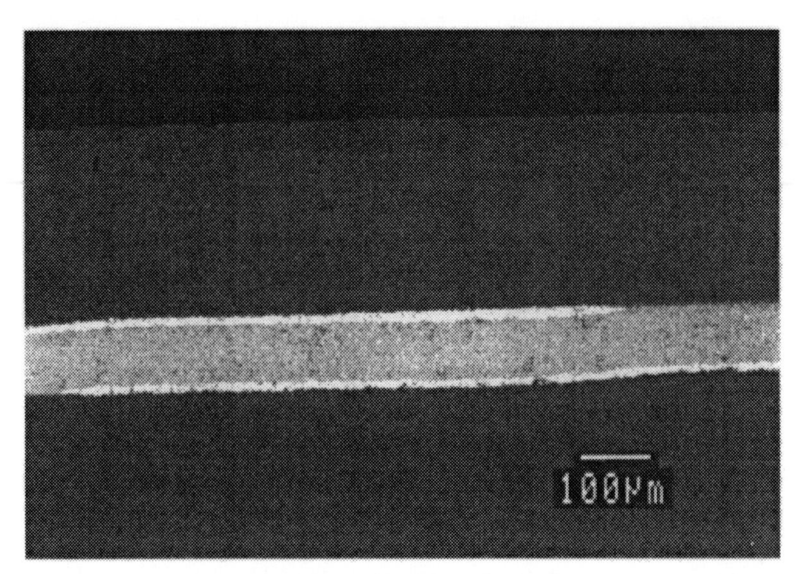

Figure 4a 100X 875°C -15 minutes

Figure 4B 500X 875°C - 15 minutes

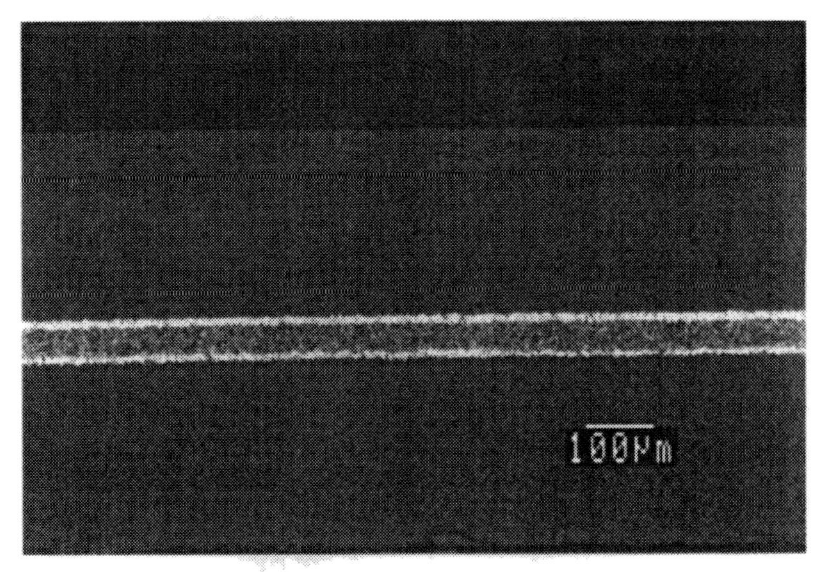

Figure 5A 100X 875°C 15 minutes

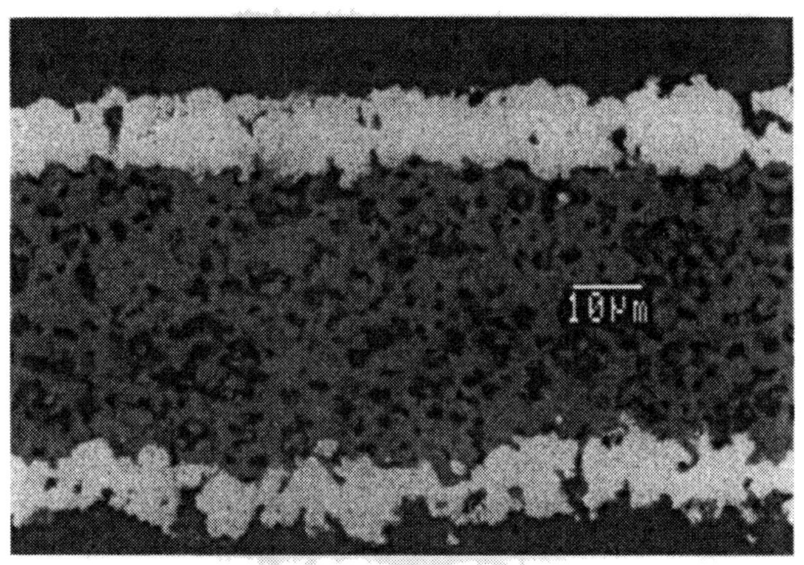

Figure 5B 500X 875°C 15 minutes

combinations are shown in Figure 6 . Again, good bonding is seen between the various components. It is more clearly shown in Figure 6B, which is at higher magnification than Figure 6A.

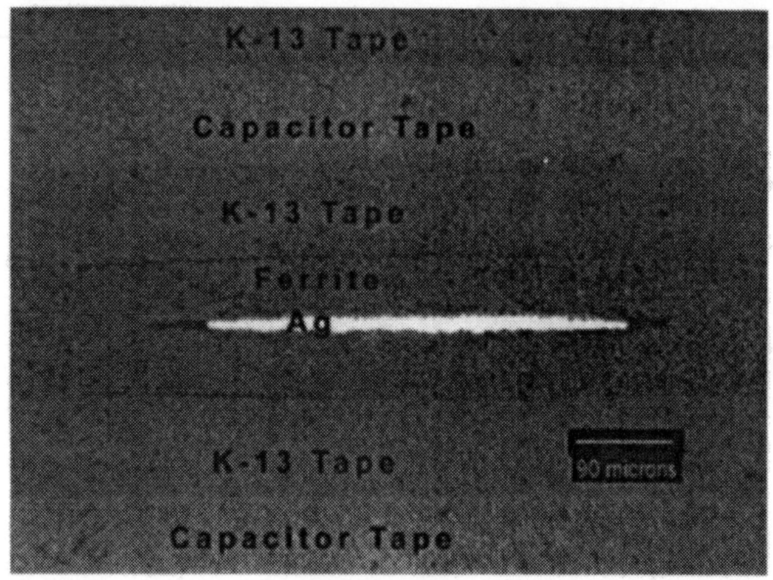

Figure 6A-100X

Figure 6B 500X

SUMMARY

A materials system was developed for telecommunication applications consisting of:

- LTCC tape with better loss than FR-4 and a K of 13 providing for size reduction (and potential cost reduction).
- Cofireable, low cost, low loss silver base conductors.
- Reduced cost processing through materials choice, parallel processing and peak firing temperatures below 900°C
- Capacitor dielectrics in tape and paste configurations with K values from 18 to 250 which can be embedded in K-13 LTCC tape.
- An inductance increasing (area reducing) ferrite tape that can be buried in the K-13 LTCC tape.
- A completely lead free system, including LTCC tape, low cost conductors, capacitor tapes and pastes, ferrite tape and solder with demonstrated compatibility.

REFERENCES

1) R. L. Wahlers, S. J. Stein, M. A. Stein, A. H. Feingold & P. W. Bless, "Ceramic Tapes for Wireless Applications", presented at Annual Meeting of American Ceramic Society, 2000, Cincinnati.

2) D. L. Wilcox, R. F. Huang & R. Kommusch, "The Multilayer Ceramic Integrated Circuit (MCIC) Technology; An Enabler for the Integration of Wireless Radio Function", Advancing Microelectronics, vol. 25, no. 4, pps. 13-18.

3) K. Snowdon; "Towards a Green 2000" Proceedings of IMAPS Europe, June 1999, pp. 71-77.

MICROWAVE DIELECTRIC CHARACTERIZATION OF FERROELECTRIC CERAMICS WITH SLEEVE RESONATOR TECHNIQUES

RICHARD G. GEYER, PAVEL KABOS, AND JAMES BAKER-JARVIS

National Institute of Standards and Technology
RF Technology Division, M.S. 813.01
Boulder, CO 80303

Abstract

High-Q, temperature-stable sleeve resonators may be used advantageously for variable-temperature dielectric characterization of rod-shaped ferroelectric test specimens at microwave frequencies. The nominal measurement frequency is determined by the permittivity and aspect ratio of the sleeve resonator. With several sleeve resonators having differing permittivities and aspect ratios but the same internal hole sizes, dielectric characterization of a *single* specimen over a wide frequency range may be performed using dominant TE_{011} mode structure. General theoretical considerations for dielectric sleeve resonators situated centrally in a cylindrical cavity and loaded with the specimen under test are given for arbitrary TE_{0np} resonant mode structure. This technique is used for variable-temperature dielectric characterization of $Ba_xSr_{1-x}TiO_3$ ceramic composites. Uncertainty analyses are provided that illustrate the influence of sample partial electric filling factor, sleeve geometry and complex permittivity, conductor losses, and measured system unloaded Q-factors on the specimen's evaluated dielectric properties.

INTRODUCTION

Improved voltage-tunable dielectric constants and dielectric loss properties of ferroelectric ceramic composites have many potential uses in microwave component and system applications. Some of these applications are wide bandwidth tunable capacitors, resonators, delay lines, and phase shifters. Other applications are tunable lenses and substrates in "smart" antenna systems.

Generally, the performance of tunable ferroelectric electronic components depends on desired tunability characteristics that are consistent with minimum insertion losses. Maximum material loss is that in the unbiased state, but is both frequency- and temperature-dependent. The variation of permittivity over the operational temperature range must be known so that suitable temperature control is maintained for the ferroelectric device with tuning. Hence an accurate method for characterizing the real permittivity and dielectric loss properties of ferroelectric materials at specified operational frequency and temperature ranges is needed.

Use of a specimen as a dielectric resonator is a commonly used technique for accurate complex permittivity evaluation of low-loss materials [1-17]. Dielectric resonator techniques, when applicable, generally provide higher accuracies than other resonator methods because specimen partial electric energy filling factors for appropriately chosen mode structure and practically-sized samples are greater. Specimen partial electric energy filling factors are often 30 to 100 times those in cavity measurement systems [18-20]. A number of different dielectric resonator system configurations have been developed in recent years, each with their specific advantages for either constrained specimen sizes, high real permittivity, low-loss characteristics; or even high real permittivity, relatively high-loss characteristics. We will address the use of a dielectric sleeve resonator system loaded with a high-permittivity rod specimen and placed in a cylindrical cavity.

Dielectric sleeve resonators may be utilized for accurate dielectric measurements at microwave frequencies. The use of sleeve resonators for dielectric characterization has several advantages. First, they can control the nominal resonant frequency at which dielectric characterization of an inserted rod specimen is performed. Thus, with the use of several low-loss ring resonators, the dielectric properties of a *single* specimen may be evaluated at several discrete frequencies. Second, if the specimen loss characteristics are not known and are relatively high, this approach often permits an accurate loss determination since the unloaded Q-factor of the composite system may be measurable when that of the specimen as a single dielectric resonator is not. Third, smaller test specimens are required at low frequencies for characterization. The placement of the loaded sleeve resonator system in a cylindrical cavity permits ready variable-temperature dielectric characterization of the rod specimen provided the dielectric characteristics of the sleeve resonator and conductive losses of the enclosing cavity have been accurately determined.

THEORY

The dielectric sleeve resonator system is illustrated in Fig. 1. The sample has complex relative permittivity $\epsilon^*_{r,spec} = \epsilon'_{r,spec} - j\epsilon''_{r,spec} = \epsilon'_{r,spec}(1 - j\tan\delta_{e,spec})$, where $\tan\delta_{e,spec}$ is the dielectric loss tangent of the specimen. The sleeve resonator is characterized by $\epsilon^*_{r,sleeve} = \epsilon'_{r,sleeve} - j\epsilon''_{r,sleeve} = \epsilon'_{r,sleeve}(1 - j\tan\delta_{e,sleeve})$. In general, the electromagnetic wave equation to be solved in cylindrical coordinates, subject to boundary conditions, is given by

$$\frac{1}{r}\frac{\partial}{\partial r}\left(r\frac{\partial F}{\partial r}\right) + \frac{1}{r^2}\frac{\partial^2 F}{\partial \phi^2} + \frac{\partial^2 F}{\partial z^2} + k^2 F = 0, \tag{1}$$

where $k = \omega\sqrt{\mu_0\epsilon_0\mu_r'\epsilon_r'}$ is the wavenumber in the medium, $k_r = \sqrt{k^2 - \beta^2}$ is the radial wavenumber, β is the z-directed propagation constant, $\omega = 2\pi f$ is the radian frequency, ϵ_0 and μ_0 are the free space permittivity and permeability (8.854×10^{-12} F/m and $4\pi \times 10^{-7}$ H/m), and F represents either the axial electric field, E_z (for TM- or E-modes), or the axial magnetic field, H_z (for TE- or H-modes). Once E_z and H_z are obtained as solutions of the wave equation, all other components can be obtained from Maxwell's equations,

$$\begin{aligned} E_r &= \frac{1}{k_r^2}\left(\frac{\partial^2 E_z}{\partial r \partial z} - \frac{j\omega\mu_0}{r}\frac{\partial H_z}{\partial \phi}\right), \\ E_\phi &= \frac{1}{k_r^2}\left(\frac{1}{r}\frac{\partial^2 E_z}{\partial \phi \partial z} + j\omega\mu_0\frac{\partial H_z}{\partial r}\right), \\ H_r &= \frac{1}{k_r^2}\left(\frac{\partial^2 H_z}{\partial r \partial z} + \frac{j\omega\epsilon_0\epsilon_r^*}{r}\frac{\partial E_z}{\partial \phi}\right), \\ H_\phi &= \frac{1}{k_r^2}\left(\frac{1}{r}\frac{\partial^2 H_z}{\partial \phi \partial z} - j\omega\epsilon_0\epsilon_r^*\frac{\partial E_z}{\partial r}\right). \end{aligned} \tag{2}$$

Typically, the TE modes are most useful in dielectric measurements because there exists no capacitive coupling between either the sleeve resonator or dielectric specimen and the upper or lower conductive end plates of the cavity. The enclosing cylindrical cavity also allows minimal dimensional changes in the positions of the upper and lower cavity end plates in variable temperature measurements that could potentially affect TE resonant mode structure in the composite dielectric resonator system. In the following, both the dielectric sleeve resonator and specimen under test are isotropic.

Permittivity Evaluation

For axisymmetric ($m = 0$) TE_{mnp} modes, where $\epsilon'_{r,spec} > \epsilon'_{r,sleeve}$, the non-zero electromagnetic field components are,

$$H_{z1} = A_1 J_0(k_{r1} r)\sin(\beta z), \tag{3}$$

$$E_{\phi 1} = \frac{j\omega\mu_0}{k_{r1}^2}\frac{\partial H_{z1}}{\partial r} = -\frac{j\omega\mu_0}{k_{r1}} A_1 J_1(k_{r1} r)\sin(\beta z), \tag{4}$$

$$H_{r1} = \frac{1}{k_{r1}^2}\frac{\partial^2 H_{z1}}{\partial r \partial z} = -\frac{\beta A_1}{k_{r1}} J_1(k_{r1} r)\cos(\beta z), \tag{5}$$

$$H_{z2} = A_2\left[J_0(k_{r2} r) + BY_0(k_{r2} r)\right]\sin(\beta z), \tag{6}$$

$$E_{\phi 2} = \frac{j\omega\mu_0}{k_{r2}^2}\frac{\partial H_{z2}}{\partial r} = -\frac{j\omega\mu_0}{k_{r2}} A_2\left[J_1(k_{r2} r) + BY_1(k_{r2} r)\right]\sin(\beta z), \tag{7}$$

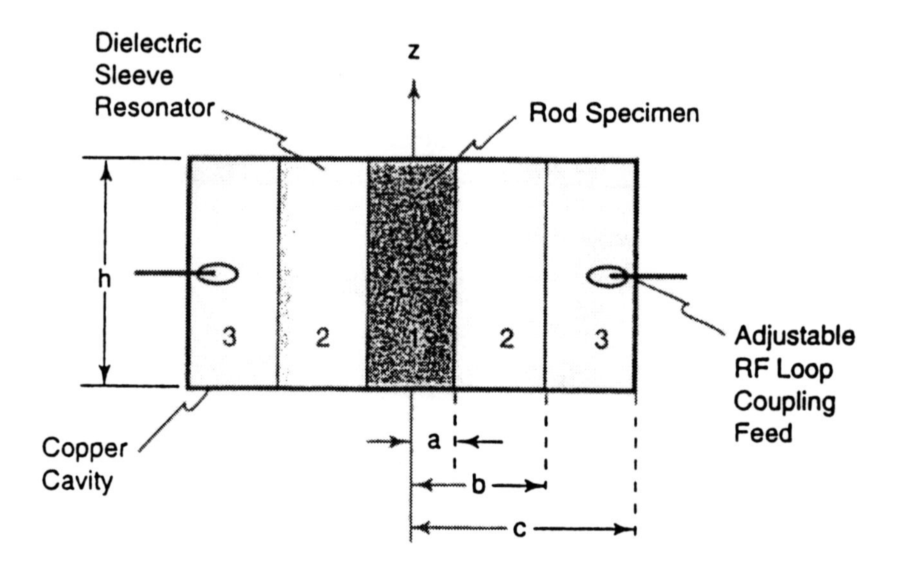

Figure 1: Dielectric sleeve resonator system for testing rod specimen with relative complex permittivity $\epsilon^*_{r,spec} = \epsilon'_{r,spec}(1 - j\tan\delta_{spec})$. Sleeve resonator has inner radius a, outer radius b, and complex permittivity $\epsilon^*_{r,sleeve} = \epsilon'_{r,sleeve}(1 - j\tan\delta_{e,sleeve})$.

$$H_{r2} = \frac{1}{k_{r2}^2}\frac{\partial^2 H_{z2}}{\partial r \partial z} = -\frac{\beta}{k_{r2}}A_2\left[J_1(k_{r2}r) + BY_1(k_{r2}r)\right]\cos(\beta z), \tag{8}$$

$$H_{z3} = A_3\left[K_0(k_{r3}r) + CI_0(k_{r3}r)\right]\sin(\beta z), \tag{9}$$

$$E_{\phi 3} = \frac{j\omega\mu_0}{k_{r3}^2}\frac{\partial H_{z3}}{\partial r} = -\frac{j\omega\mu_0}{k_{r3}}A_3\left[K_1(k_{r3}r) - CI_1(k_{r3}r)\right]\sin(\beta z), \tag{10}$$

$$H_{r3} = \frac{1}{k_{r3}^2}\frac{\partial^2 H_{z3}}{\partial r \partial z} = -\frac{\beta}{k_{r3}}A_3\left[K_1(k_{r3}r) - CI_1(k_{r3}r)\right]\cos(\beta z), \tag{11}$$

where $k_{r1}^2 = k_0^2\epsilon'_{r,spec} - \beta^2$, $k_{r2}^2 = k_0^2\epsilon'_{r,sleeve} - \beta^2$, $k_{r3}^2 = \beta^2 - k_0^2$, J_m, Y_m are Bessel functions of the first and second kind of order m, and K_m, I_m are modified Bessel functions of the first and second kind. At $z = 0$ and $z = h$, $E_{\phi 1} = E_{\phi 2} = E_{\phi 3} = 0$ so that the axial propagation constant is,

$$\beta = \frac{p\pi}{h}, \qquad p = 1, 2, 3, \ldots \tag{12}$$

At the cylindrical cavity wall $r = c$, the tangential electric field $E_{\phi 3}$ is 0, so that

$$C = \frac{K_1(k_{r3}c)}{I_1(k_{r3}c)}. \tag{13}$$

Continuity of the tangential electric and magnetic field components at $r = a$ and $r = b$ yields the following characteristic equation for TE_{0np} resonant modes [21],

$$\frac{k_{r1}J_0(k_{r1}a)J_1(k_{r2}a) - k_{r2}J_1(k_{r1}a)J_0(k_{r2}a)}{k_{r2}J_1(k_{r1}a)Y_0(k_{r2}a) - k_{r1}J_0(k_{r1}a)Y_1(k_{r2}a)} + \frac{k_{r2}J_0(k_{r2}b) - k_{r3}P_1'(b)J_1(k_{r2}b)}{k_{r2}Y_0(k_{r2}b) - k_{r3}P_1'(b)Y_1(k_{r2}b)} = 0, \tag{14}$$

where

$$P_1'(b) = -\frac{K_0(k_{r3}b) + CI_0(k_{r3}b)}{K_1(k_{r3}b) - CI_1(k_{r3}b)}. \tag{15}$$

The first zero of eq(14) for $p = 1$ corresponds to TE_{011} mode resonance for a given aspect ratio $2a/h$ of the sample, inner and outer diameters $2a$ and $2b$ of the sleeve resonator of height h, enclosing cavity diameter $2c$, and a nonmagnetic specimen real relative permittivity $\epsilon'_{r,spec}$. This equation may be used for predicting TE_{0mp} resonant frequencies for a loaded sleeve dielectric resonator or for evaluating permittivity of inserted rod specimens when identified TE_{0mp} frequencies, sleeve permittivity, and sleeve and rod specimen geometries are known.

Dielectric Loss Tangent

The relation for the dielectric loss tangent of the specimen under test is evaluated from the expression,

$$\frac{1}{Q_0} = \frac{1}{Q_s} + \frac{1}{Q_p}, \tag{16}$$

where Q_0 is the unloaded Q-factor, $Q_s = 1/(p_{e,s} \tan \delta_{e,s})$ is the sample quality factor, $p_{e,s}$ is the sample electric energy filling factor and parasitic losses Q_p^{-1} are given by

$$\frac{1}{Q_p} = p_{e,sleeve} \tan \delta_{e,sleeve} + \frac{1}{Q_c}, \tag{17}$$

with $p_{e,sleeve}$ being the electric energy filling factor of the sleeve resonator.

In general, Q_0 is given as $Q_0 = \omega_0 W/P_{tot}$, where W represents the *total* electric field energy in the resonant system and P_{tot} represents all power losses in the resonant system. The total normalized power losses are those dissipated in the dielectric specimen, $Q_{spec}^{-1} = p_{e,spec} \tan \delta_{e,spec}$, in the sleeve resonator, $Q_{sleeve}^{-1} = p_{e,sleeve} \tan \delta_{e,sleeve}$, and in the cavity metal walls, Q_c^{-1}, which include both cavity endplates and the cylindrical cavity wall. In order to determine individual electric field energy filling factors in specimen and sleeve, we must determine the electric field energy stored in each region relative to the total electric field energy in the resonant system.

After some manipulation, the partial electric field filling factor in the specimen can be shown to be [21],

$$p_{e,spec} = \left(1 + \frac{\epsilon'_{r,sleeve}}{\epsilon'_{r,spec}}\left[N_2^2\left(\frac{k_{r1}}{k_{r2}}\right)^2\frac{\bar{I}_2}{\bar{I}_1} + \frac{1}{\epsilon'_{r,sleeve}}N_2^2N_3^2\left(\frac{k_{r1}}{k_{r3}}\right)^2\frac{\bar{I}_3}{\bar{I}_1}\right]\right)^{-1}. \quad (18)$$

Similarly, the electric field partial filling factor in the sleeve resonator is

$$p_{e,sleeve} = \left(1 + \frac{\epsilon'_{r,spec}}{\epsilon'_{r,sleeve}}\left[\frac{1}{N_2^2}\left(\frac{k_{r2}}{k_{r1}}\right)^2\frac{\bar{I}_1}{\bar{I}_2} + \frac{1}{\epsilon'_{r,spec}}N_3^2\left(\frac{k_{r2}}{k_{r3}}\right)^2\frac{\bar{I}_3}{\bar{I}_2}\right]\right)^{-1}, \quad (19)$$

where

$$\bar{I}_1 = \int_0^a rJ_1^2(k_{r1}r)dr, \quad (20)$$

$$\bar{I}_2 = \int_a^b r[J_1(k_{r2}r) + BY_1(k_{r2}r)]^2dr, \quad (21)$$

$$\bar{I}_3 = \int_b^c r[K_1(k_{r3}r) - CI_1(k_{r3})]^2dr, \quad (22)$$

$$B = \frac{k_{r1}J_0(k_{r1}a)J_1(k_{r2}a) - k_{r2}J_1(k_{r1}a)J_0(k_{r2}a)}{k_{r2}J_1(k_{r1}a)Y_0(k_{r2}a) - k_{r1}J_0(k_{r1}a)Y_1(k_{r2}a)}, \quad (23)$$

$$N_2 = \frac{J_0(k_{r1}a)}{J_0(k_{r2}a) + BY_0(k_{r2}a)}, \quad (24)$$

and

$$N_3 = \frac{J_0(k_{r2}b) + BY_0(k_{r2}b)}{K_0(k_{r3}b) + CI_0(k_{r3}b)}. \quad (25)$$

Hence the dissipative losses in the specimen and dielectric sleeve resonator are

$$\begin{aligned} Q^{-1}_{spec} &= p_{e,spec}\tan\delta_{e,spec} \\ &= \left(1 + \frac{\epsilon'_{r,sleeve}}{\epsilon'_{r,spec}}\left[N_2^2\left(\frac{k_{r1}}{k_{r2}}\right)^2\frac{\bar{I}_2}{\bar{I}_1} + \frac{1}{\epsilon'_{r,sleeve}}N_2^2N_3^2\left(\frac{k_{r1}}{k_{r3}}\right)^2\frac{\bar{I}_3}{\bar{I}_1}\right]\right)^{-1} \cdot \\ &\quad \cdot\tan\delta_{e,spec}, \end{aligned} \quad (26)$$

and

$$\begin{aligned} Q^{-1}_{sleeve} &= p_{e,sleeve}\tan\delta_{e,sleeve} \\ &= \left(1 + \frac{\epsilon'_{r,spec}}{\epsilon'_{r,sleeve}}\left[\frac{1}{N_2^2}\left(\frac{k_{r2}}{k_{r1}}\right)^2\frac{\bar{I}_1}{\bar{I}_2} + \frac{1}{\epsilon'_{r,spec}}N_3^2\left(\frac{k_{r2}}{k_{r3}}\right)^2\frac{\bar{I}_3}{\bar{I}_2}\right]\right)^{-1} \cdot \\ &\quad \cdot\tan\delta_{e,sleeve}. \end{aligned} \quad (27)$$

Conductor losses can also be derived [21],

$$Q_c^{-1} = \frac{R_s}{G}, \tag{28}$$

where G is the geometric factor given by,

$$G = \frac{\omega^3 \mu_0^2 \epsilon_0 \epsilon'_{r,spec} h^3 T_1}{4\pi^2 p^2 T_2 + h^3 (k_{r1} N_2 N_3)^2 T_3 / \bar{I}_1}, \tag{29}$$

R_s is the measured surface resistance of the metal comprising the cavity endplates and sidewall and

$$T_1 = 1 + \frac{\epsilon'_{r,sleeve}}{\epsilon'_{r,spec}} \left[N_2^2 \left(\frac{k_{r1}}{k_{r2}}\right)^2 \frac{\bar{I}_2}{\bar{I}_1} + \frac{1}{\epsilon'_{r,sleeve}} N_2^2 N_3^2 \left(\frac{k_{r1}}{k_{r3}}\right)^2 \frac{\bar{I}_3}{\bar{I}_1} \right], \tag{30}$$

and

$$T_2 = 1 + N_2^2 \left(\frac{k_{r1}}{k_{r2}}\right)^2 \frac{\bar{I}_2}{\bar{I}_1} + (N_2 N_3)^2 \left(\frac{k_{r1}}{k_{r3}}\right)^2 \frac{\bar{I}_3}{\bar{I}_1}, \tag{31}$$

and

$$T_3 = \left[K_0(k_{r3} c) + C I_0(k_{r3} c) \right]^2 . \tag{32}$$

The integrals $\bar{I}_1$, $\bar{I}_2$, and $\bar{I}_3$ may be evaluated numerically or in closed form [21].

Dielectric Sleeve Resonator Without Sample

The first step in experimental evaluation of specimen permittivity and loss is to determine the real permittivity and dielectric loss of the sleeve resonator. For this case $k_{r1}^2 = k_{r3}^2 = k_c^2 = \beta^2 - k_0^2$ and $k_{r2}^2 = k_0^2 \epsilon'_{r,sleeve} - \beta^2$. An exactly analogous procedure as that used above for a loaded sleeve resonator may be employed in deriving the eigenvalue equation for the dielectric sleeve resonator without specimen loading, positioned centrally in a cylindrical cavity. For the empty sleeve resonator case, the eigenvalue equation becomes,

$$\frac{k_{r2} I_1(k_c a) J_0(k_{r2} a) + k_c I_0(k_c a) J_1(k_{r2} a)}{k_{r2} I_1(k_c a) Y_0(k_{r2} a) + k_c I_0(k_c a) Y_1(k_{r2} a)} - \frac{k_c \tilde{P}_1(b) J_1(k_{r2} b) - k_{r2} J_0(k_{r2} b)}{k_c \tilde{P}_1(b) Y_1(k_{r2} b) - k_{r2} Y_0(k_{r2} b)} = 0, \tag{33}$$

where

$$\tilde{P}_1(b) = \frac{K_0(k_c b) + \tilde{C} I_0(k_c b)}{K_1(k_c b) - \tilde{C} I_1(k_c b)}, \tag{34}$$

and

$$\tilde{C} = \frac{K_1(k_c c)}{I_1(k_c c)}. \tag{35}$$

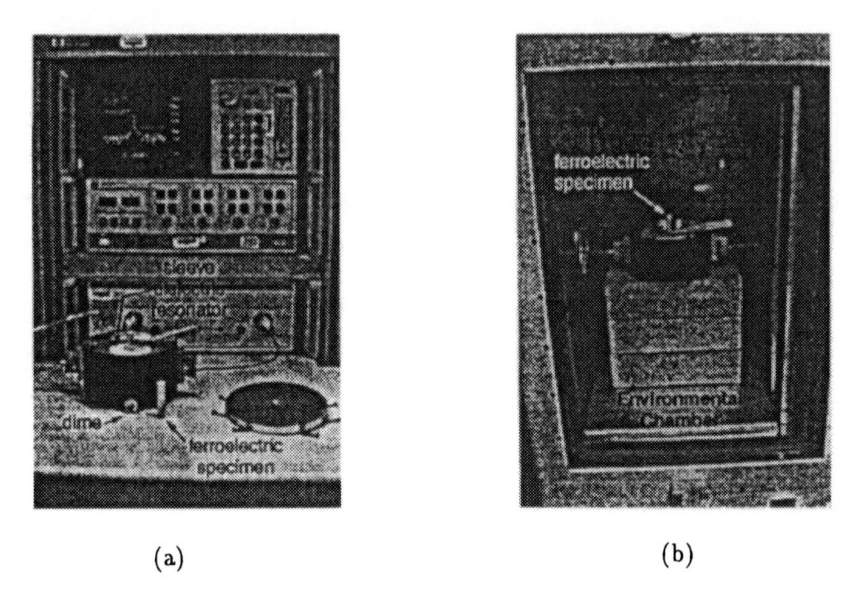

(a) (b)

Figure 2: (a) Dielectric sleeve resonator system for performing dielectric measurements at nominal 1 GHz and (b) sleeve resonator system in environmental chamber.

MEASUREMENTS

An example of a dielectric sleeve resonator system for performing variable-temperature measurements on high-permittivity ferroelectric materials at UHF frequencies close to 1 GHz is shown in Fig. 2. A low-loss, commercially available dielectric sleeve resonator was fabricated and placed inside a metal cavity for variable-temperature measurements. Although any TE_{0mp} resonant mode could be used, only the dominant, most easily identified TE_{011} mode results are reported here. The permittivity and dielectric loss tangent of the sleeve resonator were first determined. Then the permittivity and dielectric loss tangent of a rod specimen are evaluated by measuring the differences in resonant frequency and unloaded Q-factors for the TE_{011} mode with specimen inserted versus without insertion. Adjustable rf coupling loops are used so that the measured (loaded) Q-factor is essentially the same as the unloaded Q-factor for the 2-port system in transmission. Typical diameters and heights of rod-shaped specimens are 10 mm and 45 mm for the measurements reported here. The entire resonant system is then placed into an environmental chamber illustrated in Fig. 2.

The average time needed to attain $\pm 1°C$ temperature stability is 2 hours for each measurement temperature. Use of easily identified TE_{011}-mode structure allows no capacitive coupling between the specimen and the cavity end plates. Hence

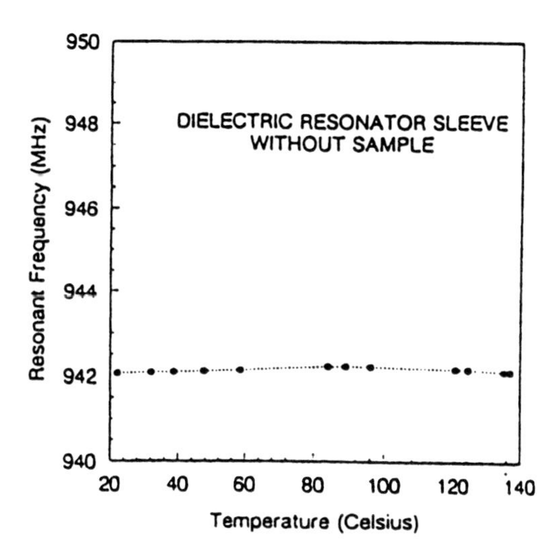

Figure 3: Resonant frequency of dielectric sleeve resonator as function of temperature without sample insertion.

repeatable measurements can be performed with disassembly and reassembly of the cavity.

Variable-temperature measurements were first made at UHF frequencies (942 MHz) on the dielectric sleeve resonator without sample insertion. These measurements, taken between 20°C and 140°C are shown in Figs. 3 and 4. The sleeve resonator has a real relative permittivity and dielectric loss tangent of 37.1±0.2 and $3.24\times10^{-5}\pm2\times10^{-5}$, respectively, at 20°C. The internal and external diameters of the sleeve are 10 mm and 60 mm and the height of the sleeve was 45 mm. The diameter of the cavity was 100 mm.

Although the sleeve resonator exhibited a temperature coefficient with respect to resonant frequency, τ_f, that was close to 0 ppm/°C, the measured unloaded Q-factor of the empty dielectric sleeve resonator decreases as temperature increases over this temperature range, almost by 20%, and must be taken into account in measuring dielectric losses of the ferroelectric specimens. It is also important to note that the surface resistance of the metal comprising the cavity was measured over the temperature range using an oriented sapphire dielectric rod resonator dimensioned for TE_{011} resonance at 10 GHz. For other measurement frequencies the surface resistance is scaled proportionally to the square root of frequency. Measured resonant frequencies and unloaded Q-factors with various ferroelectric sample in-

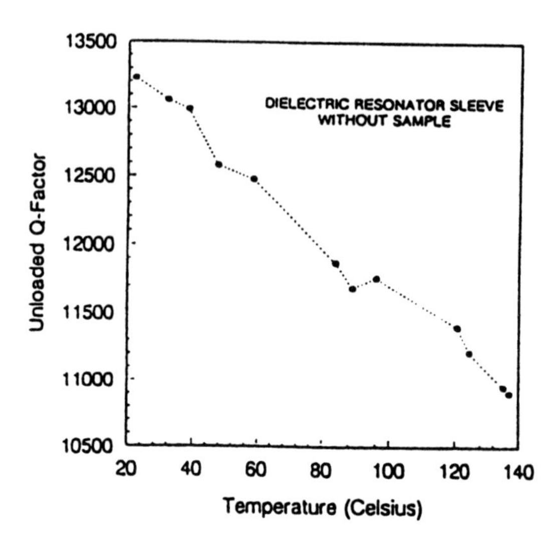

Figure 4: Unloaded Q-factor of dielectric sleeve resonator system as function of temperature without sample insertion.

sertions are given in Figs. 5 and 6. As temperature increases for the specimens tested, the resonant frequency shift decreases and unloaded Q-factor increases. This occurs because dielectric measurements are being performed at temperatures more distant from the Curie temperature that defines the ferroelectric/paraelectric state of the tested specimen. For the specimens reported here, Curie temperatures were less than -20 °C. Evaluated real relative permittivities and dielectric loss tangents for these example materials are shown in Figs. 7 and 8. In Fig 8 the variable-temperature and variable-frequency conductor losses were taken into account, as were the variable- temperature characteristics of the sleeve resonator.

UNCERTAINTY ANALYSES

The rms relative uncertainty in specimen real permittivity evaluation, assuming negligible gaps between sample and sleeve and negligible uncertainties in the height of specimen and sleeve, is given by

$$\frac{\Delta\epsilon'_{r,spec}}{\epsilon'_{r,spec}} = [(S_b^{\epsilon'_{r,spec}}\frac{\Delta b}{b})^2 + (S_c^{\epsilon'_{r,spec}}\frac{\Delta c}{c})^2 + (S_{\epsilon'_{r,sleeve}}^{\epsilon'_{r,spec}}\frac{\Delta\epsilon'_{r,sleeve}}{\epsilon'_{r,sleeve}})^2 + (S_f^{\epsilon'_{r,spec}}\frac{\Delta f}{f})^2]^{1/2}, \quad (36)$$

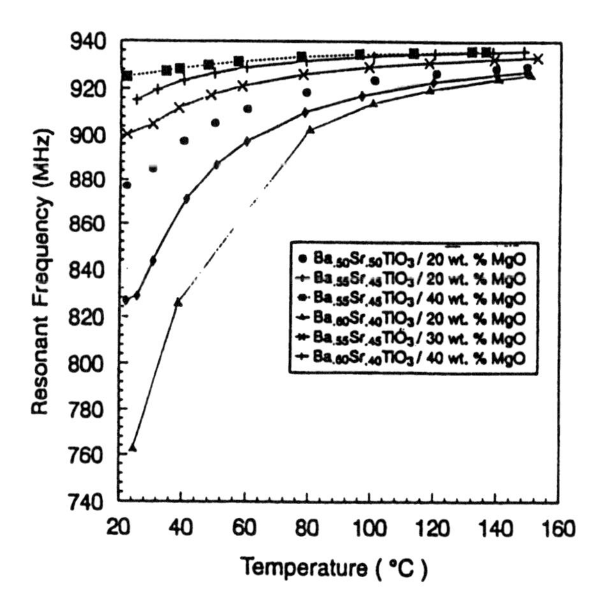

Figure 5: Resonant frequency of dielectric sleeve resonator system as function of temperature with sample insertion.

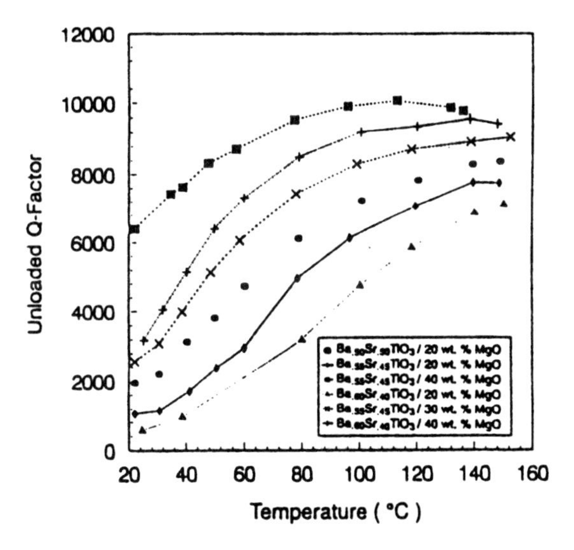

Figure 6: Unloaded Q-factor of dielectric sleeve resonator system as function of temperature with sample insertion.

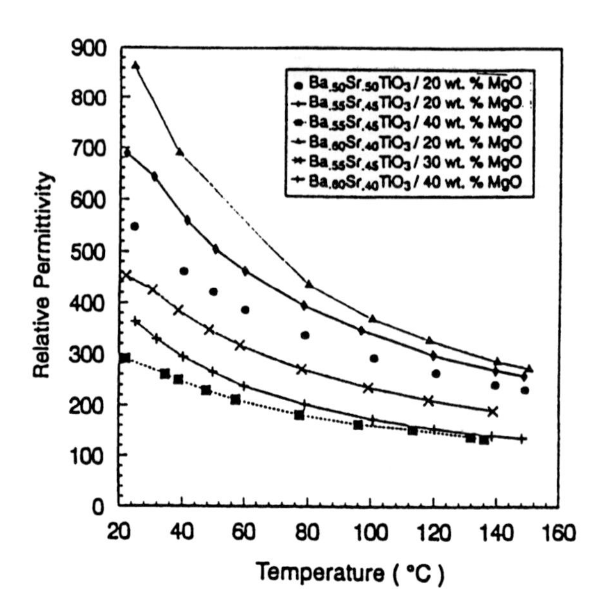

Figure 7: Computed real permittivity of rod specimen as a function of temperature.

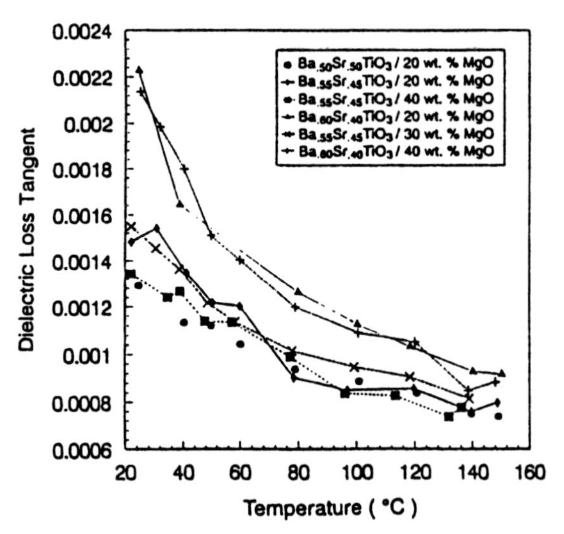

Figure 8: Computed dielectric loss tangent of rod specimen as a function of temperature.

where the weight sensitivity functions of the respective relative uncertainties in b, c, $\epsilon'_{r,sleeve}$, and f are given by

$$S_x^{\epsilon'_{r,spec}} = \frac{x}{\epsilon'_{r,spec}} \frac{\partial \epsilon'_{r,spec}}{\partial x}. \tag{37}$$

The rms relative uncertainty in specimen dielectric loss tangent may similarly be given in terms of the relative uncertainties of the sample and sleeve partial electric filling factors, the surface resistance of the enclosing cavity, the geometric factor, the loss tangent of the sleeve resonator, and the system unloaded Q- factor; that is,

$$\begin{aligned} \frac{\Delta \tan \delta_{e,spec}}{\tan \delta_{e,spec}} &= [(S_{p_{e,spec}}^{\tan \delta_{e,spec}} \frac{\Delta p_{e,spec}}{p_{e,spec}})^2 + (S_{p_{e,sleeve}}^{\tan \delta_{e,sleeve}} \frac{\Delta p_{e,sleeve}}{p_{e,sleeve}})^2 \\ &+ (S_{R_s}^{\tan \delta_{e,spec}} \frac{\Delta R_s}{R_s})^2 + (S_G^{\tan \delta_{e,spec}} \frac{\Delta G}{G})^2 \\ &+ (S_{\tan \delta_{e,sleeve}}^{\tan \delta_{e,spec}} \frac{\Delta \tan \delta_{e,sleeve}}{\tan \delta_{e,sleeve}})^2 + (S_{Q_0}^{\tan \delta_{e,spec}} \frac{\Delta Q_0}{Q_0})^2]^{1/2}. \end{aligned} \tag{38}$$

The uncertainty in the diameter of the sleeve resonator and real permittivity of the sleeve resonator dominate the uncertainty in the specimen permittivity for dimensional uncertainties of 0.01 mm, sleeve permittivity uncertainty of 0.3%, and resonant frequency uncertainty of 1 kHz. Total relative uncertainties in evaluated specimen permittivity for various values of test specimen permittivity are given in Fig. 9. To decrease the uncertainty in measured $\epsilon'_{r,spec}$ for $\epsilon'_{r,spec} < 300$, specimen diameters should be increased so that $p_{e,spec}$ is greater than 0.1.

Total uncertainties in dielectric loss tangent are dominated by uncertainties in measured unloaded Q-factor, sample partial electric energy filling factor, sleeve resonator loss tangent and cavity conductor losses. Total relative uncertainties in specimen loss tangent for the given sleeve resonator system as a function of test specimen permittivity are given in Fig. 10.

SUMMARY

Low-loss dielectric sleeve resonators can be used for accurate variable-temperature dielectric property measurements of high-permittivity ferroelectric materials, even at temperatures close to the Curie temperature. Use of multiple sleeve resonators or higher-order resonant modes permit wider bandwidth characterization of the same specimen. For $Ba_xSr_{1-x}TiO_3$ ceramic composites, permittivity and dielectric loss tangent both significantly change with temperature where tuning is likely to be employed. As x increases, both permittivity and dielectric loss predictably increase, although less rapidly at temperatures further from the Curie temperature. Hence

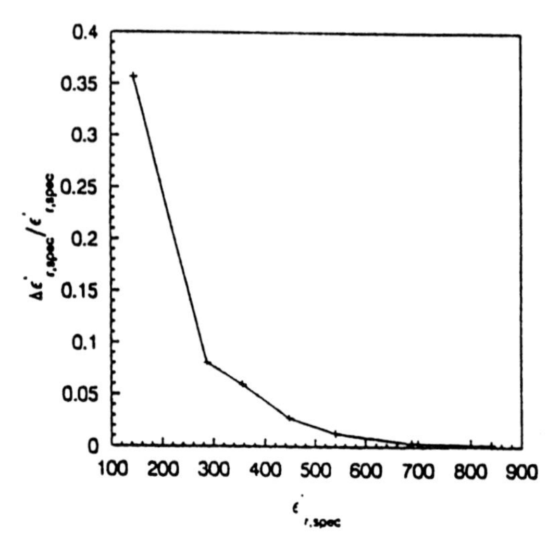

Figure 9: Total uncertainty in measured specimen real permittivity as a function of specimen permittivity for geometry and relative uncertainties given in text.

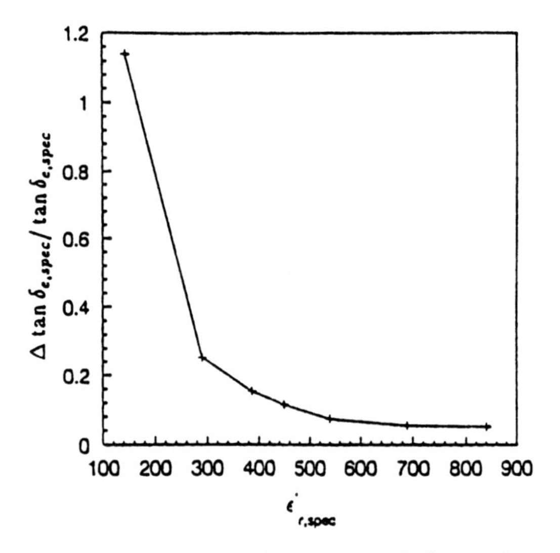

Figure 10: Total relative uncertainty in specimen dielectric loss tangent as function of test specimen permittivity.

accurate measurements of the variable-temperature properties at the operational frequency are useful for prediction of ferroelectric device figure-of-merit and any necessary temperature control. When use is made of sleeve dielectric resonators, care must be exercised in their design to ensure sufficient specimen electric energy filling factors for prescribed measurement uncertainties in permittivity and dielectric loss tangent.

ACKNOWLEDGMENTS

We wish to acknowledge the U.S. Army Research Laboratory for providing test specimens. We also thank J. Krupka for useful conversations regarding this work and J. DeRiso of Trans-Tech Inc. for supplying the sleeve resonator used in test measurements.

REFERENCES

[1]B.W. Hakki and P.D. Coleman, "A dielectric resonator method of measuring inductive capacities in the millimeter range," *IEEE Trans. Microwave Theory Tech.*, **8** 402-410 (1960).

[2]W.E. Courtney, "Analysis and evaluation of a method of measuring the complex permittivity and permeability of microwave insulators," *IEEE Trans. Microwave Theory Tech.*, **18** 476-485 (1970).

[3]D. Kajfez and P. Guillon, *Dielectric Resonators*, Dedham, MA: Artech House, 1986.

[4]Y. Kobayashi, N. Fukuoka, and S. Yoshida, "Resonant modes for a shielded dielectric rod resonator," *Electron. and Commun. (Japanese)*, **64-B** 46-51 (1981).

[5]Y. Kobayashi, Y. Aoki, and Y. Kabe, "Influence of conductor shields on the Q-factors of a TE_0 dielectric resonator," *IEEE MTT-S Int. Microwave Symp. Dig.*, 281-284 (1985).

[6]S. Vigneron and P. Guillon, "Mode matching for determination of the resonant frequency of a dielectric resonator placed in a metallic box," *IEE Proc., Pt. H*, **134** [2] 151-155 (1987).

[7]J. Krupka, "Resonant modes in shielded cylindrical and single-crystal dielectric resonators," *IEEE Trans. Microwave Theory Tech.*, **37** 691-697 (1989).

[8] *Dielectric Resonators-A Designer Guide to Microwave Dielectric Ceramics*, Trans-Tech Inc. Pub. No. 50080040, Rev. 2, 1990.

[9] M.E. Tobar and A.G. Mann, "Resonant frequencies of higher order modes in cylindrical anisotropic resonators," *IEEE Trans. Microwave Theory Tech.*, **39** 2077-2083 (1991).

[10] Y. Kobayashi and T. Senju, "Resonant modes in shielded uniaxial-anisotropic dielectric rod resonator," *IEEE Trans. Microwave Theory Tech.*, **41** 2198-2205 (1993).

[11] J. Krupka, R.G. Geyer, and D. Cros, "Measurements of permittivity and dielectric loss tangent of low-loss dielectric materials with a dielectric resonator operating on higher-order $TE_{0\gamma\delta}$ modes," *Proc. of MIKON 94 X Int. Microwave Conf.*, **2** 567-572 (1994).

[12] J. Krupka, R.G. Geyer, M. Kuhn, J.H. Hinken, "Dielectric properties of single crystals of Al_2O_3, $LaAlO_3$, $NdGaO_3$, $SrTiO_3$, and MgO at cryogenic temperatures," *IEEE Trans. Microwave theory Tech.* **42** 1886-1890 (1994).

[13] J. Krupka, D. Cros, M. Aubourg, and P. Guillon, "Study of whispering gallery modes in anisotropic single-crystal dielectric resonators," *IEEE Trans. Microwave Theory Tech.*, **42** 56-61 (1994).

[14] R.G. Geyer and J. Krupka, "Microwave dielectric properties of anisotropic materials at cryogenic temperatures," *IEEE Trans. Instr. and Meas.*, **44** 329-331 (1995).

[15] J. Krupka, K. Derzakowski, A. Abramowicz, M. Tobar, and R.G. Geyer, "Measurements of the complex permittivity of extremely low loss dielectric materials using whispering gallery modes," *IEEE MTT-S Int. Microwave Symp. Digest*, **III** 1347-1350 (1997).

[16] R.G. Geyer, C. Jones, and J. Krupka, "Microwave characterization of dielectric ceramics for wireless communications," in *Advances in Dielectric Ceramic Materials, Trans. Am. Ceram. Soc.*, **88** 75-91 (1998).

[17] J. Krupka and R.G. Geyer, "Loss-angle measurement," in *Wiley Encyclopedia of Electrical Electron. Engineering*, **11** 606-619 (1999).

[18] R.J. Cook, "Microwave cavity methods," in *High Frequency Dielectric Measure-*

ments, (J. Chamberlain and G.W. Chantry, ed.): IPC Science and Tech. Press, Guildford, U.K., 12-27 (1973).

[19]E. Ni and U. Stumper, "Permittivity measurements using a frequency-tuned microwave TE_{01} cavity resonator," *IEE Proc., Pt. H*, **132** 27-32 (1985).

[20]E. Vanzura, R.G. Geyer, and M. Janezic, "The NIST 60- millimeter diameter cavity resonator: performance evaluation for permittivity measurements," *Natl. Inst. Stand. Technol. Tech. Note 1354*, 1993.

[21] R.G. Geyer, P. Kabos, and J. Baker-Jarvis, "Dielectric Sleeve Resonator Techniques for Microwave Complex Permittivity Evaluation," to be published (2001).

FIELD DEPENDENCE OF THE DIELECTRIC PROPERTIES OF BARIUM STRONTIUM TITANATE SINGLE CRYSTALS

D. Garcia*, R. Guo and A. S. Bhalla

Materials Research Laboratory, The Pennsylvania State University, University Park, PA, 16802

ABSTRACT

In this work, BST single crystals were grown by the Laser Heated Pedestal growth technique and the dielectric permittivity and loss were analyzed as functions of temperature and the dc bias (electric field) for low frequencies. Dielectric dispersion phenomenon was also performed at high frequencies (10^6-10^9 Hz). High tunability factors, 40-60% (for 10kV/cm), were measured at room temperature for the crystals with compositions around Ba/Sr=65/35 ($T_c \approx 20^oC$). Nevertheless, relatively lower dielectric losses in paraelectric phase lead to the higher tunability figure of merit for the crystal with T_c<RT.

INTRODUCTION

Barium strontium titanate (BST) solid solutions have been subject of many studies based upon their high potential for technological applications as pyroelectric thermal imaging sensors, multi-layer capacitors and dynamic random access memories (DRAM) (1-3). New interest in such materials is focused on their dielectric properties under dc-bias electric field to be used as tunable elements in microwave circuits (4-6).

Most of the processing routes encountered for BST material aimed to produce bulk ceramic or thin/thick films (1-6). The former represents an

* Permanent address: Department of Physics, Federal University of Sao Carlos, Sao Carlos, SP. E-mail address: ducinei@psu.edu or ducinei@power.ufscar.br

inexpensive route for the commercial production and the later, in spite of the higher cost, provides the needed microelectronic integration. However, besides the unquestionable importance of polycrystalline materials, tailoring of properties and phenomenological studies are still dependent on the studies in the single crystal form. This fact is reinforced by recent results on the improvement of electromechanical properties of complex ferroelectric perovskites through crystallographic engineering (7). Nevertheless, BST based systems are difficult to obtain in good quality crystalline form due to the high melting point and the lack of any compositional congruency (8). The recent work showed that it is possible to obtain high quality single crystals of $BaTiO_3$-$SrTiO_3$ solid solution compositions by the Laser Heated Pedestal Growth technique (LHPG) (9-10).

This report is part of a ongoing work on the characterization and properties of BST single crystal fibers grown by the LHPG technique. Particularly in this paper, the focus was to study the effect of the bias field dependence on dielectric behavior of $(Ba_{1-x}Sr_x)TiO_3$ ferroelectric single crystals, within the compositions showing ferroelectric-paraelectric phase transition closer to the room temperature. The goal of this research, however, is to explore the ferroelectrics with desirable characteristics for microwave tunable capacitor applications, and which also can be processed with reproducible physical properties. Features of dielectric tuning under bias field, high frequency dependence and some recent improvements on the fiber growth affecting the dielectric properties are discussed.

EXPERIMENTAL PROCEDURE

Ceramic rods of barium strontium titanate were prepared by the conventional solid state reaction method to be used as seed and feed for the growth. Details of ceramic processing are described in reference 10. As reported earlier (9,10), Laser Heated Pedestal Growth technique was employed to grow the BST single crystal fibers. The growth conditions, including fine-tuning of molten zone temperature, seed/feed-pulling rates, feedrod alignment, have been continuously improved and optimized, favoring the better properties of the single crystals, especially those related to the compositional distribution. Single crystal fibers ~15mm in length and 0.9mm in diameter could be grown. The as grown fibers were transparent and of light brown color. More details on the BST growth conditions and equipment can be found in the references 10 and 11, respectively.

BST fibers were cut, polished, annealed at low temperature to release mechanical stress and gold-electroded on the circular cross-section areas for the dielectric measurements. Low frequency field-dependent dielectric data was obtained using an automated system, which included an HP4284A LCR-meter and a TREK high voltage supply. Dielectric constant and loss versus electric field (up to 10kV/cm) measurements were taken at small temperature intervals during cooling, from +40°C to 0°C , at 10 kHz. The period of a complete electric field cycle was ~100s. For the field dependent measurement, a blocking circuit was used. High frequency dielectric measurements were carried out in an HP4291A RF impedance analyzer, at room temperature.

The dc-biasing effect in dielectric permittivity was quantified by the tunability factor, D_t, defined at a specific temperature and field as:

$$D_t(E,T)=[1-\varepsilon_r(E,T)/\varepsilon_r(0,T)], \quad (1)$$

where $\varepsilon_r(0,T)$ and $\varepsilon_r(E,T)$ are the dielectric constants under zero and non-zero dc-bias field.

RESULTS AND DISCUSSION

Lowering of phase transition temperatures occurs for the three known phase transformations (cubic-tetragonal, tetragonal-orthorhombic and orthorhombic-rhombohedral), which are found in BaTiO3-based systems (12), when strontium content is increased in the BST. These results are displayed in figure 1 that also shows the low frequency dielectric constant as a function of temperature for $(Ba_{1-x}Sr_x)TiO_3$ single crystal fibers. Compositions with

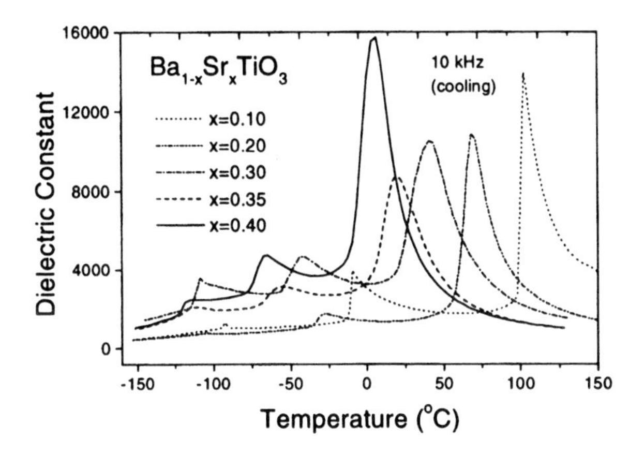

Figure 1. Temperature dependence of the dielectric constants for $Ba_{1-x}Sr_xTiO_3$ single crystal fibers during cooling runs at constant cooling rate of 4°C/min.

$0.30 \leq x \leq 0.40$ show paraelectric-ferroelectric phase transition around room temperature and will be the focus of this study. The crystal orientation in the growth direction (here the measurement direction) is close to <110> for $x<0.30$ and <100> for $x>0.30$, as observed from X-ray diffraction analysis of the sample surfaces. High dielectric constants at T_c and sharp phase transitions could be observed for all compositions. This is a result of a systematic optimization process of the growth conditions, especially for the Sr concentration $x=0.40$. After many growth attempts and the fine tuning of the molten zone temperature and pulling rate, fibers with higher crystal quality were grown. Comparison between the dielectric properties of crystals from the first batch and those from the recent batch showed the differences resulting from the growth optimization, as displayed in figure 2. New BST single crystal fibers, with $x=0.40$, showed the maximum dielectric constant two times higher than the former one (figure 2a). However the dielectric losses were not reduced at the phase transition temperature, but they were improved at temperatures above T_c (figure 2b).

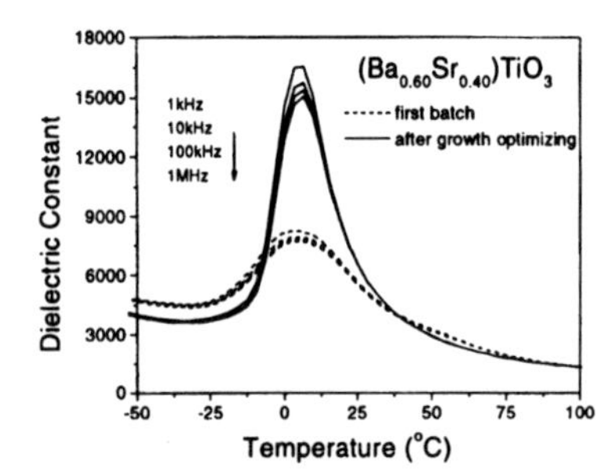

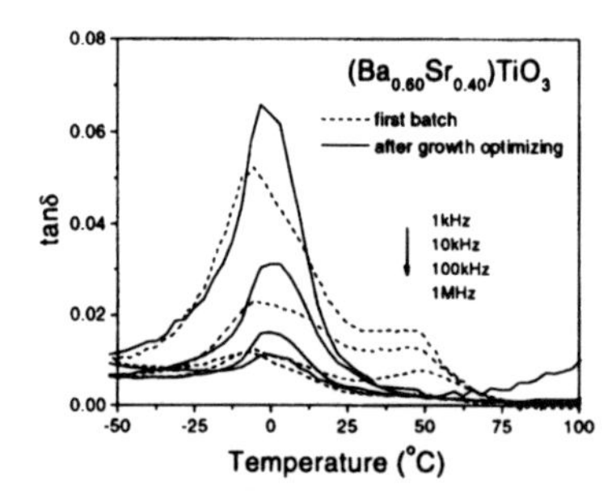

Figure 2. Temperature dependence of (a) dielectric constant and (b) loss factor for the crystal grown in the earlier attempt and after optimizing the growth condition for $Ba_{0.60}Sr_{0.40}TiO_3$ single crystal fibers.

Figure 3 and 4 show the results for the isothermal measurements of the dielectric constant and loss factor (tanδ) under bias field conditions, respectively, for BST single crystals with $x=0.30$, 0.35 and 0.40. The paraelectric-ferroelectric phase transition temperatures, 4°C, 18°C and 42°C, respectively, are practically covered in the temperature range of the measurement. The results shown are for the first electric field cycle. From figure 3, it can be seen that the changes in dielectric constant under field are higher for the composition where the phase transition is closer to the measurement temperature. Linear field dependence and small hysteresis effects occur when the crystals are in paraelectric phase, as expected. This can be observed clearly for BST with x=0.40. The dielectric losses decrease with the field, but the levels (and hysteresis behavior) stay high at temperatures just below T_c (figure 4). Thus, in the measured temperature range and from the point of view of losses, the crystal compositions x=0.35 and 0.40 show the most promising characteristics for the tunable applications

The values of dielectric tunability factor D_t, defined in equation 1, are plotted as a function of dc-bias field in figure 5. The determination of D_t was made using the data points of $\varepsilon_r[T,0]$ in the end of bias field cycle, to avoid any influence from the hysteresis behavior of the first quarter of the cycle. The level of tunability factors and the nature of the curves are related to the temperature (T-Tc). The closer the phase transition, the highest was the tunability; reaching

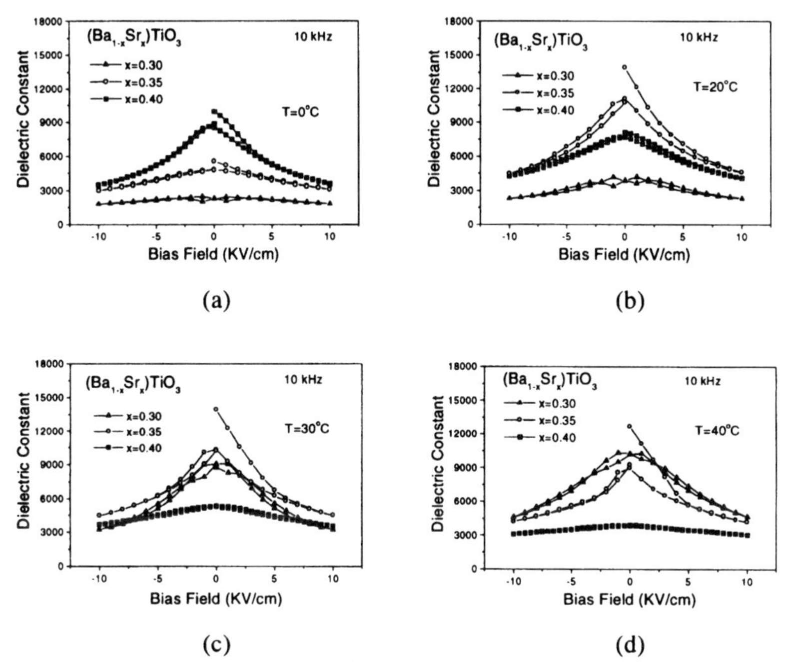

Figure 3. Field dependence of dielectric constant ε_r at (a) 0°C, (b) 20°C, (c) 30°C and (d) 40°C for BST single crystal fibers (first cycle curves).

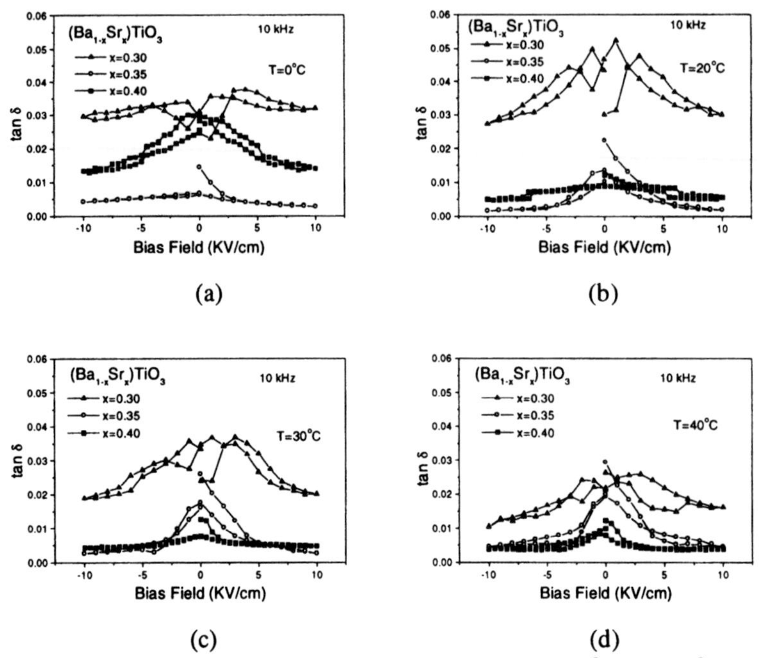

Figure 4. Field dependence of dielectric loss tanδ at (a) 0°C, (b) 20°C, (c) 30°C and (d) 40°C for BST single crystal fibers (first cycle curves).

60% at 10kV/cm and $T \approx T_c$ for the compositions studied. Linear and non-hysteretic behavior can be observed at $T=T_c+25°C$ (figure 5d) for BST crystal with x=0.40, but D_t shows a significant decrease (~20%). Comparison of the dielectric features between the different compositions, at 20°C, is summarized in the Table I, where the tunability figures of merit $K=D_t/\tan\delta$, were calculated. The tunability level at room temperature is highest for BST crystal with x=0.35, due to the proximity of T_c, but the lower dielectric loss brings comparable figure of merit for x=0.40 composition.

Figure 6a gives the RF-frequency dependence of dielectric constant and loss factor for the BST single crystals with x=0.30 and 0.40, at room temperature. Strong dispersion of dielectric constant is observed over 100MHz, accompanied with a peak in dielectric loss around 380MHz, and approaching very high tanδ levels. The frequency, where the loss peak occurs, can be characterized as

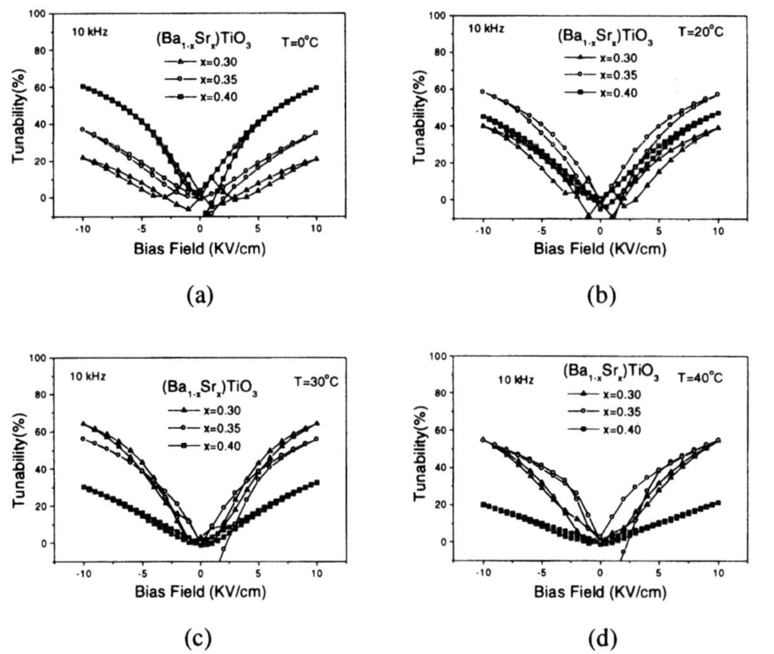

Figure 5. Field dependence of dielectric tunability D_t at (a) 0°C, (b) 20°C, (c) 30°C and (d) 40°C for BST single crystal fibers (first cycle curves).

Table I. Dielectric properties, tunability factor D_t and figure of merit, K, for the $(Ba_{1-x}Sr_x)TiO_3$ single crystal fibers, at 20°C.

Composition	ε_r*	tanδ*	D_t (at 10kV/cm)	K
x=0.30	3856	0.043	0.39	9.1
x=0.35	10790	0.010	0.57	57.0
x=0.40	7761	0.009	0.47	52.2

*At zero field, after first cycle is completed.

a relaxation frequency for this system. Similar results were found by Kazaoui et al. for ceramics of BST system (13). Figure 6b shows the influence of the bias-field on the high-frequency dielectric constant and respective tunability factor for BST with x=0.40, at room temperature and 500MHz. Only the low field levels could be used due to the limit of bridge internal bias-voltage supply (40V) but, making a linear extrapolation of the data points, a value of ~27% for tunability at 5kV/cm could be obtained. On comparing with the data of figure 5b, high-frequency tunability values were the same as obtained in the low-frequency measurements.

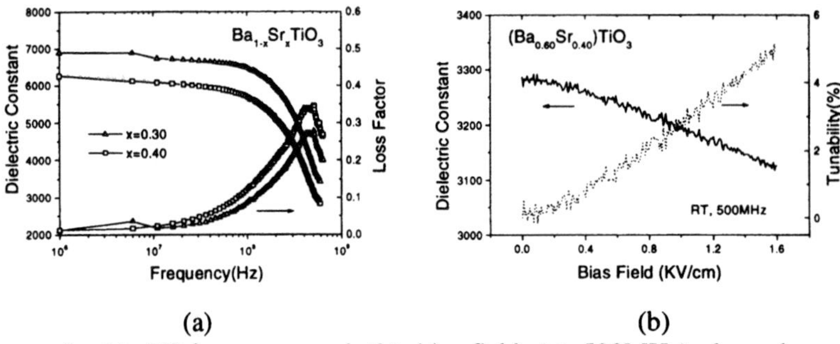

(a) (b)

Figure 6. (a) RF-frequency and (b) bias-field (at 500MHz) dependence of dielectric properties for BST single crystal fibers, at room temperature.

SUMMARY

High quality BST single crystals were grown by the LHPG technique under the optimized conditions. BST single crystals, 0.20<x<0.40, presented high permittivity, high dielectric tunability and relatively low loss around room temperature. At $T<T_c$, hysteretic behavior under bias field was observed in the ferroelectric phase. However, tunability factor remains the same in the high frequency measurements. The dielectric dispersion identified at ~ 400MHz may interfere the device performance operating at this frequency.

ACKNOWLEDGEMENTS

This work was supported by a grant from DARPA, under contract no. DABT63-98-1-002. D. Garcia acknowledges the support of FAPESP (Brazilian agency). The authors are grateful also to Dr. Petr Hana for helping in the dielectric measurements with the bias field.

REFERENCES

(1) R.W. Whatmore, P.C. Osbond, and N.M. Shorrocks, *Ferroelectrics*, **76**, 351(1987).

(2) T. Itoh, S. Tashiro, and H. Igarashi, *Jpn. J. Appl. Phys.*, **32**, 4261 (1993).

(3) A.I. Kingon, S.K. Streiffer, C. Basceri, and S.R. Summerfelt, *Mater. Res. Bull.*, **21**, 18 (1995).

(4) J.-W. Liou and B.S. Chiou, *J. Am. Ceram. Soc.*, **80**[12] 3093-99 (1997).

(5) C.L. Chen et. al, *Appl. Phys. Lett.*, **75**(3), 412 (1999).

(6) J. Im et al., *Appl. Phys. Lett.*, **76**(5), 625 (2000).

(7) S.-E. Park et. al, Proceedings of the SPIE – The International Society for Optical Enginnering, vol. 3675, pp.2-9 (1999).

(8) K.-H. Hellwegge, A.M. Hellwegge, ed., "Numerical Data and Functional Relationship in Science and Technology New Series", vol.3 (Ferro- and Anti-Ferroelectric Substances), Springer-Verlag, Berlin-Heidelberg,-New York, 1969.

(9) D. Garcia, R. Guo, and A. S. Bhalla, *Materials Letters*, **42**, 136 (2000).

(10)D. Garcia, R. Guo, and A.S. Bhalla, in *Electronic Ceramic Materials and Devices*; Ceramic Transactions, vol. 106, 175 (2000).

(11)J. Yamamoto and A.S.Bhalla, *Mater. Res. Bull.*, **24**, 761 (1989).

(12)F. Jona and G. Shirane, "Ferroelectric Crystals", Dover Publications, Inc., New York, 1993.

(13) S. Kazaoui et. al, *Ferroelectrics*, **135**, 85(1992).

ELECTRIC FIELD DEPENDENCE OF DIELECTRIC BEHAVIOR OF $(Sr_{1-x}Pb_x)TiO_3$

Y. Somiya, Ruyan Guo, A. S. Bhalla and L. E. Cross
Materials Research Institute, The Pennsylvania State University,
University Park, PA16802

ABSTRACT

The need for high frequency dielectric materials has increased due to the big demand of tunable communication devices. In this paper, $(Sr_{1-x}Pb_x)TiO_3$ compositions are selected as the potential materials for such devices.

Dielectric properties are measured without or with electric DC bias on $(Sr_{0.7}Pb_{0.3})TiO_3$. As expected with an increase of applied electric field, dielectric constants decrease and transition temperature increases along with high tunability values and low dielectric loss.

Dielectric behavior of diluted $(Sr_{0.7}Pb_{0.3})TiO_3$ is also investigated and (i) increase in transition temperature (ii) low dielectric constant (iii) low loss tangent at room temperature (iv) less change in transition temperature under DC bias are observed.

INTRODUCTION

Microwave dielectrics are in demand for wireless communication systems. In order to decrease number of components, there is a need for filters which have roles of both transmitter and receiver. Materials for these filters are required to have field dependence properties and one of the easiest ways to satisfy the demand is to find materials dielectric constant which show electric DC bias dependence. Furthermore, low dielectric loss tangents are considered to be necessary for the microwave dielectric materials. Several materials have been proposed for such applications.

In this paper, properties of one of the potential materials are described and results of diluted materials with nonferroelectric material, MgO, are also explained to seek possibility of using composite at microwave frequency since pure materials have high dielectric constant.

Regarding selection of materials, it should be an important consideration that preparation method is easy, simple and reproducible and that the transition temperature of the material is below devices' operating temperatures. This is because of the obvious reasons that the dielectric constants have typical hysteresis behavior in ferroelectric region under electric field. Nomura and Sawada have studied $SrTiO_3$: $PbTiO_3$ solid solution system and found the easy formation of good homogeneous composition[1)]. Therefore, lead strontium titanate(SPT) and specifically composition $(Sr_{0.7}Pb_{0.3})TiO_3$ (with Tc < room temperature) has been selected for the present study.

EXPERIMENTS

$(Sr,Pb)TiO_3$ compositions were prepared from high purity strontium carbonate, lead oxide and titanium oxide by mixing a suitable ratios of Sr:Pb:Ti :: 0.7:0.3:1. The raw materials were ball milled with media and liquid, and then calcinated. The reactant powder was pressed in a pellet form using a cold isostatic press and sintered at 1350°C.

Composites were prepared by addition of MgO to calcinated powders, the mixture was ball milled with media/ liquid and then procedure similar to the process for $(Sr, Pb)TiO_3$ was followed.

Phases of samples were identified by their X-ray diffraction (XRD) patterns using Cu $K\alpha$ at room temperature(PDIV, SCINTAG, Inc). XRD patterns were analyzed by Profit, program developed by Toraya [2)].

Dielectric properties were measured on the gold electrode coated sample using LCR meter (4284A: Hewlett Packard) without or with a blocking circuit, under the constant applied DC electric field conditions in cooling cycles except thermal hysteresis measurement. LCR meter and oven were controlled by the computer interfaces, *etc* and change in dielectric constant were measured on small samples under various DC bias. Dielectric tunability was obtained by projections of the plot that showed dielectric constant vs. temperature under various DC bias field.

Thermal expansion coefficients of samples, which were cut similar length to the reference, 6.54mm fused silica, were measured by the dilatometer (Theta Industries, inc.). These measurements are desirable to assess the physical strains in the samples.

Microstructures were studied using the gold sputtered fractured surfaces by scanning electron microscope (SEM: S-3500N, Hitachi).

RESULTS AND DISCIUSSION

(1) Pure $(Sr_{0.7}Pb_{0.3})TiO_3$

Well characterized samples of $(Sr_{0.7}Pb_{0.3})TiO_3$ (SPT) were used for this study. Figure 1 shows the dielectric response measured in cooling and heating cycles and dielectric behavior has less frequency dependence. Transition temperature of SPT in cooling cycle is ~12°C and difference of transition temperature between cooling and heating cycle is ~ 1°C. Dielectric loss tangents are about 5×10^{-3} at the maximum in the temperature range from 40 °C to 0 °C.

The blocking circuit was used for the measurements under DC bias and impedance range of LCR meter was selected to avoid discontinuous of loss tangent due to change of impedance range for measurements of capacitance. If capacitance of samples are changed from one condition to the another one of LCR meter, steps generally occur. Dielectric constant was not much affected by the blocking circuit, but loss tangent showed some difference between measurement without or with the blocking circuit. Low dielectric constant generally show relatively more variation due to the mismatch of impedance.

Figure 2 shows dielectric response of SPT ceramic as a function of temperature and electric DC bias at 10KHz. As applied DC bias increases, dielectric constants of $(Sr_{0.7}Pb_{0.3})TiO_3$ ceramic becomes lower, transition temperature increases and loss tangent peak around transition temperature is suppressed. Loss tangent around room temperature is below 5×10^{-3} even though loss tangent includes some error due to the measurements with the blocking circuit. Figure 3(a) shows dielectric constant of $(Sr_{0.7}Pb_{0.3})TiO_3$ ceramic as a function of electric DC bias and figure 3(b) dielectric loss tangent of SPT ceramic as a function of electric DC bias measured from 0KV/cm to 15KV, then from 10KV/cm to –15KV/cm and finally from –10KV/cm to 0KV/cm. These data are projections of dielectric properties vs. temperature at different DC bias field. SPT shows hystersis under lower DC bias below 2.5 KV/cm. This could be caused by remnant charges in the sample. Maximum value of dielectric constant is close to that of dielectric constant at heating cycle, so after appropriate annealing the sample could be recovered to initial maximum value of dielectric constant.

Tunability is calculated using dielectric constant under 0KV/cm in the first run (figure 4). K_{E1} expresses dielectric constant at electric DC bias E1and K_{E2} means dielectric constant at electric DC bias E2.

$$\text{Tunability}(\%)=\frac{K_{E1}-K_{E2}}{K_{E1}}\times 100$$

Tunability at 30°C, roughly about 15 °C above transition temperature, is 60% at 15KV/cm.

(2) Diluted $(Sr_{0.7}Pb_{0.3})TiO_3$ by addition of MgO

In order to explore materials which have reasonable dielectric constant, low dielectric loss tangent and suitable transition temperature (below room temperature), the composites are investigated and the results are compared with the ceramics of $(Sr_{0.7}Pb_{0.3})TiO_3$ sintered in a similar way to the composites.

Dielectric response of $(Sr_{0.7}Pb_{0.3})TiO_3$: MgO composite with 50: 50 composition by weight is shown in figure 5. Some changes in the transition temperature from the ceramics, reduction of dielectric constant and broadening of dielectric constant peak are observed. Regarding frequency dependence, composite samples do not show much dependence, but frequency dependence is noticed in ferroelectric region. Figure 6 shows the thermal hysteresis of SPT : MgO composite. The change in transition temperature is not obvious and thermal hysteresis is minimal.

Figure 7(a) shows a typical X-ray diffraction (XRD) pattern of SPT : MgO composite with 50 : 50 composition by weight after sintering, while figure 7(b) shows a XRD pattern of $(Sr_{0.7}Pb_{0.3})TiO_3$ powders mixed with MgO before sintering. The peaks corresponding to only SPT and MgO are observed. From dielectric response data, the transition temperature of the composite is shifted to above room temperature, but the structure change is not reflected in the XRD pattern. FWHM (full width at half maximum) of sintered SPT pattern in the composite is similar to that of corresponding SPT peaks before sintering. However, FWHM of MgO peaks in the sintered composite becomes narrower than that in the composite before sintering. Figure 7(c) is a typical XRD pattern of sintered SPT and figure 7(d) shows a XRD pattern of calcinated SPT powders. In this case, narrowing of FWHM of SPT peaks are observed. To understand changes in FWHM, XRD patterns are analyzed [2)] (figure 8(a) change in FWHM of SPT and 8(b) change in FWHM of MgO).

Change in FWHM is caused by several reasons such as changes in the grain sizes and stresses. Accordingly, microstructures of samples were investigated by SEM. The microstructure of sintered SPT: MgO composite, sintered SPT, and calcinated SPT powders are shown in figure 9(a)-9(c). Grain sizes of the calcinated powders are below a few micrometer, but grain sizes of the sintered SPT are over 10 micrometer. Grain size of SPT: MgO composite is below 10 micrometer, mostly on the order of a few micrometer. Although pure SPT grains grow during sintering, growth of SPT is restricted in the composite due to the MgO grain growth. There is always a possibility of small Mg doping on the B-site in SPT and this may cause some change of transition temperature. In addition, some stresses in the samples can alter transition temperature. In order to get some insight, the thermal expansion behavior of the samples was measured.

Thermal expansion coefficient is calculated using the results of thermal expansion to investigate change in transition temperature of composite. Figure 10(a) shows thermal expansion and thermal expansion coefficient of SPT : MgO composite and figure 10(b) shows the measurements on pure SPT. The transition temperature of SPT: MgO composite is not very sharp, whereas thermal expansion coefficient of SPT abruptly decreases around transition temperature and then increases. Thermal expansion coefficient of composite is similar to that of SPT above transition temperature of SPT. SPT itself shrinks around transition temperature. The composite SPT may be under the tensile stresses and therefore, transition temperature of SPT in the composite may be having some differences.

The results of field dependence of dielectric constants are shown in Figure 11. Since the measured capacitance of sample was small (below 10pF), therefore, measurements with the blocking circuit introduced little errors. Figure 11 (a) shows the dielectric response of composite as functions of temperature and electric DC bias at 10KHz from 0 to 20KV/cm. A small hysteresis is observed in case of SPT : MgO composite. Loss tangent does not change with the DC bias. Effect of DC bias on dielectric constant is shown in figure 11(b) and the tunability of the composite is plotted in figure 11(c). The preliminary data is interesting and the work related to the improvement of SPT : MgO composites is in progress.

SUMMARY

Dielectric behavior of $(Sr_{0.7}Pb_{0.3})TiO_3$ (SPT)was investigated and the results summarized as follows: (i) Dielectric response of SPT showed low frequency dependence, (ii) SPT showed small thermal hysteresis; difference of the transition temperature between cooling and heating cycles was $\sim 1^{o}C$ (iii) dielectric response was suppressed under the DC bias field as expected (iv) hysteresis was observed under low DC bias field after the samples were gone through the higher DC bias field cycle and this may be due to the effect of remaining charges on electrodes, (v) high dielectric tunability was observed.

For the SPT: MgO composite with 50 : 50 composition by weight, the results are summarized as follows: (i)the composite showed reduction of the dielectric constant and some frequency dependence, particularly in ferroelectric region, (ii) the composite did not show much thermal hysteresis of the transition temperature, (iii) reduction in grain growth of SPT in the composite was observed because of the MgO grain growth (iv) increase in transition temperature of SPT in the composite was observed. It may be due to the small substitution of Mg on B-site and the effect of small tensile in the samples. (v) dielectric loss tangent stayed below 5×10^{-3} at around room temperature, and (vi) dielectric tunability at ~ room temperature was about10 % at 20KV/cm.

ACKNOWLEDGEMENT

This study is sponsored by DARPA under contract No. DABT 63-98-1-002.

REFERENCES

1) S. Nomura and S.Sawada, "Dielectric Properties of Lead•Strontium Titanate", *J. Phys. Soc. Japan*, **10**, 108(1955).

2) H. Toraya, "Whole-Powder-Pattern Fitting Without Reference to a Structural Model: Application to X-ray Powder Diffractometer Data", *J. Appl. Cryst.*,**19**, 440(1986).

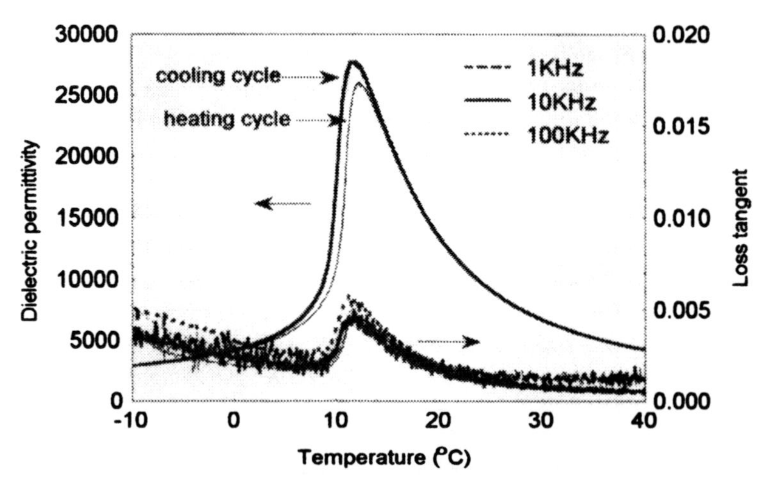

Figure 1 Dielectric response of $(Sr_{0.7}Pb_{0.3})TiO_3$ ceramic as functions of temperature and frequency in cooling and heating cycles

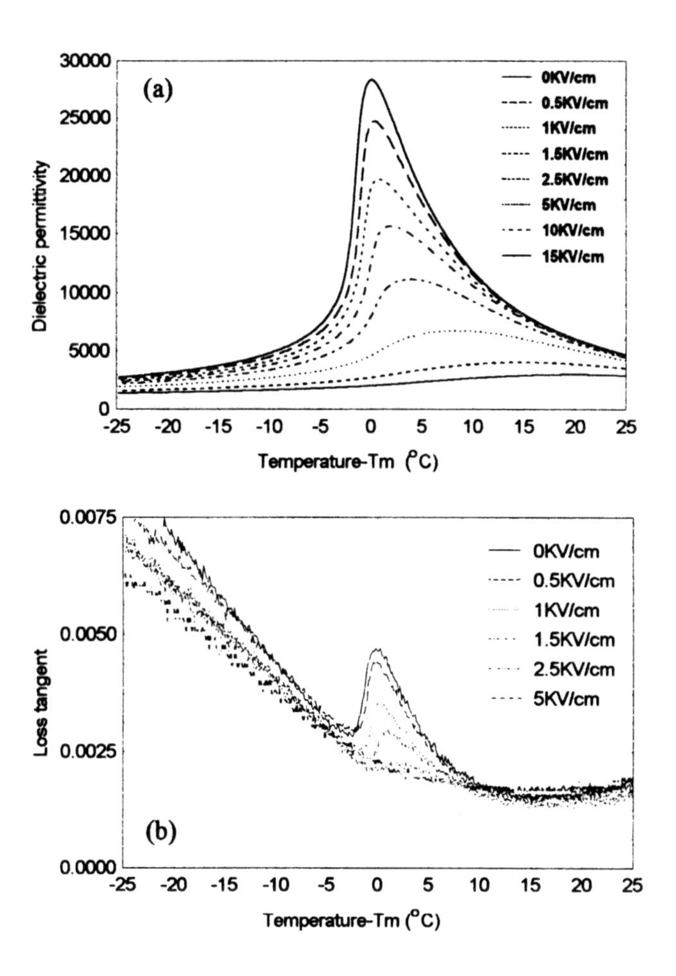

Figure2 Dielectric response of $(Sr_{0.7}Pb_{0.3})TiO_3$ ceramic as functions of temperature and applied electric DC bias field at 10KHz ; (a) dielectric permittivity and (b) dielectric loss tangent

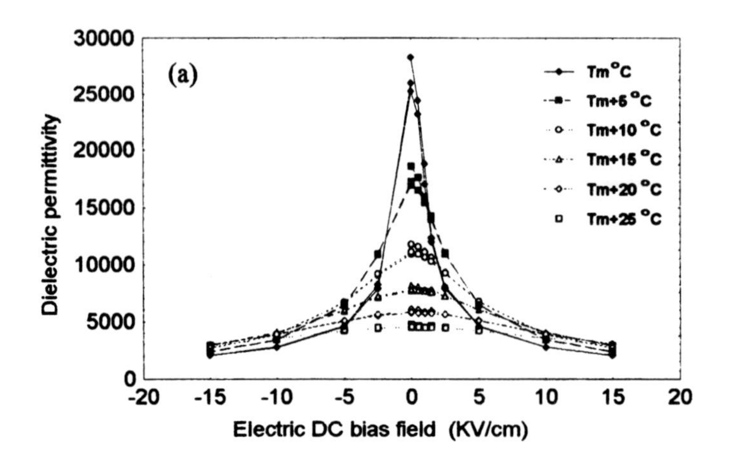

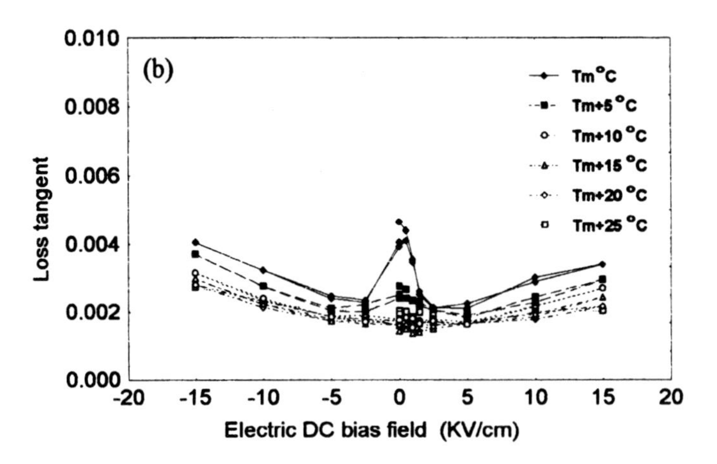

Figure 3 Dielectric response of $(Sr_{0.7}Pb_{0.3})TiO_3$ ceramic as a function of temperature and electric DC bias (0 – 15 – 0 – -15 – 0KV/cm) at 10KHz ; (a) dielectric permittivity and (b) dielectric loss tangent

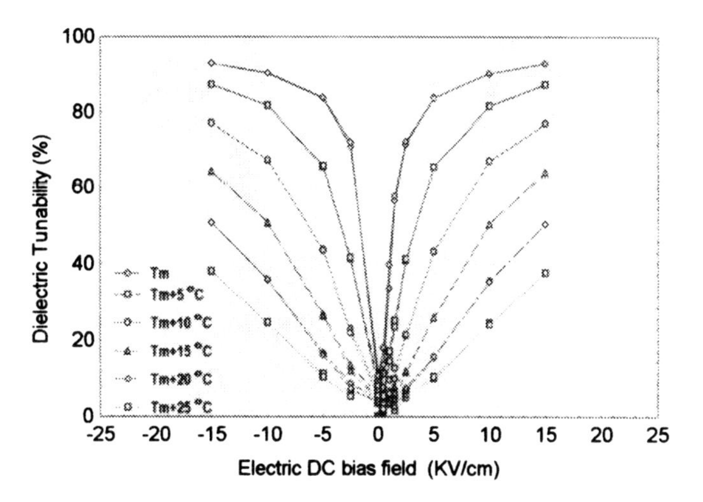

Figure 4 Dielectric tunability of $(Sr_{0.7}Pb_{0.3})TiO_3$ ceramic under DC bias field as a function of temperature at 10KHz

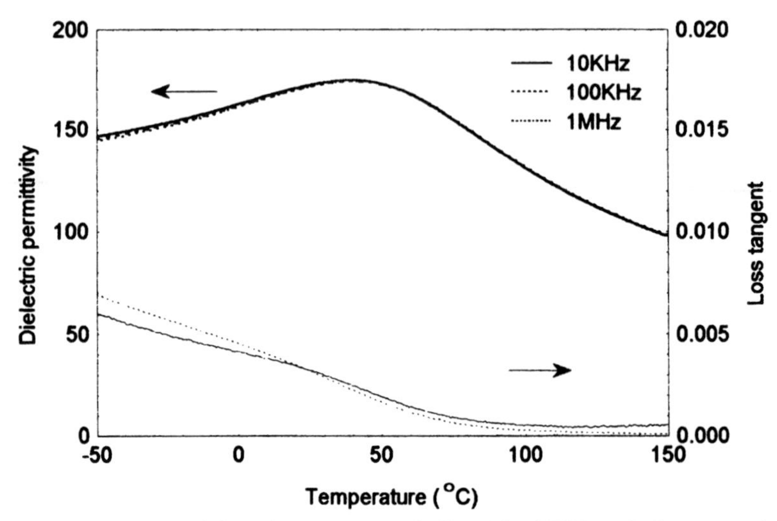

Figure 5 Dielectric response of $(Sr_{0.7}Pb_{0.3})TiO_3$: MgO composite as a function of temperature and frequency in cooling cycle

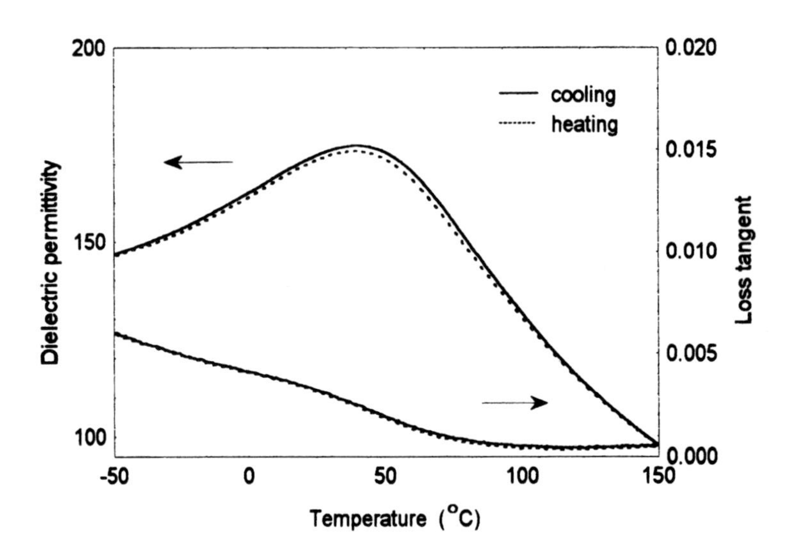

Figure 6 Dielectric response of $(Sr_{0.7}Pb_{0.3})TiO_3$: MgO composite as a function of temperature at 10KHz in cooling and heating cycles

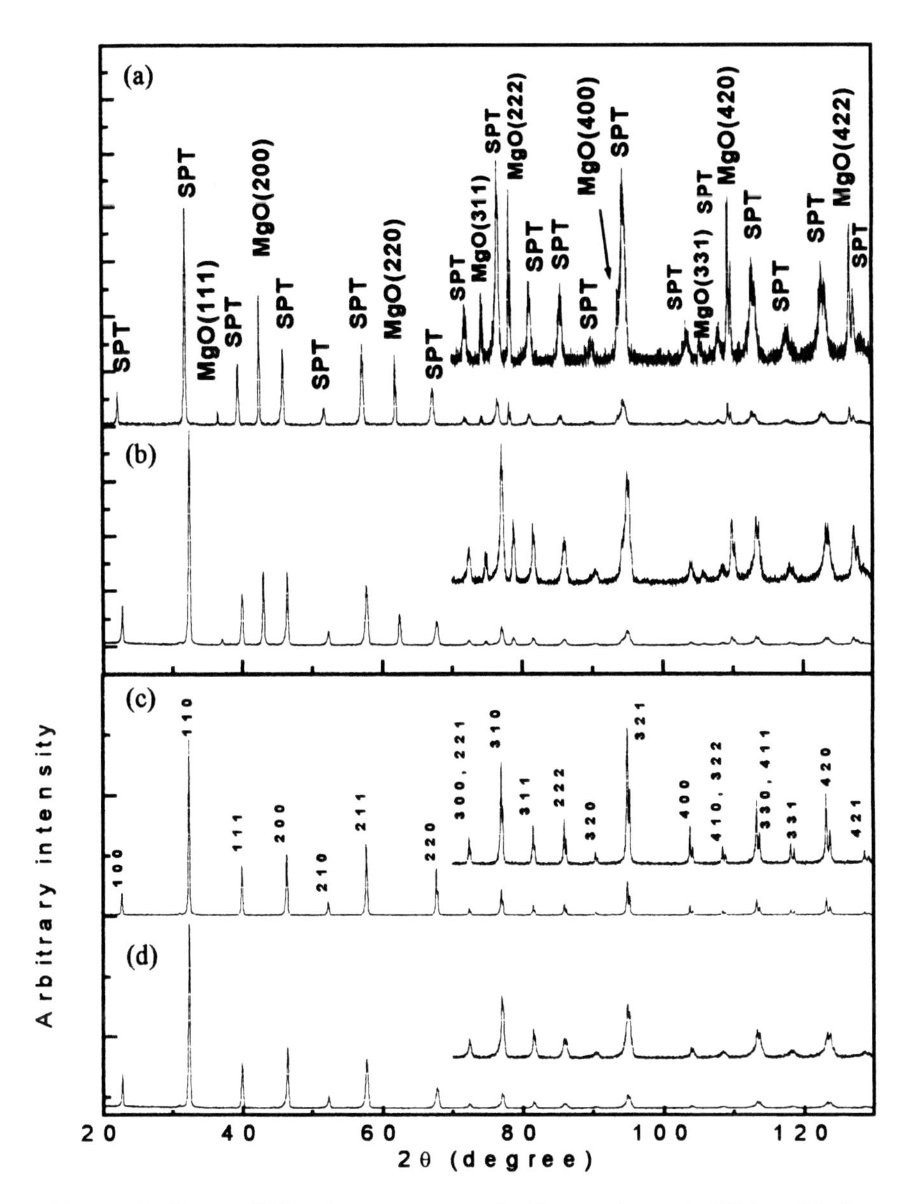

Figure 7 X-ray diffraction patterns of (a) as sintered SPT : MgO composite, (b) mixture of calcinated SPT and MgO powders, (c) sintered SPT and (d) calcinated SPT powders(SPT : $(Sr_{0.7}Pb_{0.3})TiO_3$)

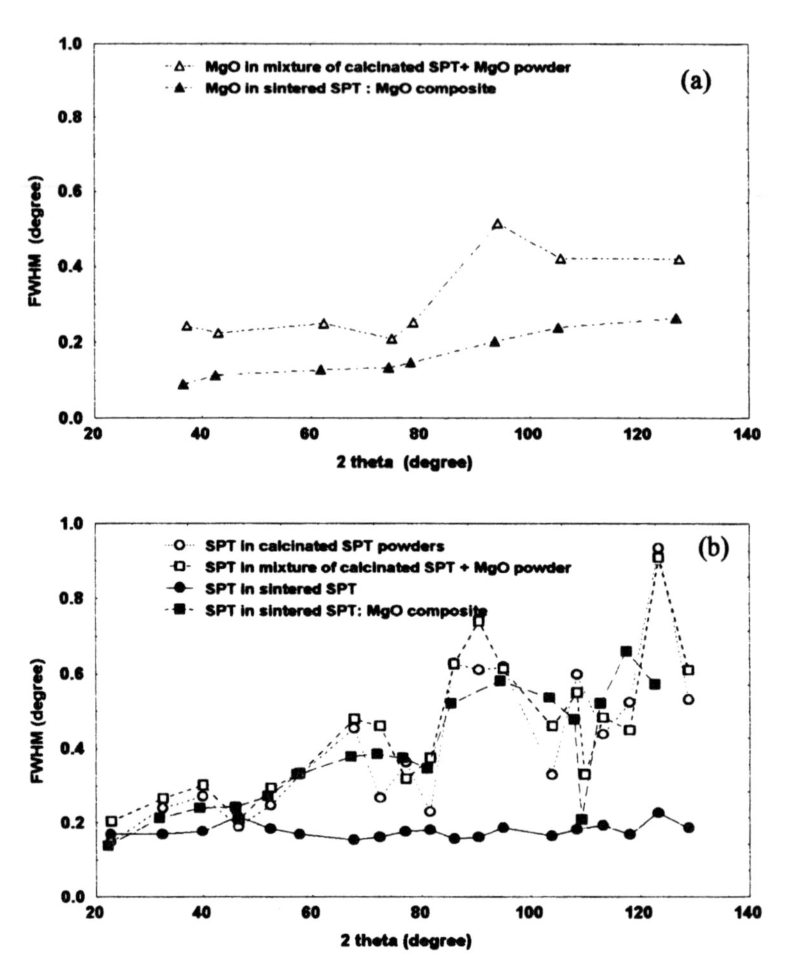

Figure 8 Change in FWHM (full width at half maximum) of (a) SPT, $(Sr_{0.7}Pb_{0.3})TiO_3$ and (b) MgO after sintering

(a)

(b)

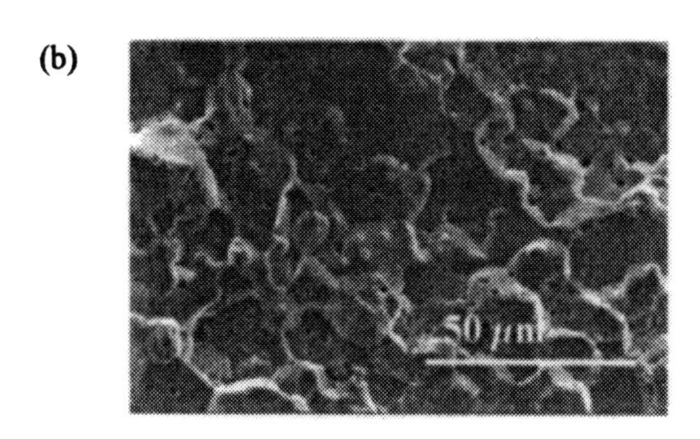

(c)

Figure 9 Scanning electron microscope images of (a)sintered SPT: MgO composite and (b)sintered SPT and (c)calcinated SPT ; SPT=$(Sr_{0.7}Pb_{0.3})TiO_3$

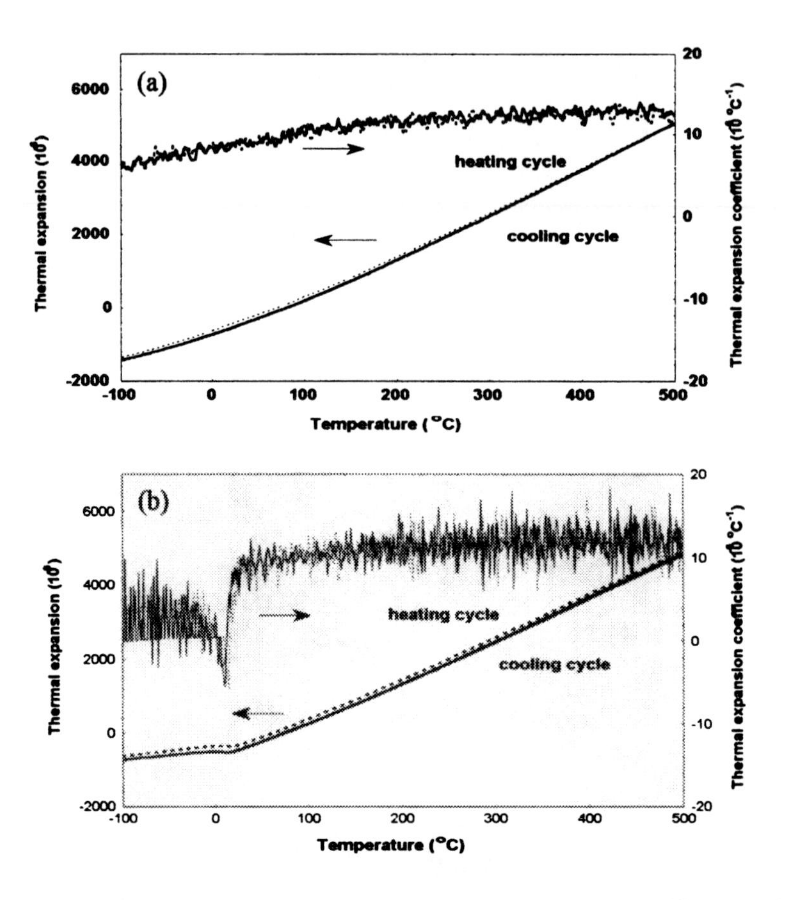

Figure 10 Thermal expansion and thermal expansion coefficient of (a) SPT : MgO composite and (b) SPT as a function of temperature in cooling and heating cycles; SPT=$(Sr_{0.7}Pb_{0.3})TiO_3$

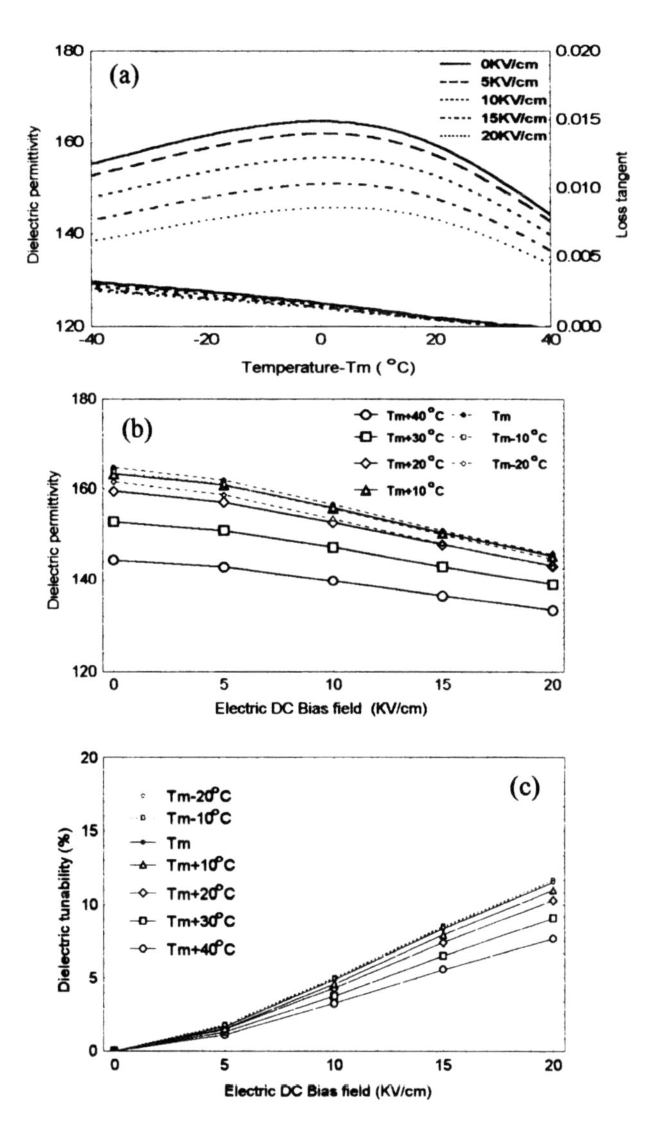

Figure 11 Dielectric response of SPT : MgO composite under DC bias fieldas a function of temperature(0-20KV/cm) at 10KHz; (a) dielectric permittivity vs. temperature, (b)dielectric permittivity vs. DC bias field and dielectric tunability vs. DC bias field
(SPT= $(Sr_{0.7}Pb_{0.3})TiO_3$)

LATTICE DYNAMICS AND DIELECTRIC PROPERTIES OF FERROELECTRIC THIN FILMS FOR FREQUENCY AGILE DEVICES

X. X. Xi, A. A. Sirenko, I. A. Akimov, A. M. Clark, and J.-H. Hao
Department of Physics
The Pennsylvania State University
University Park, PA 16802

INTRODUCTION

The most critical materials issue facing the frequency-agile electronics using ferroelectric thin films is the higher loss tangent in the thin films [1, 2]. Also, a larger tunability is often accompanied by a higher loss [3], and the dielectric constant tends to be significantly reduced in thin films compared to in single crystals [4, 5]. Strain [6, 7] and defects such as oxygen vacancies [8, 9] have been suggested as important materials parameters influencing these thin film properties. Buffer layers [10, 11] and oxygen pressure during the deposition [6, 7] have been used to control strain in the films. Epitaxial lift-off using sacrificial layers [12, 13] and Crystal Ion Slicing [14] have been used to relieve the strain in the films. Oxygen annealing [15, 16] and doping [17, 18] have been investigated to reduce oxygen vacancies in the ferroelectric thin films. Understanding the mechanisms of dielectric properties in thin film is important for these intense efforts to improve the thin film properties and the performance of the frequency-agile electronics [1, 19].

Lattice dynamics is of central importance for ferroelectrics [20]. The ferroelectric distortion involves the same iconic movement as the vibration of a zone-center transverse optical phonon mode, the "soft mode". The soft mode has a low frequency which decreases towards zero as the temperature is lowered to the Curie temperature T_c, and the ferroelectric phase transition represents the freezing of the soft mode. In the paraelectric phase, according to the

Lyddane-Sachs-Teller (LST) relation,

$$\frac{\epsilon(0)}{\epsilon(\infty)} = \prod_{j}^{N} \frac{\omega_{LOj}^2}{\omega_{TOj}^2}, \tag{1}$$

where $\epsilon(0)$ and $\epsilon(\infty)$ are the static and the high frequency dielectric constants, and ω_{LOj} and ω_{TOj} are the frequencies of the longitudinal and transverse optical phonon modes, respectively. The frequencies of the higher optical modes exhibit no sizeable temperature dependence, therefore $\epsilon(0)$ is inversely proportional to the square of the soft-mode frequency, ω_{TO1}:

$$\epsilon(0) \propto \frac{1}{\omega_{TO1}^2}. \tag{2}$$

The electric-field induced hardening of the soft-mode frequency is the origin of the electric-field dependence of $\epsilon(0)$ [21]. The dielectric loss is related to the damping of the soft mode through multiple-phonon processes in the centrosymmetric crystals, and dominated by the quasi-Debye contribution in the non-centrosymmetric crystals [22].

This paper describes our research to support the materials efforts in frequency agile electronics with fundamental understandings of the tunability and loss mechanisms in thin films. Our focus is to investigate the lattice dynamical properties, in particular the soft-mode behaviors, in thin films. We focus on thin films of $SrTiO_3$ (STO), which is a prototype, albeit incipient, ferroelectric materials [23]. The results of our lattice dynamical studies in STO films provide important insights on the issues described above. It is our goal to extend these studies to thin films of $(Ba,Sr)TiO_3$ (BST), which are more widely used for frequency-agile devices [24, 6, 25].

METAL-OXIDE BILAYER RAMAN SCATTERING

In spite of its critical importance, there is very little information available on lattice dynamics in ferroelectric thin films. Phonon spectra in STO bulk crystals have been investigated using infrared spectroscopy [26], neutron scattering [27], Raman scattering [21], and hyper-Raman scattering [28]. The neutron scattering technique is limited to the bulk material since it usually requires several grams of sample. Hyper-Raman scattering utilizes very high excitation laser power density which will damage thin film samples. It is difficult to carry out Raman and infrared measurements of phonons in transparent thin films: the excitation light goes through the film into the substrate, whose signal dominates the resultant spectrum. To overcome this problem, we have

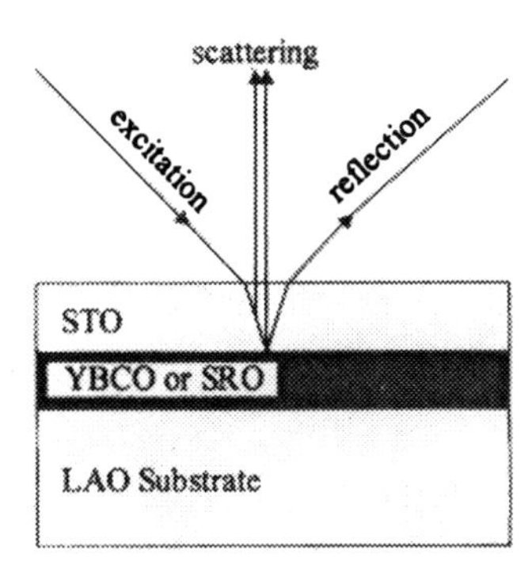

Figure 1: A schematic of metal-oxide bilayer Raman scattering. The conducting layer (such as SRO) reflects the laser beam so that it cannot reach the substrate, allowing the study of Raman scattering from the STO thin film.

used a metal-oxide bilayer structure, schematically shown in Fig. 1 [29]. In this structure, a conducting metal oxide layer, $SrRuO_3$ (SRO) in our studies, was deposited between the STO layer and the substrate, which blocks the signal from the substrate. In contrast to growing on reflective substrates such as Si or Pt-coated Si [30], or on substrates with low Raman or infrared activities at the frequency of interest, such as Al_2O_3, $KTaO_3$, MgO, and fused quartz, [31, 32], the metal-oxide bilayer technique ensures high quality epitaxial growth of the ferroelectric thin films, which is important for probing intrinsic thin film properties. The bilayer geometry allows the Raman scattering, far-IR ellipsometry, and low-frequency dielectric measurements to be performed on the same sample. In Fig. 2 the Raman spectra of three STO films in the STO/SRO bilayer structure on $LaAlO_3$ substrate are shown. The Raman signals from the substrate are absent, those from the SRO buffer layer are weak (denoted by the stars), and those from the STO films are clearly observed [29].

SOFT MODE HARDENING IN STO THIN FILMS

The soft mode frequency of films was measured by far-infrared ellipsometry [33]. This newly developed technique combines ellipsometry with the high brightness of the synchrotron radiation, and provides a powerful capability to measure vibrational properties with high reliability and accuracy. In Fig. 3 the square of the soft-mode frequency, ω_{TO1}^2, is plotted together with the inverse dielectric constant, $1/\epsilon(0)$, for a 2 μm-thick STO film and a single crystal. As

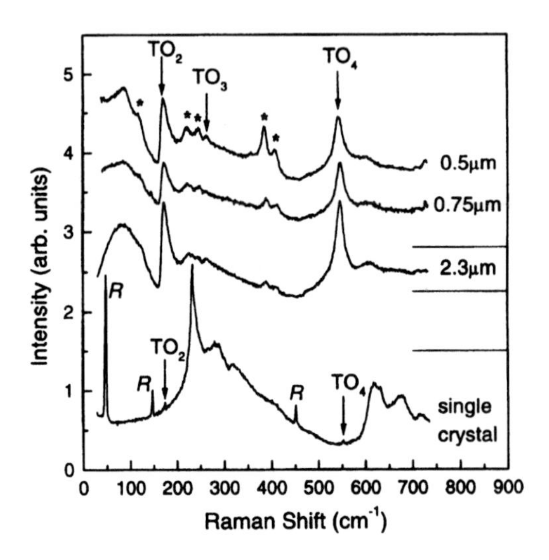

Figure 2: Normalized Raman spectra of STO films and a single crystal measured at $T = 5\,\mathrm{K}$. The stars denote the SRO Raman lines, R the structural modes, and the arrows the zone center $\mathrm{TO}_{2,3,4}$ phonons.

seen in the figure, the soft mode frequency in the thin film is higher than that in the crystal, which leads to a lower dielectric constant as shown by Fig. 3(b). Quantitative analysis shows that in the entire temperature range the LST relation between the measured optical phonon frequencies and the static dielectric constant is maintained with an accuracy of better than 10%.

The observation of the soft-mode hardening in STO films is important for such high-storage-density-capacitor applications of ferroelectrics as in DRAM. It is often thought that the significant reduction in the static dielectric constant ϵ_0 in thin films arises at least to some extent from an interfacial "dead layer" which has a low dielectric constant [5, 34, 35]. In our $2\,\mu$m-thick sample, the interfacial "dead layer" effect is negligibly small, but the dielectric constant is still well below the value in bulk single crystals [36]. Our infrared result shows that when the interfacial dead layer effect is not significant, the lower dielectric constant in STO thin films than in the bulk is due to the soft-mode hardening.

SYMMETRY BREAKING IN STO FILMS

As shown in Fig. 2, we have clearly observed Raman scattering signal from

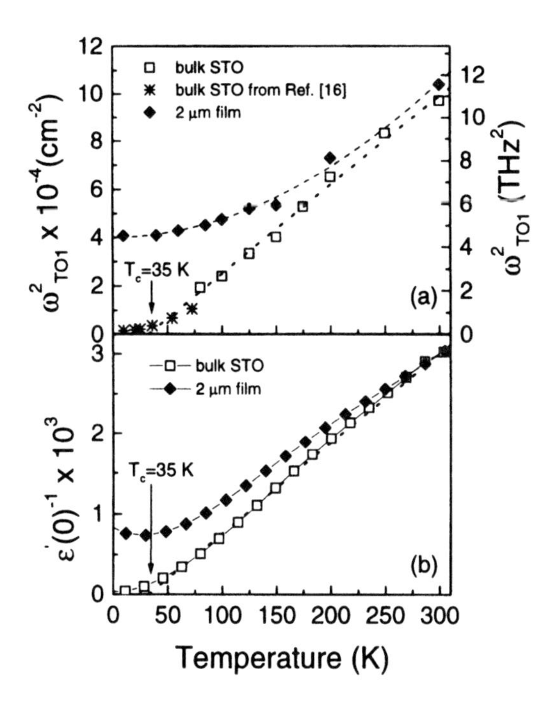

Figure 3: (a) ω_{TO1}^2 and (b) $1/\epsilon(0)$ vs. temperature for a 2 μm-thick STO film and an STO single crystal.

the STO films using the MOB-RS technique [29]. Bulk STO crystals have a centrosymmetric structure and the zone-center optical phonons are of odd parity, therefore they are not Raman active [37]. The results Fig. 2, however, show strong first-order peaks from the optical phonons, which can be seen up to room temperature. This indicates a lowering of the crystal symmetry in the STO films, the breaking of inversion and/or translation symmetries. In Fig. 4, the enlarged spectra of the polar $TO_{2,4}$ and non-polar TO_3 peaks for the 2.3 μm film are displayed for $T = 5\,K$. The TO_2 peak is strongly asymmetric and exhibits a Fano profile, which indicates a coherent interference between the polar TO_2 phonon with a continuum of excitations in STO films [38]. In doped STO and KTO single crystals, similar asymmetric line shape has been attributed to the polarization fluctuations in the defect-induced micro polar regions [39, 40, 41]. The Fano effect in our STO films can similarly result from the interaction of the TO_2 phonons with the polarization fluctuations in such local polar regions.

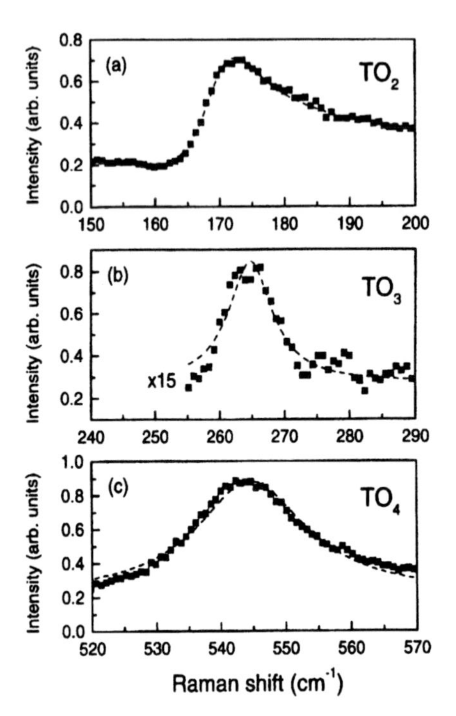

Figure 4: (a) The Fano-like TO_2 peak for a 2.3 μm STO film at T = 5 K. (b,c) The symmetric peaks for the TO_3 and TO_4 phonons, respectively.

Concerning the origin of the local polar region, it is impossible to rule out the existence of impurities: even in samples of highest purity, acceptor-type impurities have been detected [42]. The defect chemistry studies show, however, that the existence of these impurities, as well as cation off-stoichiometry, often results in oxygen vacancies in the titanates [42]. It was shown by Uwe *et al.* that when a nominally pure STO single crystal is made oxygen deficient, ferroelectric micro-regions are induced in it [43]. Both theory and experiment show that the singly-charged oxygen vacancies give rise to dipole centers [44, 45]. In thin films, besides cation off-stoichiometry and acceptor-type impurities, oxygen vacancies can also result from insufficient oxygenation during the deposition process. Based on these considerations, we conclude that the local polar regions in the STO films are most likely caused by the oxygen vacancies.

The existence of local polar regions have been independently observed using time-resolved confocal scanning optical microscopy by Hubert *et al.* [19]. The microwave dielectric response of a BST thin film was measured locally, which shows ferroelectric nanodomains whose size-dependent relaxation frequencies lead to strong dielectric dispersion at mesoscopic scales.

ELECTRIC-FIELD INDUCED SOFT-MODE HARDENING

In STO single crystals, the dielectric nonlinearity vanishes above $T \sim 80\,\mathrm{K}$ [46] while in thin films it remains non-zero to very high temperature [11]. To find the mechanism of the dielectric nonlinearity in STO thin films, we have performed Raman scattering under external electric field. The measured sample consisted of a $0.2\,\mu$m-thick transparent conducting indium-doped tin oxide (ITO) top electrode deposited on a STO/SRO bilayer structure. The ITO top electrode allows the application of electric field normal to the film plane during the Raman measurements. The temperature dependence of the TO_1 phonon frequency, ω_{TO1}, for various external electric fields is shown in Fig. 5(a). The soft mode frequency increases when an electric field is applied, and the electric field induced soft-mode hardening is observed in the entire temperature range of the measurement. This is different from bulk crystals where the mode hardening vanishes above $T \sim 80\,\mathrm{K}$ [47].

Bulk STO crystals have a TiO_6 cubic-to-tetragonal octahedral-rotation phase transition at 105 K [48]. As a result, R-point phonon modes become visible in the Raman spectrum below the phase transition temperature. In Fig. 5(b), the normalized R-mode intensity is shown as a function of temperature. The structural transition temperature in our films is found to be $120 \pm 5\,\mathrm{K}$, higher than that in the bulk at 105 K. This observation is in a quantitative agreement with the x-ray diffraction result. Below the cubic-to-tetragonal structural phase transition temperature, the soft mode split into two branches of different symmetries ("A" and "E" symmetry). The electric-field dependence of $1/\omega_{TO1}^2$ for the A-branch is found to be consistent with that of $\epsilon(0)$, indicating that the mechanism for dielectric nonlinearity in films is, as in the bulk, the field induced hardening of the soft mode. The different dielectric nonlinearity in thin films is a reflection of the different soft-mode behavior from that in bulk crystals [49].

ROLES OF LOCAL POLAR REGIONS

The symmetry lowering, and the suggestions based on it that local polar regions exist in STO thin films, may hold the key to the understanding of the lattice dynamical and dielectric properties of ferroelectric thin films. Fleury and Worlock [21, 47] first pointed out that the soft-mode frequency can be expressed as a function of lattice polarization, P, based on Devonshire's expansion of Gibb's free-energy density and the LST relation. Ignoring tensor

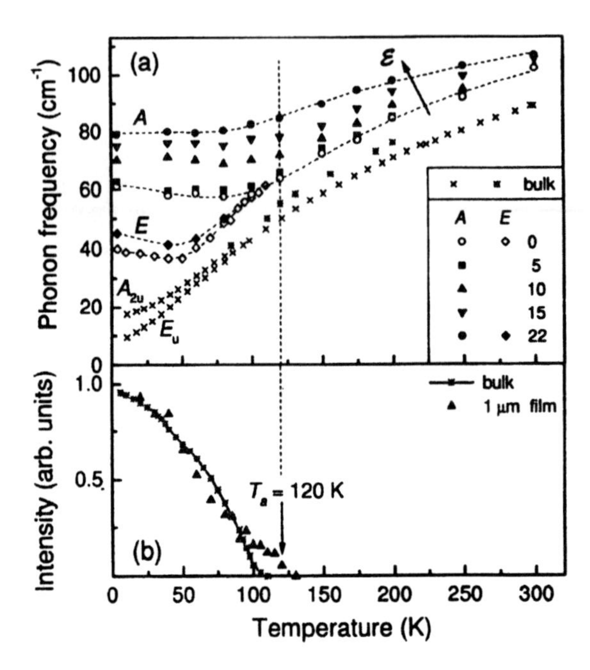

Figure 5: (a) Soft mode frequency vs. temperature for different external electric field, given in 10^4 V/cm. (b) Intensity of the structural R modes as a function of temperature.

indices, the Gibbs free-energy density, G, is expanded in a power series of P as

$$G(T, P) = G(T, 0) + \frac{1}{2}\chi(T)P^2 + \frac{1}{4}\xi(T)P^4 + \frac{1}{6}\zeta(T)P^6 + \cdots, \tag{3}$$

where χ is the electric susceptibility, and ξ and ζ are expansion coefficients. The macroscopic electric field, E, is the derivative of G with respect to P:

$$E = \frac{\partial G}{\partial P} = \chi(T)P + \xi(T)P^3 + \zeta(T)P^5 + \cdots, \tag{4}$$

and the second derivative gives the inverse dielectric constant:

$$\frac{1}{\epsilon_0(T)} \approx \frac{1}{\epsilon_0(T) - 1} = \frac{\partial E}{\partial P} = \frac{\partial^2 G}{\partial P^2} = \chi(T) + 3\xi(T)P^2 + 5\zeta(T)P^4 + \cdots. \tag{5}$$

The approximation $\epsilon_0(T) \approx \epsilon_0(T) - 1$ is used because $\epsilon_0(T) \gg 1$. Rewriting Eq. (2) as

$$\omega_{TO1}^2(T) = \frac{A}{\epsilon_0(T)}, \tag{6}$$

where A is a constant, and combining Eqs. (5) and (6), one obtains

$$\omega_{TO1}^2(T, P) = \omega_{TO1}^2(T, 0) + A[3\xi(T)P^2 + 5\zeta(T)P^4 + \cdots]. \tag{7}$$

In general, P can be either spontaneous polarization, P_s, or polarization induced by an external electric field. This was how Fleury and Worlock explained the dielectric nonlinearity, which is determined by the higher-order terms in Devonshire's expansion. Ideal STO single crystal is an incipient ferroelectric and therefore there is no spontaneous polarization and $P = 0$ in the absence of external electric field. However, when local polar regions exist in STO thin films, $P_s \neq 0$ in those regions, and therefore the P^2, P^4, ... terms' contributions to the soft-mode frequency are not zero. This will cause the soft-mode frequency to be higher than in the ideal single crystal. Neglecting the higher-order terms, the soft-mode hardening can be expressed as

$$\omega_{TO1}^2(T, P_s) - \Omega_{TO1}^2(T) \propto \langle P_s^2(T) \rangle, \tag{8}$$

where $\Omega_{TO1}^2(T)$ is the soft-mode frequency of the ideal bulk STO crystal, and $\langle P_s^2(T) \rangle$ is the mean square of polarization averaged over the length scale comparable to the wavelength of the soft mode phonon. The higher the density of the local polar regions, the larger is $\langle P_s^2(T) \rangle$ and thus the soft mode frequency of the STO thin films. Similar approach has been used by Vendik it et al. [50] and by Vogt to explain the hardening of the soft modes in Li-doped bulk $KTaO_3$ [51].

The existence of local polar regions will increase dielectric loss in the thin films. It has been shown that three main mechanisms of intrinsic dielectric loss exist in crystals [22]: three-quantum loss, four-quantum loss, and quasi-Debye loss. The first two mechanisms involve multi-phonon scattering processes, while the quasi-Debye mechanism involves the relaxation of the phonon distribution function and can occur only in non-centrosymmetric crystals. The quasi-Debye loss is much larger in magnitude than the multi-phonon losses. In ideal bulk crystals of STO, which is centrosymmetric, only the three- and four-quantum losses are important, which are very low. However, if the local polar regions exist as in the case of STO thin films, these regions are non-centrosymmetric, which is evidenced by the observation of first-order Raman peaks, and the quasi-Debye loss will become possible. Depending on the density of the local polar regions, this could dramatically increase the dielectric loss in STO thin films.

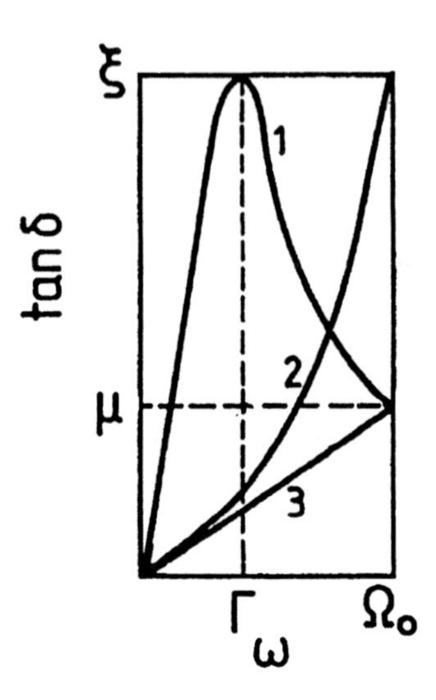

Figure 6: Intrinsic loss in ferroelectrics: (1) quasi-Debye, (2) three-quantum, and (3) four-quantum mechanisms. Ω_0 and Γ are soft mode frequency and damping. The microwave frequency is much smaller than Ω_0. From Tagantsev, Ref. [22].

CONCLUSIONS

The lattice dynamical properties are important for a fundamental understanding of the tunability and loss mechanisms in thin films. Through Raman scattering and infrared ellipsometry measurements, we found that the soft mode in STO thin films is harder than that in bulk crystals, consistent with the reduced dielectric constant following the LST relation. The symmetry lowering in the STO films, evidenced by the observation of the symmetry-forbidden first-order optical phonon peaks in Raman scattering and the Fano asymmetry in the TO_2 peak, suggests the existence of the local polar regions likely due to oxygen vacancies. Based on the Devonshire's model and the LST relation, the existence of local polar regions can give rise to the harder soft-mode frequency. It can also explain the higher dielectric loss in STO films because the quasi-Debye loss, which is much larger than the multi-phonon losses, but is active only in non-centrosymmetric crystals, can also contribute to the dielectric loss due to the symmetry lowering in STO thin films.

REFERENCES

[1]X. X. Xi, Hong-Cheng Li, Weidong Si, A. A. Sirenko, I. A. Akimov, J. R. Fox, A. M. Clark, and Jianhua Hao, "Oxide thin films for tunable microwave

devices", *J. Electroceramics* 4, 407 (2000).

[2]F. A. Miranda, G. Subramanyam, F. W. Van Keuls, R. R. Romanofsky, and J. D. Warnerand C. H. Mueller, "Design and development of ferroelectric tunable microwave components for $Ku-$ and $K-$band satellite communication systems", *IEEE Tran. Micro. Theo. and Techn.* 48, 1181 (2000).

[3]M. J. Dalberth, R. E. Stauber, J. C. Price, C. T. Rogers, and David Galt, "Improved low frequency and microwave dielectric response in strontium titanate thin films grown by pulsed laser ablation", *Appl. Phys. Lett.* 72, 507 (1998).

[4]D. E. Kotecki, J. D. Baniecki, H. Shen, R. B. Laibowitz, K. L. Saenger, J. J. Lian, T. M. Shaw, S. D. Athavale, C. Cabral, Jr., P. R. Duncombe, M. Gutsche, G. Kunkel, Y. J. Park, Y. Y. Wang, and R. Wise, "(Ba,Sr)TiO_3 dielectrics for future stacked-capacitor (DRAM)", *IBM J. Res. Develop.* 43, 367 (1999).

[5]C. Zhou and D. M. Newns, "Intrinsic dead layer effect and the performance of ferroelectric thin film capacitors", *J. Appl. Phys.* 82, 3081 (1997).

[6]W. J. Kim, W. Chang, S. B. Qadri, J. M. Pond, S. W. Kirchoefer, D. B. Chrisey, and J. S. Horwitz, "Microwave properties of tetragonally distorted $(Ba_{0.5}Sr_{0.5})TiO_3$ thin films", *Appl. Phys. Lett.* 76, 1185 (2000).

[7]W. J. Kim, , H. D. Wu, W. Chang, S. B. Qadri, J. M. Pond, S. W. Kirchoefer, D. B. Chrisey, and J. S. Horwitz, "Microwave dielectric properties of strained $(Ba_{0.4}Sr_{0.6})TiO_3$ thin films", *J. Appl. Phys.* 88, 5448 (2000).

[8]A. A. Sirenko, I. A. Akimov, J. R. Fox, A. M. Clark, Hong-Cheng Li, Weidong Si, and X. X. Xi, "Observation of the first-order raman scattering in $SrTiO_3$ thin films", *Phys. Rev. Lett.* 82, 4500 (1999).

[9]Oleg Tikhomirov, Hua Jiang, and Jeremy Levy, "Direct observation of local ferroelectric phase transitions in $Ba_xSr_{1-x}TiO_3$ thin films", *Appl. Phys. Lett.* 77, 2048 (2000).

[10]Q. X. Jia, A. T. Findikoglu, D. Reagor, and P. Lu, "Improvement in performance of electrically tunable devices based on nonlinear dielectric $SrTiO_3$ using a homoepitaxial $LaAlO_3$ interlayer", *Appl. Phys. Lett.* 73, 897 (1998).

[11]H. -C. Li, Weidong Si, Alexander D. West, and X. X. Xi, "Near single crystal-level dielectric loss and nonlinearity in pulsed laser deposited $SrTiO_3$ thin films", *Appl. Phys. Lett.* 73, 190 (1998).

[12]M. M. Eddy, R. Hanson, M. R. Rao, and B. Zuck, "Oxide epitaxial lift-off (OELO)", *Mater. Res. Soc. Symp. Proc.* 474, 365 (1997).

[13]M. J. Dalberth, J. C. Price, and C. T. Rogers, "Epitaxial lift-off of strontium titanate thin films and the temperature dependence of the low frequency dielectric properties of the films", *Mater. Res. Soc. Symp. Proc.* 493, 371

(1998).
[14]M. Levy, R. M. Osgood, Jr., R. Liu, L. E. Cross, G. S. Cargill III, A. Kumar, and H. Bakhru, "Fabrication of single-crystal lithium niobate films by crystal ion slicing", *Appl. Phys. Lett.* 73, 2293 (1993).
[15]W. T. Chang, James S. Horwitz, Adriaan C. Carter, Jeffrey M. Pond, Steven W. Kirchoefer, Charles M. Gilmore, and Douglas B. Chrisey, "The effect of annealing on the microwave properties of $Ba_{0.5}Sr_{0.5}TiO_3$ thin films", *Appl. Phys. Lett.* 74, 1033 (1999).
[16]C. M. Carlson, T. V. Rivkin, P. A. Parilla, J. D. Perkins, D. S. Ginley, A. B. Kozyrev, V. N. Oshadchy, and A. S. Pavlov, "Large dielectric constant ($\epsilon/\epsilon_0 > 6000$) $Ba_{0.4}Sr_{0.6}TiO_3$ thin films for high-performance microwave phase shifters", *Appl. Phys. Lett.* 76, 1920 (2000).
[17]H. Chang, C. Gao, I. Takeuchi, Y. Yoo, J. Wang, P. G. Schultz, X.-D. Xiang, R. P. Sharma, M. Downes, and T. Venkatesan, "Combinatorial synthesis and high throughput evaluation of ferroelectric/dielectric thin-film libraries for microwave applications", *Appl. Phys. Lett.* 72, 2185 (1998).
[18]P. C. Joshi and M. W. Cole, "Mg-doped $Ba_{0.6}Sr_{0.4}TiO_3$ thin films for tunable microwave applications", *Appl. Phys. Lett.* 77, 289 (2000).
[19]Charles Hubert, Jeremy Levy, E. J. Cukauskas, and Steven W. Kirchoefer, "Mesoscopic microwave dispersion in ferroelectric thin films", *Phys. Rev. Lett.* 85, 1998 (2000).
[20]W. Cochran, "Crystal stability and the theory of ferroelectricity", *Advan. Phys.* 9, 387 (1960).
[21]J. M. Worlock and P. A. Fleury, "Electric field dependence of optical-phonon frequencies", *Phys. Rev. Lett.* 19, 1176 (1967).
[22]A. K. Tagantsev, "Mechanisms of dielectric loss in microwave materials", in "MRS symposium proceedings: Materials issues for tunable RF and microwave devices", Q. X. Jia, F. A. Miranda, D. E. Oates, and X. X. Xi, eds. (Materials Research Society), vol. 603, 221 (2000).
[23]K. A. Müller and H. Burkard, "$SrTiO_3$:an intrinsic quantum paraelectric below 4 K", *Phys. Rev. B* 19, 3593 (1979).
[24]P. Padmini, T. R. Taylor, M. J. Lefevre, A. S. Nagra, R. A. York, and J. S. Speck, "Realization of high tunability barium strontium titanate thin films by rf magnetron sputtering", *Appl. Phys. Lett.* 75, 3186 (1999).
[25]B. H. Park, J. Lee, E. J. Peterson, X. Zeng, W. Si, X. X. Xi, and Q. X. Jia, "Effects of very thin strain layers on dielectric properties of epitaxial $Ba_{0.6}Sr_{0.4}TiO_3$ films", *Appl. Phys. Lett.* submitted.
[26]J. L. Servoin, Y. Luspin, and F. Gervis, "Infrared dispersion in $SrTiO_3$ at high temperature", *Phys. Rev. B* 22, 5501 (1980).

[27]J. D. Axe, J. Harada, and G. Shirane, *Phys. Rev. B* 1, 1227 (1970).
[28]H. Vogt, "Refined treatment of the model of linearly coupled anharmonic oscillators and its application to the temperature dependence of the zone-center soft-mode frequencies of $KTaO_3$ and $SrTiO_3$", *Phys. Rev. B* 51, 8046 (1995).
[29]Vladimir I. Merkulov, Jon R. Fox, Hong-Cheng Li, Weidong Si, A. A. Sirenko, and X. X. Xi, "Metal-oxide bilayer raman scattering in $SrTiO_3$ thin films", *Appl. Phys. Lett.* 72, 3291 (1998).
[30]E. Ching-Prado, J. Cordero, R. S. Katiyar, and A. S. Bhalla, "Raman-spectroscopy and x-ray-diffraction of $PbTiO_3$ thin-film", *J. Vac. Sci. Tech. A* 14, 762 (1996).
[31]L. H. Robins, D. L. Kaiser, L. D. Rotter, P. K. Schenck, G. T. Stauf, and D. Rytz, "Investigation of the structure of barium-titanate thin-films by Raman-spectroscopy", *J. Appl. Phys.* 76, 7487 (1994).
[32]I. Fedorov, J. Petzelt, V. Zelezny, G. A. Komandin, A. A. Volkov, K. Brooks Y. Huang, and N. Setter, "Far-infrared dielectric response of $PbZr_{1-x}TiO_3$ thin ferroelectric-films", *J. Phys.-Cond. Mat.* 7, 4313 (1995).
[33]A. A. Sirenko, C. Bernhard, A. Golnik, A. M. Clark, Jianhua Hao, Weidong Si, and X. X. Xi, "Soft-mode hardening in $SrTiO_3$ thin films", *Nature* 404, 373 (2000).
[34]C. Basceri, S. K. Streiffer, A. I. Kingon, and R. Waser, "The dielectric response as a function of temperature and film thickness of fiber-textured (Ba, Sr)TiO_3 thin films grown by chemical vapor deposition", *J. Appl. Phys.* 82, 2497 (1997).
[35]C. T. Black and J. J. Welser, "Electric-field penetration into metals: Consequences for high-dielectric-constant capacitors", *IEEE Trans. Electron Devices* 46, 776 (1999).
[36]H. -C. Li, Weidong Si, Alexander D. West, and X. X. Xi, "Thickness dependence of dielectric loss in $SrTiO_3$ thin films", *Appl. Phys. Lett.* 73, 464 (1998).
[37]W. G. Nilsen and J. G. Skinner, "Raman spectrum of strontium titanate", *J. Chem. Phys.* 48, 2240 (1968).
[38]U. Fano, "Effects of configuration interaction on intensities and phase shifts", *Phys. Rev.* 124, 1866 (1961).
[39]S. K. Manlief and H. Y. Fan, "Raman spectrum of $KTa_{0.64}Nb_{0.36}O_3$", *Phys. Rev. B* 5, 4046 (1972).
[40]U. Bianchi, W. Kleemann, and J. G. Bednorz, "Raman scattering of ferroelectric $Sr_{1-x}Ca_xTiO_3$, $x = 0.007$", *J. Phys.: Condens. Matter* 6, 1229 (1994).
[41]J. Toulouse, P. DiAntonio, B. E. Vugmeister, X. M. Wang, and L. A.

Knaus, "Precursor effects and ferroelectric microregions in $KTa_{1-x}Nb_xO_3$ and $K_{1-y}Li_yTaO_3$", *Phys. Rev. Lett.* 68, 232 (1992).

[42]R. Waser and D. M. Smyth, "Defect chemistry, conduction, and breakdown mechanism of perovskite-structure titanates", in "Ferroelectric thin films: synthesis and basic properties", Carlos Paz de Araujo, James F. Scott, and George W. Taylor, eds. (Gordon and Breach Publishers, Amsterdam), 47 (1996).

[43]H. Uwe, H. Yamaguchi, and T. Sakodo, "Ferroelectric microregion in $KT_{1-x}Nb_xO_3$ and $SrTiO_3$", *Ferroelectrics* 96, 123 (1989).

[44]S. A. Prosandeyev, A. V. Fisenko, A. I. Riabchinski, I. A. Osipenko, I. P. Raevski, and N. Safontseva, "Study of intrinsic point defects in oxides of the perovskite family: I. Theory", *J. Phys.: Condens. Matter* 8, 6705 (1996).

[45]I. P. Raevski, S. M. Maksimov, A. V. Fisenko, S. A. Prosandeyev, I. A. Osipenko, and P. F. Tarasenko, "Study of intrinsic point defects in oxides of the perovskite family: II. Experiment", *J. Phys.: Condens. Matter* 10, 8015 (1998).

[46]J. Hemberger, P. Lunkenheimer, R. Viana, R. Böhmer, and A. Loidl, "Electric-field-dependent dielectric constant and nonlinear susceptibility in $SrTiO_3$", *Phys. Rev. B* 52, 13159 (1995).

[47]P. A. Fleury and J. M. Worlock, "Electric-field-induced Raman scattering in $SrTiO_3$ and $KTaO_3$", *Phys. Rev.* 174, 613 (1968).

[48]G. Shirane and Y. Yamada, "Lattice-dynamical study of the 110 °K phase transition in $SrTiO_3$", *Phys. Rev.* 177, 858 (1969).

[49]I. A. Akimov, A. A. Sirenko, A. M. Clark, Jianhua Hao, and X. X. Xi, "Electric-field induced soft-mode hardening in $SrTiO_3$ films", *Phys. Rev. Lett.* 84, 4625 (2000).

[50]O. G. Vendik, L. T. Ter-Martirosyan, and S. P. Zubko, "Microwave losses in incipient ferroelectrics as functions of the temperature and the biasing field", *J. Appl. Phys.* 84, 993 (1998).

[51]H. Vogt, "Stiffening and splitting of the soft mode of $KTaO_3$ induced by doping with Li", *Ferroelectrics* 202, 157 (1997).

KEYWORD AND AUTHOR INDEX

Lightning Source UK Ltd.
Milton Keynes UK
UKOW02n0636301014

240830UK00001B/50/P